Empiricist Research on Teaching

Philosophy and Education

VOLUME 4

Series Editors:

C. J. B. Macmillan
College of Education, The Florida State University, Tallahassee

D. C. Phillips
School of Education, Stanford University

Editorial Board:

Richard J. Bernstein, *Haverford College*
David W. Hamlyn, *University of London*
Richard J. Shavelson, *U.C.L.A.*
Harvey Siegel, *University of Miami*
Patrick Suppes, *Stanford University*

The titles published in this series are listed at the end of this volume.

Empiricist Research on Teaching

A Philosophical and Practical Critique
of its Scientific Pretensions

by

JOHN H. CHAMBERS

KLUWER ACADEMIC PUBLISHERS
DORDRECHT / BOSTON / LONDON

Library of Congress Cataloging-in-Publication Data

```
Chambers, John H., 1939-
   Empiricist research on teaching : a philosophical and practical
critique of its scientific pretensions / by John H. Chambers.
      p.    cm. -- (Philosophy and education)
   Includes index.
   ISBN 0-7923-1848-X
   1. Education--Research.  2. Teaching.  I. Title. II. Series.
LB1026.C47  1992
370'.7'8--dc20                                          92-17197
```

ISBN 0-7923-1848-X

Published by Kluwer Academic Publishers,
P.O. Box 17, 3300 AA Dordrecht, The Netherlands.

Kluwer Academic Publishers incorporates
the publishing programmes of
D. Reidel, Martinus Nijhoff, Dr W. Junk and MTP Press.

Sold and distributed in the U.S.A. and Canada
by Kluwer Academic Publishers,
101 Philip Drive, Norwell, MA 02061, U.S.A.

In all other countries, sold and distributed
by Kluwer Academic Publishers Group,
P.O. Box 322, 3300 AH Dordrecht, The Netherlands.

Printed on acid-free paper

All Rights Reserved
© 1992 Kluwer Academic Publishers
No part of the material protected by this copyright notice may be reproduced or
utilized in any form or by any means, electronic or mechanical,
including photocopying, recording or by any information storage and
retrieval system, without written permission from the copyright owner.

Printed in the Netherlands

For my Maria

TABLE OF CONTENTS

Acknowledgments   xi

Introduction   1
I   The Many Kinds of Theory   7
    Cluster 1 : Theory Contrasted with Fact   10
    Cluster 2 : Theory Contrasted with Practice   10
    Cluster 3 : Theory as Evolving Explanation   11
    Cluster 4 : Practical Theory   12
    Cluster 5 : Theory as Hypothesis   14
    Cluster 6 : Theory as Presupposition   16
    Cluster 7 : Normative Theory   19
    Cluster 8 : Empiricist Theory   20
    Examples of the Confusion between Different Kinds of Theories   26

II   Scientific Research   28
    Copernicus and *De Revolutionibus Orbium Coelestium*   29
    Kepler and his Three Laws of Planetary Motion   32
    Galileo's Dynamics   35
    Newton's Laws of Motion and Gravitation   38
    The Scientific Achievement of Galileo and Newton   43
    Lavoisier and the Discovery of Oxygen   44
    The Discovery by Crick and Watson of the Structure of DNA   49
    Wegener, Continental Drift and Plate Tectonics to 1991   59
    Cluster 9: Scientific Theory   67
    Through Scientific Research, Scientific Theory Improves   76
    Conclusion   92

III   The Precursors of Modern Empiricist Research on
      Teaching   93
    J.S.Mill   93
    Alexander Bain, James Sully and Others   96
    The Statistics of Francis Galton   97
    Karl Pearson   98
    Rice, Thorndike, Taylor, Ayres, Bobbitt, the Efficiency Movement   100
    Ronald Fisher   103
    W.W.Charters and A.S. Barr   106
    Conclusion   109

IV   The General Form of Empiricist Research on Teaching   111
    1 Experimental Process-Product Studies at the University of
    Canterbury (New Zealand) (1970-1973)   112
    2 Questionnaires, Observation, and Factor and Cluster Analysis:
    Bennett's (1976) *Teaching Styles and Pupil Progress* (England)   113
    3 Teachers' Use of Grading Instruments: DeCasper's
    Questionnaire Research of the Use of Georgia Eighth Grade

	Criterion-Referenced Tests of Basic Skills, by Ninth Grade Teachers	115
	4 Pre-Test, Post-Test, Control Group Experiment: Study by Anderson, Brophy and Evertson on Reading (U.S.A.)	116
	5 Research with Meta-Statistics: Gage (1978) and Walberg (1986) (U.S.A.)	118
	6 Pre-Test, Post-Test, Observational-Statistical Research: Helmke and Schrader's Practice of Mathematics During "Seatwork" (Germany)	121
	7 Performance Based Teacher Education (PBTE) (U.S.A.)	125
	8 Observational-Statistical Research: Academic Learning Time (ALT) (U.S.A.)	126
	9 Experimental-Statistical Research by Finn and Achilles: A State-Wide Study of the Effect of Size of Class (U.S.A.)(1990)	127
	Conclusion	129
V	**What is the View of Scientific Research Assumed in the Studies?**	130
	Views of Teaching Assumed in Empiricist Research	130
	Notions of What it is to be Scientific in Research	134
	Conclusion	143
VI	**Preliminary Critique of the Present Empiricist Approach**	144
	Critique of the Apologia by Gage	144
	Abstract Concepts cf. General Concepts	156
	When Do We Require Such Research Anyway?	164
	Conclusion	166
VII	**The Many Contrasts Between Scientific Research and Empiricist Research on Teaching**	167
	The Inductivist Orientation of Empiricist Research	170
	There is No Scientific Method or Art of Discovery As Such	173
	The Context of Discovery is not Isolated from Other Inquiries	179
	Because of Their Theoretical Background, Scientists are Aware of a Question, Problem, Puzzle, or Discontent, or see Anomalous or Unexpected Phenomena	180
	There is Potential Falsifiability	183
	It is Helpful to have Alternative Paradigms, Programs, Traditions, Hypotheses, Theories, if a Science is to Progress	185
	There are no Theoretically-Neutral Observations in Research	186
	Scientific Research is Open Ended and Continually Evolving	186
	Scientific Research uses Abstract Concepts Embedded in Theory	187
	The Statistics and Probabilities of Empiricist Research are not the Mathematics and Probabilities of Scientific Research	188
	Scientific Research makes use of Ratio Scales in its Measurements	192
	Scientific Experiments are Radically Different from Empiricist Experiments	193
	Serendipity Occurs in Scientific Research	197

	Scientific Research as an Enterprize has Social Aspects	198
	Conclusion	200
VIII	Acknowledging the Complexities of Teaching which are being ignored by Present-Day Empiricist Research	202
	Allowing for the Complexities: Sam Bush, Master Teacher of Woodwork	203
	Allowing for the Complexities: Don Campbell-- Teaching High School Physics	204
	Allowing for the Complexities: Vera Milz and the teaching of Reading	207
	Allowing for the Complexities: Harry, and the teaching of American History	209
	Some Conceptual Matters: Distinctions between Reason and Cause and Action and Movement	210
	Some Seminal Features of Teaching, Inadequately Construed	215
	Acting on the Intention to Achieve Learning	215
	There is a "Content" to be Acquired or Learning to be Achieved and the Teacher Must Know It	219
	There is a Particular Context Within Which Teaching is Embedded	223
	There are Learners and the Method or Manner Must be Appropriate for Them	225
	The Lack of "Validity" of Empiricist Pedagogical Research	227
IX	Empiricist Research should Acknowledge its Empiricist Status, drop its Scientific Pretensions, and become embedded in an Adequate Pedagogical Theory	229
	Embedding Empiricist Research in an Appropriate Pedagogical Theory	229
	Normative Theory in Pedagogical Theory	230
	Some Empiricist Research as Anti-Educational	232
	An Adequate Conception of Teaching	234
	Local Theories as Part of Pedagogical Theory	235
	Misunderstanding the Importance of a More Complex Empiricist Method	236
	An Example of an Improved Kind of Empiricist Research: System for the Classroom Observation of Teaching Strategies	239

Conclusion	245
References	247
Index of Names	269
Index of Topics	275

ACKNOWLEDGMENTS

Many people helped me to develop this book. I should cite first of all Professor Jonas Soltis of Columbia University, New York City, who read several earlier arrangements of the material, and aided me in organizing the argument. I also owe thanks to colleagues at the former Tasmanian State Institute of Technology, and at the University of Tasmania. Over several years I discussed issues at the interface between philosophy of science and pedagogy with Dr. Mike Degenhardt, who also alerted me to many of the pretensions of empiricist pedagogical research. Geologist, Mr.David Hannan helped me greatly with understanding of and sources on continental drift, and physicist, Dr. Bob Rogers, corrected my ideas of Newtonian theory. Philosopher, Dr. John Norris first drew my attention to the conceptually-confused research being performed on praise. Articles by and critiques of some of my work by Professor C.J.B.Macmillan of Florida State University, Tallahassee, the writings of Professor D.C.Phillips of Stanford University, and Emeritus Professor Arnold Rothstein of The City University of New York, and discussions with them over the years, and the work of many other philosophers too numerous to list, have all been extremely helpful. Professor Rothstein also commented on the final two chapters. Discussions with and articles by Professor James Garrison of Virginia Technological Institute and State University were seminal in the earlier stages, and he graciously read the entire final manuscript. Professor Harvey Siegel of The University of Miami saved me from some "philosophical howlers" and after reading two earlier versions, made the invaluable suggestion that the philosophy of science be integrated naturally into the body of the account of scientific research. Indeed, it might not be inappropriate to suggest, that it was Professor Siegel, or Professor Garrison, who really ought to have written such a book.

Through the good offices of Dr. George Cook, Vice-President of Bermuda College, the following members of the college faculty kindly read the case studies: the President of the College, Dr.A.C.Hollis Hallet (Copernicus, Kepler, Galileo and Newton), Dr. Gloria Frederick (Lavoisier), and Dr. Nicola O'Leary (Crick and Watson); Dr. Martin Brewer, formerly of The Institute of Geological Sciences, London, read the case-study of Wegener and Plate Tectonics. The generous assistance of these scholars prevented me from committing several errors.

Dr. Barbara Walton, of The College for Human Services, New York, gave me encouragement over several years. My sisters Mrs. Joy Chambers-Grundy and Mrs Coral Chambers-Garner, from the orientations of their own professional fields have selflessly given much in the way of time and energy in discussions. My wife, Maria Figueroa, also read the whole manuscript and made some valuable suggestions.

Several of these readers may not always see matters in the way I do, e.g. they may not consider that abstract concepts are as central to the nature of scientific research as I suggest. So I should emphasize that none of them is responsible for my personal views of philosophy of science, or of the nature of scientific research, or for my direct suggestions, or for what I imply for the improvement of pedagogical research.

Much of my inspiration comes from the incisive and powerful critiques of

ACKNOWLEDGMENTS

empiricist sociology in the books of the Willers: Judith Willer, 1971; David and Judith Willer, 1972; David Willer, 1965, 1987. Alan R. Tom's book, *Teaching as a Moral Craft* was especially valuable in orienting my ideas of early empiricist research.

During the almost two years in which Kluwer showed interest in publishing this book, their editor, Mr. Peter de Liefde was always helpful, positive, and encouraging.

Sincere gratitude goes to my brother-in-law, Mr. Reg Grundy, O.B.E., for his generous help over the years when this book was evolving. Many thanks also go to Miss Margaret Moore, and Mrs. Laura Gilmour, who provided valuable clerical skills.

Grateful acknowledgement is made to the following persons and publishers for allowing selected quotations from the publications listed below.

To the Scottish Council for Research in Education for: J.L. Powell, 1985, *The Teachers' Craft*.

To the American Educational Research Association and Macmillan Publishing Company for the following articles from Merlin C Wittrock, (Ed.), 1986, *Handbook of Research on Teaching* (3rd Edition): Lee S. Shulman, "Paradigms and Research Programs in the Study Of Teaching: A Contemporary Perspective"; Jere E. Brophy and Thomas L. Good, "Teacher Behavior and Student Achievement"; Herbert J. Walberg, "Syntheses of Research on Teaching".

Copyright 1986 by the American Educational Research Association. Reprinted by permission of the publisher.

To Teachers College Press. Reprinted by permission of the publisher from Gage, N. L., *The Scientific Basis of the Art of Teaching*. New York: Teachers College Press, c 1978 by Teachers College, Columbia University. All rights reserved. Selected quotations.

To London and Buttenwieser for: James D. Watson, *The Double Helix*.

To Longman Publishing Group for selected quotations, from *Educational Research: An Introduction* (5th Edition), by Walter R. Borg and Meredith Damien Gall. Copyright 1989 by Longman Publishing Group.

To Holt, Rinehart and Winston for: Fred N. Kerlinger, 1986, *Foundations of Behavioral Research* (3rd Edition).

An earlier version of parts of Chapter 1 first appeared in, *Educational Foundations*, Vol 4 No 4, Fall 1990. I am grateful to the publisher Alan H. Jones and the Caddo Gap Press for permission to use this material.

J.H.C.,
Isabela, Puerto Rico
February, 1992.

INTRODUCTION

For perhaps one hundred years, there has been a pervasive misunderstanding in the minds of many of the people who have performed research on teaching. *In attempting to be scientific they have been empiricist.* Moreover, because they have used only a very restricted kind of empiricist method, their results have been commonplace. Until the difference between scientific and empiricist activity is recognized and *until empiricist research as presently practised is radically modified* this disappointing situation must continue.

The researchers themselves have called their methods "empirical", and it is that term which I shall use to describe such research in the next few pages. By "empirical" they meant observational, experimental and quasi-experimental research of all kinds, statistical and correlational research, *ex post facto* research, triangulation, cluster analysis, factor analysis, questionnaires, Process-Product research[1], and so on. Such approaches are of course not mutually-exclusive, but overlap, and are used in combination in various complicated ways. All such approaches claim to inform us about the nature of the world of teaching; and to draw their conclusions they all use one or another kind of statistical method at some stage. As already mentioned, practitioners of these approaches have attempted to be scientific and have usually claimed the kudos of scientific status for their work. Indeed, they often misleadingly use the words, "empirical" and "scientific" interchangably.

Such research has a massive literature, as well as legions of enthusiastic users and proselytisers. There are thousands of schools of education and national and private institutes of pedagogical research around the world, whose faculty apply such methods in their own research and who instruct teachers and apprentice researchers in the various designs, and in the panoply of statistical techniques which are adopted in the establishment of the initial data, the analysis of the resulting data, and the drawing of the conclusions. Prestigious government and quasi-government institutions, such as the United States Department of Education, and its subsidiary, the (U.S.) Office of Educational Research and Improvement (formerly the National Institute of Education), many American regional educational laboratories, the Scottish Council for Research in Education, the National Foundation for Educational Research (England and Wales), The Australian Council for Educational Research, numerous significant private foundations such as the Carnegie Foundation, and the Niels Bohr Peda-

---

1. Lest it be thought I am guilty of a category mistake, I should emphasize that Process-Product research on teaching is not a research technique, but rather is the name given to a cluster of research designs with a particular focus, viz. attempts to discover improved ways of classroom teaching, by trying to assess what is contributed by the teacher and what is achieved as a result by the students. (It has also been, I believe, the most influential kind of pedagogical research).

gogical Institute, and powerful national societies such as the American Educational Research Association, and The Canadian Educational Researchers' Association, although now acknowledging the possibility of alternative approaches to research, still spend hundreds of millions of dollars yearly in pursuing or supporting studies which adopt such methods[2].

It is such approaches which have dominated the entries in such authoritative works as the *Encyclopedia of Educational Research* (Sixth Edition, 1992) and the *Handbook of Research on Teaching* (Third Edition, 1986) since their inception in 1941 and 1963 respectively, and the annual *Review of Educational Research*. Many educationists judge such research to be of immense importance. For instance, N.L.Gage in his widely-read book, *The Scientific Basis of the Art of Teaching* (1978) in alluding to the significance of such research says that teachers can, ". . make education a thing of joy and success or a matter of frustration and despair" (1978, p.13). And his confidence in his methods for making teaching a thing of joy and success is explicitly shown near the end of the book to be based mostly upon his kind of empirical approach when he writes,

> At least for the foreseeable future, research on teaching will proceed as "normal science" (Kuhn,1962); that is investigators will follow the elaborated process-product paradigm and work on cleaning up an enormous number of details in the unfinished business of the field. New variables will be identified, invented, and measured. More ingenious ways of relating the variables, especially in complex causal patterns, will be devised and exploited. Better qualitative investigations, more comprehensive correlational studies, more intensive and single-case experiments, and more comprehensive path analyses will be performed. Better meta-analyses will bring together the results of the research in more valid and interpretable ways (pp. 93-94).

A large number of journals, many of them world famous, such as the *American Educational Research Journal, Educational Researcher, Educational Psychology ,The Journal of Educational Research ,The British Journal of Educational Studies , The Australian Journal of Education* , and the *Journal of Educational Psychology* report findings of such empirical methods on teaching.

2. Re expenditures, consider this statement from Borg and Gall: "From 1957 to 1963, USOE appropriations for educational research and development averaged $20 million or less. In 1964, appropriations increased to $37 million. In the following year they increased again, to over $100 million, where they stabilized for the remainder of the decade. Federal funding . . in the 1970s and 1980s varied depending upon the administration in power. . A conservative estimate of recent expenditures would be on the order of several hundred million dollars annually" (1989, pp.10-11). To grasp the significance of such enormous figures, the reader should bear in mind that these are merely the expenditures of the U.S. federal government, and that the majority of the research carried out was of the kind critiqued in this book.

INTRODUCTION

So significant are observations, measurements and statistics thought to be that a *Journal of Educational Statistics* has been published by the American Educational Research Association since 1975, and Vol. 20, No. 1, 1985, of *Educational Psychologist* carried an article with the title, "An Analysis of Statistical Techniques Used in *The Journal of Educational Psychology* 1979-1983". There are also well-known, professional periodicals which devote all or most of their pages to the reporting of research on measurement or the discussion of testing. Amongst the better known are for instance, the *Journal of Educational Measurement, Educational and Psychological Measurement, Applied Psychological Measurement*, and *Measurement and Evaluation in Guidance*. Since 1973 there has been an annual *Mental Measurements Yearbook*, widely acknowledged by pedagogical researchers. There are also numerous newsletters which discuss issues in the measurement and statistical analysis of teaching.

Hundreds of text-books have been published, showing how to do such empirical research. Representative of the last ten years have been, Fraas, *Basic Concepts in Educational Research* (1983), Galfo, *Educational Research Design and Data Analysis* (1983), a third edition of Evans's Planning Small-Scale Research (1984), a third edition of Wiersma (1985), Slavin's *Research Methods in Education* (1984), a third edition of Kerlinger's *Foundations of Behavioral Research* (1986), a second edition of McMillan and Schumacher's *Research in Education: a Conceptual Introduction* (1988), *Complementary Methods for Research in Education*, edited by Jaeger and put out by the American Educational Research Association (AERA) in 1988 (this also discusses alternative historical, philosophical, ethnographic and case-study approaches), a fifth edition of Borg and Gall's *Educational Research: an Introduction* (1989), and *Educational Research: a Guide to the Process*, by Wallen and Fraenkel (1991). Enthusiastic claims continue to be made for the future achievements of such empirical approaches. Biddle and Anderson wrote in 1986, and the point still applies,"Moreover, accepting the broad thrust of these claims [the relationship between empirical research on teaching and improvements in education], the National Institute of Education now devotes substantial resources to research on teaching and literally hundreds of investigators are now committed to careers that involve this research field" (p.230).

To any newcomer to this information it might therefore be surprising to learn that on the other hand, there is a fast-growing disenchantment with these empirical approaches. Many published criticisms have been scathing. Consider the following, which are representative of the last decade: "What is [empirical] educational research? The testing of unworkable materials foisted upon schools by ivory-tower academics whose first-hand knowledge of the classroom is at best out-of-date and at worst non-existent?" (Nixon, 1981 p.5). Alan Tom wonders about the continuation and pertinacity of such research despite the striking fact that with it, ". .minimal progress has been made toward an applied science of education" and he points to, ". .the futility of building a science of education through the empirical study of teaching" (1984, p.15). Barrow says, "First, I do not believe that research has in fact established any. . general rules. Secondly, I believe that the reason for this is that it cannot hope to do so, rather than simply that the research has been incompetently carried out"(1984, p.145). Macmillan and Garrison make the point that the search for correlations between what teachers are observed to do and the scores their students achieve on tests is otiose, because the behavior of teachers ,". .is part of the mediating situation itself— not as a separate causal element, but as an essential part of teaching considered as a whole"(1988, p.68); Scriven, in referring to a key technique of such research argues that, "The substitution of statistical for evaluative criteria of significance is a blunder whose consequences included the trivialization of a very great deal, perhaps most, of educational research in most of this century" (1988, p. 134). Finn says, "To put it simply, our labors have not produced enough findings that Americans can use or even see the use of" (1991, p.39); and so I could continue.

There are growing claims to the effect that empirical research on teaching is bankrupt. Some of the most notable critiques have been: J. Wilson, *Philosophy and Educational Research* (1972); L.Elvin,*The Place of Commonsense in Educational Thought* (1977), G.D.Fenstermacher, "A Philosophical Consideration of Recent Research on Teacher Effectiveness" (1979); J.Wilson, *Fantasy and Commonsense in Education* (1979); W.B.Dockrell and D.Hamilton, *Rethinking Educational Research* (1980); R. Barrow, *Giving Teaching Back to Teachers* (1984); M.A.B.Degenhardt, "Educational Research as a Source of Educational Harm" (1984); R.F. Dearden, *Theory and Practice of Education* (1984); C.J.B.Macmillan and J.W.Garrison, *A Logical Theory of Teaching* (1988); and T.J. Sergiovanni, *"Science and Scientism in Supervision and Teaching"* (1989).

Those who believe that empirical research on teaching is bankrupt in its present form have begun using a wide range of alternative approaches, which are loosely termed "qualitative research"— with increasingly interesting results. And it is noteworthy that recent manuals of research methods such as the AERA volume mentioned above, acknowledge the significance of these approaches. But most supporters of empirical research on teaching say the situation is not as negative as critics suggest. Another of their replies is that, although results may have been disappointing so far, empirical research is still a very young science and it is hardly realistic to have expected better results so far: still more effort, manpower, and sophistication in observational and experimental techniques are required, and especially, further subtle developments in statistics, but above all, more money[3].

I reject emphatically the suggestions of the previous sentence. Indeed this book has grown out of the belief that, despite the widespread influence of its findings and the great political power of its practitioners and proselytes, there are fundamental misunderstandings which negate the efficacy of such empirical research— deep inadequacies and confusions which reduce the ability of this kind of research to provide truly useful findings for teachers and administrat-

---

3. Thus, empiricist researchers Shavelson and Berliner write: "What ails education research first and foremost is the federal government's failure. . to provide political leadership and financial support for education research" (1991, p.79).

ors. Nevertheless, I have admiration for its *aims* – they are both laudable and significant. As will be argued at length below and as stated in the first paragraph, the main problem is that *empirical research is not scientific as its practitioners claim, but rather it is empiricist and a specific and limited kind of empiricist at that . It is this confusion of status which has led such researchers astray* ; *and it is its specific and restricted kind of empiricist approach which has severly limited the usefulness* of the enormous amount of research which has been carried on in this way.

The book concentrates upon this central matter of the claimedly scientific status, and what follows from the fact that such research is not scientific. The argument focuses on the difference between scientific research and empiricist research. Note the contrast between "empirical" and "empiricist". From now on, the word, "empiricist" will be used in this book in distinction to the word, "scientific". What has normally been called "empirical research" on teaching is, in my view, empiricist. *Both* empiricist research and scientific research are empirical, i.e. both attempt to discover and explain the nature of the world of empirical phenomena. Nevertheless they deal with that empirical world in radically different ways. To make my point, there is a consideration of examples from astronomy, physics, chemistry, biology and the earth sciences . There is an examination of nine paradigmatic studies or clusters of studies in empiricist research on teaching. The many striking differences between scientific enterprises and such empiricist enterprises are considered and the claim by pedagogical researchers to scientific status is examined and rejected. Such research uses exceedingly sophisticated statistical methods, but that sophistication is not enough to make it scientific.

This profound confusion between the scientific and the empiricist leaves empiricist research on teaching *in a no-man's land, where it is impossible for it to be effective scientifically, or truly effective in an empiricist manner*. However, the argument of the book continues that all is not lost. It is suggested that empiricist research on teaching may be able to redeem itself in either of two ways. First, it can attempt to meet the criticism about its lack of scientific status by attempting to make itself scientific, in particular by creating truly scientific theory which amongst other things interrelates abstract concepts in propositions which can be interpreted to the world of sense experience. Given the general failure of social science to achieve such theoretical sophistication, this first suggestion, though perhaps not impossible, would however seem an unlikely one to be achieved by pedagogical researchers. Or second, and more realizable, empiricist researchers can recognize the empiricist status of their endeavor, grasp what this means for its future development, and modify its methods in ways which take into account the wide range of humanistic and normative critiques which have already been made by other writers, as well as by taking note of the suggestions made in the final chapters of the present book. In particular, I suggest that matters can be improved by embedding empiricist research in a Practical Theory of Pedagogy, one which takes into account several of the other kinds of theory inadequately used or currently ignored. In short, it is more realistic for its practitioners to be less ambitious and to see what can be done in a piecemeal way to improve the empiricist approach. I recognize that I am merely the latest person to have found such research problematic, though I believe the orientation of the criticism is original, viz. the confusion of empiricist research with scientific research.

This Introduction ought perhaps to end on an exculpatory note. The arguments are not meant to impugn the integrity, aspirations, dedication, or abilities of those who have performed the research criticized here. They are rather to point to debilitating problems which I believe to be endemic in the very approaches made use of by the researchers, and to the significant negative effects which result from misconstruing empiricist research as scientific. If empirically-minded researchers could be brought to see that their work has involved a very limited form of empiricist research, rather than scientific research (without thinking that being empiricist as such involves loss of face) they might do better work and bring to the field a richer understanding of the object of their research— teaching. That is the aim of this book.

CHAPTER I

THE MANY KINDS OF THEORY

The central thesis of this book is that empirical research on teaching is not scientific as its practitioners claim, but rather, empiricist, and that it is this confusion which leads to its present inadequacies. To show that this is so, it will first be necessary in this and the following chapter to examine some crucial aspects of scientific research.

Central to any adequate understanding both of scientific and of empiricist pedagogical research is an understanding of the multiple roles played by theory. Because the first and greatest inadequacy of pedagogical researchers' notions of scientific research involves a misunderstanding of the complex functions of theory, because theory of several kinds will be fundamental in an improved research, because such different kinds of theory introduce their own considerable complexities and confusions, and because allusions to theory will be made regularly throughout the book, it is necessary that the widely different uses and meanings of this term be distinguished.

In both everday and pedagogical discussion, the word, "theory" is used in a bewildering variety of ways. Meanings shift even as we talk. To clarify matters, I have here reduced the uses to nine "Clusters". These Clusters are not logically pure, but overlap and criss-cross in a complex manner, and represent different emphases and purposes. Such Clusters do not form a taxonomy. Even if there is a sense in which some develop out of others, the many subtle relationships are much too complex, multi-directional, and multi-dimensional to form a taxonomy.

Ignoring the different types of theory muddles thinking and prevents researchers from understanding the diverse and multitudinous effects of theory on their work. A good example of this lack of understanding is provided by the leading Process-Product researcher, Nathaniel Gage (1985). Until his paper of 1989 (with Margaret Needles), which was itself a reply to his philosophical critics, Gage had explicitly claimed to be doing research which was *non*-theoretical. But non-theoretical research, at least in the sense of Cluster 6 (explained below), and probably in other senses also, is logically-impossible (Garrison & Macmillan, 1984; Garrison, 1988).

The situation can be confusing. A colleague of mine reports that whereas he believed a particular paper at a conference on research to be highly theoretical, a questioner suggested that it entirely lacked theory. Presumably, they were using different meanings of the word, "theory". As an example of confusion, consider the different meanings used (but not distinguished amongst) over two and a quarter pages in the following quotation from Kerlinger's, *The Foundations of*

Behavioral Research (1986), probably the most popular textbook used in university courses which train empiricist researchers:

> The second stereotype of scientists is that they are brilliant
> individuals who think, spin complex *theories* . . .
> They are [regarded as] impractical *theorists* , even though
> their thinking and *theory* occasionally lead to results
> of practical significance. . . The static view . .[of science
> concentrates on]. .the present state of knowledge and adding to
> it and on the present set of laws, *theories* , hypotheses,
> and principles. . . The heuristic view in science emphasizes
> *theory* and interconnected conceptual schemata that are
> fruitful. . .The basic aim of science is *theory* . . is to explain. . .
> Such explanations are called *theories* . . . [a psychologist might
> call]. . a general explanation a *theory* of problem solving. . .
> If we accept *theory* as the ultimate aim of science [then]. .
> explanation and understanding become subaims of the ultimate
> aim. . .a *theory* is a set of interrelated constructs (concepts),
> definitions, and propositions that present a systematic
> view of phenomena by specifying relations among variables,
> with the purpose of explaining and predicting. . [italics added].

In a manner common to much research on teaching, and in educational research and the social sciences more generally, Kerlinger appears to be unaware that he has used the word, "theory" in a variety of different ways (see below) and that as a result his own account of the nature of science is deeply confused. More to the point, what does this say about the nature of such supposedly scientific research on teaching, if the writer of one of the leading textbooks is himself muddled about the variety of theoretical features of scientific research?

Amongst the numerous types of theory, the following can be distinguished. The types do not arise from any preconceived "theory of theory", but (a) from a gradual realization of the subtle confusion, garnered from many years of reading pedagogical writing and research which allude to theory in one way or another, (b) from recent discussions in philosophy of science, and as the immediate catalyst, (c) from R.F. Dearden's incisive discussion in his *Theory and Practice in Education* (1984). To imagine that there would always have to be an explicit theory of theory, a meta-theory (though there *might* have been something like one) would be an error, for it would eventually lead to an infinite regress. However, one must agree that there will necessarily be theory in the sense of Cluster 6 (Presupposition), underlying how I construe matters.

There is no pretence that the present list is exhaustive (I have, for example, written nothing about mathematical theory), or that there may not be alternative classifications. I said above that this list is not a taxonomy, though there may be a kind of continuum; certainly there is a change of emphasis between the Clusters. It is also the case that particular propositions and claims, or sets of propositions and claims may at times be classified under several of the Clusters.

THE MANY KINDS OF THEORY

The kinds included here are chosen not because they exhaust the possibilities, but because they can all be seen as having significance for pedagogical activity, study, writing and research.

In most cases, I initially provide an example drawn from some aspect of scientific research, in order to show how complex are the varieties of theory even in that paradigmatic area of acquiring knowledge, and how confusion is likely to occur even there. Then I provide an example or examples from teaching or education. Scientific Theory is discussed in detail in Chapter II. For the convenience of the reader, the following chart outlines the nine clusters and their various types.

CLUSTER 1 : THEORY CONTRASTED WITH FACT
1. as Hunch or Speculation (to be contrasted with Fact)

CLUSTER 2 : THEORY CONTRASTED WITH PRACTICE
2. as the Opposite of Practice
3. as Instructions to be Applied

CLUSTER 3 : THEORY AS EVOLVING EXPLANATION
4. as Evolving Explanation

CLUSTER 4 : PRACTICAL THEORY
5. as Practical Theory guiding a profession or art

CLUSTER 5 : THEORY AS HYPOTHESIS
6. as Model
7. as Heuristic
8. as Hypothesis

CLUSTER 6 : THEORY AS PRESUPPOSITION
9. as Ontological Presupposition
10. as Observational Presupposition

CLUSTER 7 : NORMATIVE THEORY
11. as Doctrine and Dogma
12. as Rational Normative Arguments

CLUSTER 8 : EMPIRICIST THEORY
13. as Empiricist Accounts and Explanations

CLUSTER 9 : SCIENTIFIC THEORY
14. as First-Level Scientific Theory
15. as Second-Level Scientific Theory

Cluster I : Theory Contrasted with Fact

1. A theory may merely be *a novel but undeveloped way of considering some problem or issue* : HUNCH or SPECULATION. "It's just a theory, but I think that. ." Someone may say, "I have the theory that international sport is really sublimated warfare", or, "I have the theory that schools mess up more than educate kids." Or a teacher may say,"I know it's not a fact, but I have this theory that Jimmy's getting into drugs." There is no implication here of anything detailed or scholarly. The theorists (in this sense of the word) are not claiming that what they are saying is a fact. It is *by contrast with* fact. *Its whole point* is that it is speculative, not factual.

With appropriate hard work this Type of speculative theory can develop into other Types, e.g. modified and extended, it can become theory of Type 7 (Heuristic) below: ideas whose cogency it is believed can be tested against predicted empirical outcomes. If it evolves out of Scientific Theory (Cluster 9), it becomes Type 8 (Hypothesis). Or theory of this type may develop with an emphasis on relating, understanding, or explaining presently-known facts, in which case it may become Cluster 3 (Theory as Explanation). This may itself become the source of theory Cluster 5, Type 7, and even, potentially, Type 8.

When theory is in contrast with fact, there is commonly a denigratory connotation: *mere* theory, i.e. *being* theory, is itself seen as suspect— ". . it's *only* a theory, not a fact." This disparaging usage occurs for instance, when opponents of Darwin fail to grasp that the Theory of Evolution is Cluster 9 (Scientific Theory)[1], and misconstrue it as falling under this Cluster 1 sense, saying that it is merely a theory not a fact. Such people are mostly ignorant of the variety of meanings of "theory", but some are deliberately equivocating[2].

T.W.Moore, in discussing the attitude of some teachers to theory says that theory, ". . having to do with proposals which may conflict with well-tried ways. . is often dismissed by them as 'mere theory', as opposed to common sense practice"(1974, p.2). His statement is an interesting example of the subtle way the meanings of the word can shift. For the phrase, "mere theory" would more often be used to contrast with "fact". But Moore, in a somewhat less common usage is here contrasting it with practice. His example leads to the second cluster.

Cluster 2 : Theory Contrasted with Practice

2. *In contrast to practice, or to that which can be applied* : OPPOSITE OF PRACTICE. An experienced high school teacher may argue that what student-teachers need is not more theory from university educationists and researchers, but more practice: "Your college courses are too theoretical". In this

---

1. See Chapter II, for details of Scientific Theory.
2. See, for instance, W.W.Wassinger's letter in the *New York Times*, on December 15, 1989, and the replies on 5th January 1990.

instance, the word, "theory" denotes thought or thinking, the word, "practice" denotes action or doing. Though denotatively the theory-practice contrast is emotively neutral, the use of the word, "theory" in this context, as with theory Type 1, may sometimes carry critical or derogatory connotations.

3. There is a somewhat different emphasis when *theory is seen as a collation of instructions and general principles* : INSTRUCTIONS TO BE APPLIED. A chemistry student learns the theory of analysis to be applied in the laboratory. The supervisor says to the apprentice mechanic: "The *Automobile Manual* will give you the theory, but the workshop will give you the practice." The apprentice mechanic may refer to his understanding of the contents of the automobile workshop manual as "theory" and say that he finds difficulty in making use of some of this theory when he is faced with an actual car, even though he knows that the theory is O.K. A student-teacher may say that she can understand the general principles of Behaviorist Learning Theory, but that she finds difficulty in applying this theory in her teaching.

Older textbooks in the principles of teaching usefully listed such "maxims of method" as: proceed from the known to the unknown, the simple to the complex, the whole to the part, the concrete to the abstract, the particular to the general. Such clusters of principles provide simple pedagogical examples of theory Type 3. In our own times we have had manuals of micro-teaching which have done much the same job, and those empiricist researchers who adopt the Process-Product approach would seem to be striving to produce more-specific theory of this kind[3]. Examples of the theory of pedagogical research aimed at practice are the multitude of manuals which offer "how to do it" advice, discussing such topics of empiricist research as, methods of observation, quasi-experimental design, analysis of variance, and so on.

Because Type 3 involves principles and instructions, that which is here being designated as "theory" is expected to have application in the empirical world. In this sense "theory" refers to collected thought expressed in propositions, principles and instructions and in other ways such as diagrams, flow-charts and graphs; "practice" refers to application of the propositions or principles, or to action on the basis of the instructions. There is no implication here that such theory is "impractical, or "mere". Indeed the opposite is normally implied: such theory is either the epitome of practicality, or at least believed to be so. The instructions and diagrams in the *Automobile Manual* apply to the automobile in front of the apprentice. The instructions in the *Manual of Educational Research* are to be applied by the neophyte researcher.

Cluster 3 : Theory as Evolving Explanation

4. *Theory as making some field clearer* : EVOLVING EXPLANATION. This Type is a generally inclusive, potentially checkable (by observation, commonsense argument, or logic), developing attempt at an explanatory account

---

3. See, for example, studies 1, 4 and 6 in Chapter IV.

of some field of interest or concern, or of some group of facts or data. Theory in this sense is deliberately produced in order or to show how it throws light on the matters of interest or concern, to see if such matters can be consistently and coherently fitted into it, or to see how they can be developed with its aid. It is potentially open and evolving.

There is no implication here that such theory is merely a hunch: it is not Cluster 1. It is continually developing and accumulating and grows out of an increasingly complex understanding and clarification of the field of concern to which it contributes by reflexive reflection.

In philosophy of science, the stage theory of Kuhn (1962)(discussed in Chapter II) is a good example of Theory as Explanation, as is Giere's (1988) more recent cognitive theory of the development of science. In educational scholarship, besides the theory of history of and philosophy of education, there are many other good examples of this kind of theory, especially in psychology and sociology. Three of the best known specific examples are Bruner's Theory of Instruction, Piaget's Theory of Cognitive Development, and Kohlberg's Theory of Moral Development. An excellent recent example is the "Erotetic Theory" of teaching of C.J.B.Macmillan and James W. Garrison (1988)[4]. Their theory aims at being both empirically rigorous *and* intentional, and construes teaching as the implicit or explicit asking and answering of questions about some subject matter. In showing that the key to understanding teaching is not a matter of recognizing a constant conjunction of separable As and Bs, but rather recognizing a set of logical relations, erotetic theory provides a logic of explanation which attempts to be as rich as teaching itself. The intention of Macmillan and Garrison is that erotetic theory will eventually make possible a truly scientific approach to pedagogical research. If it does, we shall have a powerful pedagogical example of theory Cluster 3 becoming theory of Cluster 9.

It is worth noting that theories such as those of Bruner, Piaget, and Kohlberg are general, in one sense of that word, but as they developed they reached a point where they did not become more precise and succinct in their integration and abilities to account for an increasing range of empirical data: in order to cover the odd cases, they became more diffuse. This feature is an important contrast to that of Scientific Theory.

Cluster 4 : Practical Theory

5. Professors at conferences such as foundations of education or of educational research can be heard to say such things as, "If teachers have superficial views of the nature of educational theory, they will have superficial notions of the significance of theory for their work." In the context of this book, one is con-

---

[4]. The theories of Bruner, Piaget, Kohlberg, Garrison and Macmillan, and so on, also form a fundamental part of Cluster 4 (see the following section). The empirical parts of such theories may equally lie in Cluster 8 (Empiricist Theory).

cerned with *Theory of Teaching, or, Pedagogical Theory* : on the present categorization such a concern will be a variety of PRACTICAL THEORY.

Whereas the function of Cluster 3 is explanation, the function of Practical Theory is to determine practice. Reasons, as well as empiricist and scientific claims enter into it ; intentions as well as causes are involved [5].

The distinguishing logic of pedagogical theory as Practical Theory is that it is a composite. It involves not just empiricist, or technical, or scientific knowledge, but additional forms of knowledge, e.g. from Cluster 7— of moral norms and oughts, of religious claims (in church schooling), as well as (from Cluster 3) historical, aesthetic, and philosophical dimensions of human relationships, psychological accounts such as Bruner's or Kohlberg's theory, and so on. In other words, it also contains examples of what has sometimes been called *an* educational theory. The interest is in what should be taught in schools, and how to teach it, and there are arguments supporting both the should and the how. The unity of pedagogical theory as Practical Theory is provided by the consistency and interrelationships of the set of principles underpinning the what and how. Implicit or explicit in all this, there will be ideas about how to act rationally in the domain of teaching: in terms of objectives, aims and goals to be reached, of principles of procedure to be followed, and of values both substantive and procedural to be maintained.

P.H.Hirst (1966 ; 1972 ; 1974) has nominated medicine, politics, navigation, and engineering as other theories having similar logical form to pedagogical theory-- to which one might add groupings such as home economics. Such Practical Theory is not to be confused with Cluster 2 (Compared with Practice) above, because it *involves* what is usually called "practice", and is not merely a matter of already-given principles to be applied.

Pedagogical theory as Practical Theory is continually being elaborated and learns from its own activities which feed back into it in a cyclical fashion. Both intellectual reflection and practical judgment are required. It certainly involves propositions, but is not merely a set of propositions. It is an on-going, living, working theory, which as such writers as Schon (1983; 1987) have pointed out, involves practice, and reflection on that practice. It is a continual reflexive evolution of means and ends-- a situation impossible on the approach of "technical rationality" (which has been assumed in such empiricist research as process-product research on teaching) which sees rational behavior as the taking of means to previously determined ends.

Its present form would seem to be a mixture of facts, rules, precepts, pedagogical traditions, professional wisdom, assumptions about the nature of learning, ideologies about the best way to teach, psychological and sociological

---

5. As I argue in Chapters VIII and IX, an improved empiricist research, a research which takes into account most of the important aspects of the pedagogical world, not just those aspects which happen to be amenable to the present observational and statistical approach, must take into account not causes alone, but reasons also. Indeed, reasons will be much more significant. The same point will also apply on several levels in any other variety of adequate Practical Theory.

claims, beliefs about the nature of persons, social norms and values, philosophical argument, general principles, socio-political presumptions, and so forth. Despite its conglomeration of different disciplines and logical kinds its point is to *help teachers and administrators decide what ought to be done in their classrooms and how it should be done*. It will therefore involve many of the other types of theory, perhaps types from every other cluster– even Cluster 1. For instance, empiricist pedagogical researchers intend their findings to be included in pedagogical theory in this sense, as theory from Cluster 2, Type 3 (Instructions to be Applied), and concomitantly, they hope, as theory from Cluster 9 (Scientific Theory). Gage, with clear intentions in both these directions, called his book, *The Scientific Basis of the Art of Teaching* (1978).

There is an important qualification. If pedagogical theory is to be Practical Theory in the sense that, say, medicine and engineering are Practical Theory, then not just anything is acceptable. At the same time that it is logically complex, *it should be logically rigorous*. To adapt some ideas from Moore (1974), insofar as it involves scientific claims, these must be checked against the relevant Scientific Theory (Cluster 9); insofar as it makes empiricist claims, these must be tested by appropriate observation; insofar as it involves norms and values, these must not be immoral, and they must be consistent; and insofar as it is a developed and extended argument, it must be non-contradictory and internally coherent.

Cluster 5 : Theory as Hypothesis

6. *Theory as* : MODEL. Sometimes in pedagogical talk, the words, "theory" and "model" are used interchangably. "Model" normally refers to the way in which an explanatory idea, system or concept from one area is used as an analogue to suggest construals or procedures in another. A model is a sort of bridge between the familiar or known and the less familiar or unknown. Consider for example: thermostatic construals of biological function; Darwin's translating of Malthus's idea of explosive growth among human populations within a limited environment to that of the biological world generally (Darwin, 1958; Mayr, 1982); the view of society as an organism; the idea that educational growth is similar to the growth in a plant or animal (Dewey, 1950; Chambers, 1989). This sort of theory is called a "model", because it extrapolates, some *particular and restricted aspects* from the one and applies them in the other, seeing what would then seem to follow for, or what insights might be suggested about the other. That which seems to follow or be suggested may become theory of Type 8 (Hypothesis), and it is hoped, eventually make a contribution to theory of Type 14 (First-order Scientific Theory).

Perhaps the most famous sequence of models in the history of scientific research was that of the nineteenth century British physicist, James Clark Maxwell, as he developed his equations of electromagnetic radiation. His sequence shows the movement from Type 6 to Type 8 (Hypothesis) and through to Cluster 9. As Oldroyd points out (1986, p. 284), Maxwell initially construed of electricity as a sort of invisible fluid. From that model, he produced theory of

Type 8 (Hypothesis), which through mathematical extrapolation, became his first inverse square law for electrostatics. Then he changed the model to rotating tubes of magnetic force, with small particles of electricity supposedly occupying the interstices between the tubes. Using this model, he developed further theory of Type 8, and then the equations to account for the propagation of electromagnetic waves. By then he had a set of equations which could be used to describe and predict the behavior of electromagnetic radiation. Thus the equations themselves finally became the most succinct account of the Scientific Theory of electromagnetic radiation (Cluster 9).

Sometimes theory as model is more literal, as occurred in the research of Crick and Watson on DNA , to be discussed in Chapter II, where they built an actual 3-D model of the molecular structure.

7. *Hopeful hypothetical idea* : HEURISTIC[6]. A common example of this kind of theory is provided by the the professor of the course, *Empirical Research IN1992* , when he says, "You require a theory or hypothesis, if you are to do empirical research on teaching." Thus, in discussing the testing of the effects of praise, Winne writes, "Note that the variety of this line of reasoning is limited to cases in which the researcher's theory is not supported" (1987, p.10). Slavin suggests that any scientific investigation begins (sic) with the statement of a, "theory or hypothesis", i.e., ". . a formalized 'hunch' about the relationship between one or more variables." Slavin writes that such a theory or hypothesis might be, "Use of daily mental arithmetic drills will increase the mathematics performance of fifth graders more than daily written drills [will] " (1984, p.11).

Here "theory" (despite the kudos-seeking use of the terms, "hypothesis" and "scientific investigation" by such writers) is merely referring to an heuristic for proceeding and for establishing *empiricist* knowledge. It is not an hypothesis in the scientific sense, for it does not derive from Scientific Theory (Cluster 9), nor is it involved with the powerful implications, interrelationships, and additional derivations of such Theory.

Theory as Heuristic is used particularly in the social sciences and empiricist pedagogical research, which, as just emphasized, unlike the following type, Type 8 (Hypothesis), lack supporting theory from Cluster 9. Where then does such theory as Heuristic come from? Heuristic suggestions, such as those by Slavin or Winne, listed above, or by Gage in his misnamed, *The Scientific Basis of the Art of Teaching* , have to come from theory in one of the other senses developed here-- from Cluster 3 (Explanation), or Cluster 7 (Normative Theory), or just, as is usually the case, from earlier empiricist pedagogical research, and ordinary, pedagogical experience, with its generalizations and commonsense (i.e. from Empiricist Theory). The point is to see if the Heuristic can be supported by way of observational-statistical, or experimental-statistical testing[7].

6. I owe this use of the term to a suggestion in a conversation in 1988 with Dr. James W.Garrison.
7. Theory as Heuristic is also different from Type 1 (Hunch or Speculation), because "theory" is not here being used in contrast to fact. The concern is not that such a theory is not a fact-- the empiricist researchers are not interested in its logical status.

8. *Theory as* : HYPOTHESIS. Theory in this sense *grows out of a background of Scientific Theory* (Cluster 9). It is a scientist's deep and detailed understanding of his material which in some way prompts the hypothesis, and which gives the hypothesis some chance of being acceptable. An hypothesis must not merely show imagination, but be apt and scientifically informed. Thus, late in 1952, Crick and Watson, in their evolving attempts to develop the correct account of the structure of DNA, might have said, "At this stage our theory runs as follows: DNA is a double helical structure, which consists of identical base sequences, held together by hydrogen bonds between pairs of identical bases-- upon that supposition, we are basing all our present work," In 1965, a theory which grew out of earth science theory was that there would be discovered in Brazil, a continuation of a geological formation across the Atlantic in Ghana. This theory was found to be correct and so made a contribution to the earth sciences more generally, i.e. became Cluster 9 theory[8].

Another easily understood example was the theory that there might be a small, fast-moving planet, referred to as "Vulcan", between Mercury and the Sun. This was an attempt to explain with the aid of Newtonian Gravitational Theory (Cluster 9), why the planet Mercury precesses at perihelion[9]. Consistency within Newtonian theory was the only reason why such a theory as this (Type 8: Hypothesis) could arise at all. Though there was a claimed sighting of "Vulcan", this theory has not found general empirical support.

Here "theory" refers to a presumably reasonable scientific speculation which is to be tested by experiment, or by seeing whether its intimations are in some other way borne out, while the hypothesis remains consonant with the general explanations and implications of that branch of Scientific Theory (Cluster 9).

Cluster 6 : Theory as Presupposition

9. *Theory as* : ONTOLOGICAL PRESUPPOSITION. The fact that scientific research can itself only proceed upon the basis of one or more ontological presuppositions, e.g. that an experiment repeated tomorrow will be essentially the same as one performed today, is often forgotten. Scientists may be quite unaware of these, but they logically must be there. They affect the way explanations develop, the kinds of questions asked, and the matters they will construe as problems; moreover, by suggesting particular kinds of thinking and not others, they will help scientists to inquire in some directions, but restrict or entirely prevent inquiry in other directions.

Does the order of the universe proceed upon the basis of "substances" which are determined in a teleological manner and reproduce qualitatively, or is the basis always some variety of quantitative change of matter in motion ? Was the

---

8. For both examples, see discussions in Chapter II.
9. The closest point (perihelion) of Mercury's orbit to the sun slowly moves around the sun, so that the orbit is actually a series of loops such as a point on the wheel of a trick circus bicycle might trace out. This phenomenon is known as precession at perihelion. (See also the discussion on p 184.)

English physicist, W.Thomson (later Lord Kelvin) correct in assuming that when,". . you can measure what you are speaking about and express it in numbers you know something about it; but when you cannot measure it, when you cannot express it in numbers, your knowledge is of a meagre and unsatisfactory kind" (Thomson, 1891, p.80). Was the American empiricist researcher, E.L.Thorndike justified in assuming that if X exists, then it exists in some amount and if X exists in some amount then it can be measured? Or did Cervantes, Boccaccio, Dostoevski, Shakespeare and Ts'ao Hsueh-ch'in show that they knew some unmeasurable things much better than Kelvin and Thorndike could ever conceive of?

There was a long controversy between Cartesians and Newtonians in the seventeenth and eighteenth centuries over whether it was acceptable to appeal to action at a distance-- which is of course presupposed by Newton's theory of universal gravitation. Another significant scientific example is the derivation of Galileo's Theory of Circular Inertia which was eventually developed into Newton's Theory of Rectilinear Inertia[10]. Galileo's idea of circular inertia followed in a presupposed and quite unexamined manner from the ancient and medieval idea that circular movement was both the natural and the most perfect movement. The theory of conservation of matter is both historical and contemporary theory of this kind. It is assumed in all chemical experiments and activities, and implicitly and explicitly played a significant role in Lavoisier's revolutionary work with oxygen[11]. The presuppositions of stabilism are a splendid modern example: for the great majority of earth scientists, stabilism functioned as the unquestioned *that into which* all evidence had to be fitted and *that with which* all hypotheses had to be made consistent. Right into 1966 stabilist papers were being produced-- at the precise moment when decisive evidence for continental drift was being produced by scientists in the very same geological institutions[12].

Modern social scientists and pedagogical researchers who search for causes rather than reasons, or who study observable movements and ignore human intentions and actions are implicitly using a Behaviorist presuppositional theory of this kind. The idea of empiricist researchers that, for example, it makes sense both to isolate aspects such as praise from a total teaching context[13], and also to combine statistically the praise of different teachers in different classrooms at different times is also based on, largely unexamined, Theory as Ontological Presupposition.

Later in the book, I argue that thought experiments are more important in scientific research than most people realize, and that pedagogical researchers who would wish to be scientific in their practises should pay more attention to them.

10. See Chapter II.
11. See Chapter II.
12. See Chapters II and VII for discussion of the theoretical (Cluster 9) and social dimensions respectively.
13. See Chapter VI for a discussion of this inappropriate "atomistic" reduction.

Thought experiments rely upon presuppositional theory: in order to proceed with them, scientists just have to presume some ontological certainties.

10. *Theory as Presupposition by which persons are enabled to "observe-as"*: OBSERVATIONAL PRESUPPOSITION. Observations require mastery of particular structures of symbolization, belief, or knowledge, which make it possible for the initiate to see things which others cannot. A good introduction to this idea is provided by a quotation from Hanson, who first drew our attention to this phenomenon in a scholarly manner, "We simply see what the time is; the surgeon simply sees a wound to be septic; the physicist sees the X-ray tube's anode overheating" (1972a, p.17). Initiates "observe-as": they observe the figures and the position of the hands as the time, they observe the shape, color and texture of the wound as septic, they observe the interior of the tube as anode and they observe the color and brightness of the anode as overheating. Those who have not mastered such theoretical understanding cannot *observe* such aspects of the universe. It is not a matter of interpreting differently, it is a matter of observing differently, of *observing-as* .

Here "theory" means something like interrelated set of epistemological presuppositions, or orienting principles, beliefs and knowledge, which we bring to our having of experiences, and which moreover make them possible. Commonsense experience and understanding are also presupposed here. That we are born into a complex culture and slowly grow up in one, actively and passively acquiring its artifacts of conceiving the world, hides the fact of the conceptualized theoretical (in this sense) nature of all observation.

It should be pointed out that serendipitous discovery[14], such as Galvani's jerking frogs' legs, or Roentgen's glowing barium platino-cyanide screen, presumes "observing-as". The researchers' conceptualizations are embedded in a way of observing. Because of this, they grasp that there is something different, or new, or odd, or anomalous, or unexplained. People who lack such theory necessarily miss the phenomenon.

Some educationists are so irremediably immersed in their behaviorism or critical theory, that they can construe classrooms only in ways revealed, or made possible, or presumed by the theory. For much of the twentieth century the words, "educational research" *meant* statistical (i.e. empiricist) studies. Non-observational, or non-experimental, or non-statistical work was just *not observed as* research. Because of the presence of theory Types 9 and 10, relatively simple empiricist views, such as those of contemporary Process-Product research on teaching are necessarily circumscribed and question begging.

Theory in these Cluster 6 senses would also seem to be a logically necessary prerequisite for theory in all of the other senses. In these senses of Theory as Ontological or Observational Presupposition, all human activity, including scientific research is as Hanson (1972a) puts it, *theory-laden*[15].

14. See Chapter VII.
15. It is argued at the end of the following chapter, that this feature does not make scientific research unrescuably subjective, as too many scholars, particularly sociologists, unfortunately still want to claim.

Cluster 7 : Normative Theory

11. *Subjective Norms* : DOCTRINE AND DOGMA. It would be inappropriate not to recognize the immense significance in human lives, of various kinds of religious, moral and political positions which draw upon greater or less consistently and cogently argued theories of norms and values. Such normative theory functions as an unchallengable given. As clear cases one may cite, say, Naziism and Mormonism, the one based upon false claims to scientific premises to do with race, the other upon abnormal empirical premises to do with the discovery of golden tablets in a hill near Palmyra, New York state, tablets which Mormons claim, after having been deciphered by Joseph Smith, just dissolved out of existence. But other well-known political, moral and religious positions also show highly debatable premises and/or unchallengable axiomatic first principles.

The theory of doctrines and dogmas also underpins many pedagogical positions. These may be religious, or quasi-religious, or politically inspired, as in Thomist bases of Roman Catholic schooling (Maritain, 1955, 1963), in the argument about divine unfolding in Frobel's theory (Frobel, 1887; Lilley, 1967), and in the Marxist-Leninist-Maoist foundations of Communist schooling (Makarenko, 1951, 1954); or they may be secular, as for instance A.S. Neill's quasi-Freudian theories (Neill, 1944, 1962/1968).

12. *Normative theory with a rational basis* : RATIONAL NORMATIVE ARGUMENT. By "rational normative argument" is meant any clearly argued value theory whose proponents allow it to evolve under the pressure of rigorous criticism, and strive valiantly for internal consistency and logical and, where appropriate, empirical check and falsification in principle. I have in mind political theories such as liberal-democracy, the theories of justice of, say, Rawls (1972) or Nozick (1974), and normative ethical theories such as Utilitarianism and Kantianism.

I also have in mind the kind of developed argument which may be termed *an educational theory*, such as that of Rousseau's individualism (1762/1966), or James Mill's Utilitarian theory (Moore, 1974), or John Dewey's theory of democratic education (1950). Each of these thinkers offers guidance on producing a certain sort of person and/or a certain sort of society. Each says something about the nature of learners. Each discusses to some extent the ancient question of what sorts of knowledge are most worthwhile[16]. Each theory provides a developed body of principles and values for teachers to follow. Each gives us arguments to show why his conclusions are *worthy* of putting into practice. And each, ". . involves a persuasive attempt to secure acceptance, not only of pedagogical recommendations, but of the social aims written or implied in the theory" (Moore, 1974, p.47). Such theoretical positions may often be construed under Cluster 3, and are also of considerable significance in the development of

---

16. For an ancient view of this fundamental pedagogical issue see Aristotle, 1953; for a contemporary one, see Degenhardt, 1982.

the kind of composite pedagogical theory, designated Practical Theory, above.

Arguments for the various fads and fashions of pedagogy: project method, I.T.A., open plan schools, open education, team teaching, teaching machines, programmed learning, behavioral objectives, mainstreaming, bilingual education, multiculturalism, computer literacy, and all the others which endlessly displace one another from year to year, show aspects of Types 11 and 12, as do the arguments in favor of present-day empiricist pedagogical research.

Obviously it is often difficult to distinguish clearly between types 11 and 12. Where I see A.S. Neill's writing almost exclusively as a grossly confused version of Type 11, some observers might see his work as Type 12; and some aspects of such writers as Rousseau, Mill and Dewey are doctrinal in the Type 11 sense. Consider also recent so-called critical theory in education. The writings of many modern critical theorists of education, as demonstrated by their intolerant reaction to criticism of their own work, falls into type 11[17]. The writings of other critical theorists, who welcome critique, and whose accounts evolve in response to it, falls into type 12.

Cluster 8 : Empiricist Theory

It is now time to examine that Cluster of theory, Empiricist Theory, which, as with Scientific Theory forms a fundamental category in the argument of this book.

First, the terminological point made in the Introduction should be repeated. The word, "empiricist" is used here to contrast with the word, "scientific". It is argued that both empiricist activity and research and scientific activity and research are empirical. In being empirical, propositions of Empiricist and Scientific theory can be contrasted with, say, logical/mathematical, or moral, or artistic/aesthetic, or religious, or philosophical propositions, or propositions about our own and other persons' minds. Both Empiricist and Scientific Theory attempt to describe and to explain the world of empirical phenomena. But as will be argued in this section and in other places, they deal with the empirical world in significantly different ways.

For most of our history, most of mankind's empirical knowledge has been empiricist. The builders of the Pyramids, of the Great Wall of China, of Greek temples and Roman aqueducts were empiricist, not scientific. The same is true of, say, the Australian aboriginal makers of boomerangs, the mediaeval smiths of Nigeria, the Chinese discoverers of the magnet, gunpowder and rockets, the ancient Peruvian makers of bronze and of copper-arsenic alloys, and the classic swordmakers of Japan.

These people worked with *general concepts defined by* observed sets of objects, and their instructions and explanations were empiricist *generalizations* – propositions which involved the general concepts and whose acceptability was established inductively *by trial and error from their observations and experience* .

---

17. See for instance Henry Giroux's (1988) intemperate, ad hominem response to analytic criticism of his work by Francis Schrag.

Such people generalized on the basis of their experience, and embodied their generalizations in rules of thumb, procedures and traditions. Those who were literate wrote down these procedures and rules as a sequence of interrelated propositions. *It is bodies of such propositions, their oral expression, and the concomitant tacit knowledge (Polanyi, 1958) connected with them, which I am calling, "Empiricist Theory"* .

Consider the invention of gunpowder. This was not a scientific development (as is so often claimed), but came as a byproduct of China's chimeral empiricist quest for a pill for longevity for the emperors. Sun Simiao, a noted alchemist/physician of the Tang Dynasty wrote the following prescription for making immortality pills:

> Place two ounces of ground sulphur and two ounces of ground saltpeter in a pan for frying. Ignite three gleditsia pods and throw them in to make the mixture burst into flame. When the flame dies down add three jin (1.5 kilograms) of wood and the same quantity of charcoal and fry again. Remove from the fire when the charcoal has been reduced by one third (*China Science*, 1983, pp.18-19).

This is actually a recipe for gunpowder! Or consider the intriguing empiricist theory for the making of swords in Japan, which has been going on to the present day in the same way since around 800 A.D. As Bronowski points out, such swordmaking, as with all metallurgy until modern times, surrounds itself with ritual-- for good reason. If there is no written language or accurate chemical formulae, ". . then you must have *a precise ceremonial* which fixes the sequence of operations so that they are exact and memorable" (1973, p.131). The ceremonial fixes the order of the empiricist features to be observed in the production.

In Europe, Medieval masons and builders had a sort of "empiricist technology" which used general concepts and was again built up through trial and error, but except for their use of elementary arithmetic and geometry, they had no abstract concepts-- the kind of concepts involved in scientific research, and in the modern scientific technology which is based upon it. They had craft wisdom, traditions, custom, rule of thumb. Verticals were fixed by a plumbline, horizontals by a plumbline joined to a right-angle. Thus, while most of the arches and vaults in their Gothic cathedrals held successfully, some such as Beauvais at 150 feet did not (Bronowski,1973, p.110). In 1284, it came crashing down. Yet as Bronowski points out, the empiricist design at Beauvais may well have been sound; probably the ground shifted under the weight of the building. The problem was that the builders could not calculate any of the stresses. The only guides and the only "explanations" such masons and builders had were empiricist generalizations, such as that this *sort* of arch, using this *kind* of stone with this *volume* and *shape* , at this *approximate height* and at *about this width*, had been observed *generally to hold in the past*. It is revealing that mediaeval builders never again attempted to achieve the height of Beauvais. Their *empiricist experience told them* (falsely) that they had reached the limit.

Yet it has been precisely the confusion of such empiricist "technology" and scientific technology based upon the laws and theories of science that has continually occurred in history textbooks and continues to be made by intelligent laymen and by most social scientists—including most pedagogical researchers. As Voltaire accurately and perceptively described the situation in 1734, "It is to a mechanical instinct, which exists in most men, that we owe all the skills, and not to a sound philosophy"(Voltaire,1980, p.59).

In like empiricist manner, "The brewer learned from long experience the conditions, not the reasons, of success" (Tyndall quoted in Conant,1947, p.41); and even today, despite great progress in scientific analysis, there is still a component of the empiricist in beer brewing. The same empiricist approaches apply with varying degrees of precision in much of what people do in everyday life to this day, from that of the master builder applying his arithmetical rules of thumb in using suitable timber beams, to that of the housewife cooking at 270 C in preparing her dinner. For instance, the general empiricist rule used by carpenters for deciding how thick a floor joist should be, is as follows. Divide the length of the space to be spanned, by the width of the timber in inches, and add one inch. Thus if eight feet are to be spanned with basic two-inch timber, the width of the floor joists should be $(8 \div 2) + 1 = 4+1 = 5$ inches; so we need joists 2 by (at least) 5 inches, by 8 feet. This is an empiricist rule which has been developed after the coming of the scientific revolution and after people had become used to elementary arithmetic and measuring in their everyday lives. To the extent that the rule makes use of arithmetic, to that extent is it beginning to use an intellectual tool which is crucial to scientific research. But that does not turn the rule into a scientific formula. For the arithmetic provides merely a veneer of precision: although the arithmetic is precise, the resulting guidance it provides for the joists is only "more or less" and approximate, as are other empiricist rules and principles. With longer spans, both the rule (and the resulting joists!) break down, and the carpenter knows this— through trial and error, customary usage, and craft wisdom. To turn the rule into a scientific formula, we need to introduce abstract concepts embedded in Scientific Theory-- the conceptions of scientific engineering with its stresses and strains and Hooke's Law and Young's Modulus.

Empiricist concepts are *defined by* observable entities in the empirical world, or by observation-based operational definitions, not by abstractions. Both the empiricist concept and the generalization which involves the concept build up as more cases are observed. Empiricist propositions are *general* and *relatively imprecise*. Consider as examples the concepts used by Helmke and Schrader in their study which is discussed in Chapter IV, concepts such as "seatwork", "monitoring", "recall question", "teachers' diagnostic competence", "supportive contacts", and their propositions (generalizations) such as, "Achievement was highest when both frequency of teachers' supportive contacts with students and their diagnostic competence were above average".

Empiricist knowledge is gained through generalizing on the basis of sensation and experience of observables. The advent of modern statistics has also made possible the development of an alternative, complex, and systematic (in one sense of that word) variety of empiricist knowledge. But the logical status of the generalizations remains the same. Although the statistics provide decision-making techniques not possible with commonsense, and thus help to produce generalizations in one kind of specific, systematic, empiricist manner, unavailable prior to their development, this does not change the logical status.

Not merely human beings, but animals also can develop empiricist knowledge (though I do not see how they could produce Empiricist Theory). This is perhaps most clearly so in the case of chimpanzees, who can even learn the use of rudimentary tools. But human beings have a vast advantage in that they can communicate their empiricist knowledge to one another by the use of the propositions of language, write these down, make them consistent, test them through experience in the empirical world, and thus produce Empiricist Theory.

The most extreme philosophical accounts of empiricist knowledge have been those of the so-called British Empiricists, such as Locke, Berkeley and Hume, who argued that we gain all our knowledge on a sensory basis. Locke insisted that the most that can be achieved is a collection of generalizations about the association and succession of phenomena. At best such generalizations are probable, not necessary. As Hume says, ". . all our ideas or more feeble perceptions are copies of our impressions or more lively ones" (1956, p.17). In the twentieth century this philosophical tradition has been taken up by logical positivists such as Ayer and Carnap and by logical empiricists such as Russell. What is not as often realized however, and what is significant for the present book is that the same tradition finds expression in the systematic work of Mill in the nineteenth century, and the late nineteenth century, early twentieth century British statistical school of Galton, Pearson and Fisher. It is significant because these four persons are the founding fathers of most of the complex empiricist methodology employed in present-day empiricist pedagogical research[18].

According to Hume, knowledge of causal relations cannot be the result of *a priori* reasoning, but rather the notion of cause is merely the result of repeated observation of one object or event followed by another. Hume defined a cause as an, ". .object followed by another, where all the objects similar to the first are followed by objects similar to the second," (1956, p.82). Knowledge of cause is thus a matter of sensation and habit. Moreover, while, because of our direct sensations we may be certain that *this* X is followed by this Y, the question whether X is *generally* followed by Y cannot be certain because we cannot observe the future connections. Thus Hume seems to be claiming that what he construes as science begins with sense impressions and can involve only those concepts which are somehow constructed out of sense data. Causal knowledge is a knowledge of contingent association of two classes of events. Statements about causes are formed through induction by enumeration.

This view can be seen to provide a method, i.e. it provides both a way of investigating empirical phenomena, and a sort of explanation of the connections discovered. To investigate the causes we categorize the objects or events X and

---

18. Their thought will be considered further in Chapter III.

the objects or events Y, *on the basis of their observed similarities*. Again, the relationship between X and Y is either observed naturally, or observed through some kind of manipulation. Moreover, having observed a psychologically sufficient number of instances of X being constantly conjoined with Y, we say X causes Y. Our explanation then is that seemingly *being so constantly conjoined* X is the cause of Y. This is the logic of Empiricist Theory. It is thus also the logic of present-day empiricist research on teaching. However, in empiricist research *the connection or "constant conjunction" being based upon statistical tests, or statistical correlations, is therefore considerably looser— but of the same logical type.*

As we shall see in the following chapter, *explanation* by way of empiricist connection is different from explanation using Scientific Theory. Empiricist explanation is the result of observation, or of statistical manipulation of propositions which involve observational concepts. Empiricist Theory is concerned with an interconnected and experientially-consistent body of empiricist generalizations and explanations, including claimed observed causal connections, and claimed calculated statistical relationships, between one proposition involving observational concepts and a second proposition involving other observational concepts. As Willer and Willer nicely express the matter in their short, but devastating critique of empiricist sociology, Scientific Theory and Empiricist Theory differ,

> .. in the way they transcend particulars in individual cases. [Empiricist Theory] transcends particulars by generalization, in which a name is applied to a set of similar objects, forming an [empiricist] category defined by those objects included within it . . and then is related to another, similarly defined, [empiricist] category through an observational process (italics added) (1972, p.22).

Consider the following empiricist explanation of my cleaning the bottom of a swimming-pool. Suppose I am having increasing difficulty in using my pool-pole to clean the bottom of the swimming pool, the farther I have to reach towards the middle of the pool. The difficulty arises because the pole seems to be becoming more and more bent, and I have to make adjustments for the seeming shortness and bend in the pole. I have known of other manifestations of this phenomenon since childhood and after discussing it with other people who have had similar experiences, have concluded that it may be generalized. So I expected this to occur because sticks do appear to bend in water, and the farther one pushes them away the more bent they appear to become. This is after all a pool of water, and I expect sticks or poles to appear increasingly bent the farther away they are poked into water. This is why I am having trouble cleaning the bottom. "Poles, like other rigid objects appear to be bent when placed in pools of water," and, "The farther away poles are pushed, the more bent they appear" are propositions of Empiricist Theory.

It should be noted how this explanation and these propositions use empiricist connections. All the concepts used refer to observables: "water", "pole", "pool", "bent", "farther away". These concepts name a group of similar objects or features to form an empiricist category *which is defined by that set*. The objects or features of observation define the category to which the empiricist concept refers. The connection between propositions linking the terms ("pole" and "pool" connected to "water", "bent", and "farther away") is made because I and others have drawn a conclusion as a result of repeated observations of the phenomenon. The explanation is a general causal claim about sticks and water, and I apply it to the present particular case. It is logically the same type as, say, our explanation of why a red billiard ball is caused to move by its being struck by a white billiard ball. This kind of explanation is typical of those which we offer for everyday happenings. Such explanations are founded upon commonsense observation and observed repetitions. Even were we to attempt more precision and to make the situation seemingly more sophisticated by having a random sample of people push the pole, and then made use of statistical correlations linking particular movements and particular positions of the pole, the epistemological status of the explanation would not be changed. Moreover, and most significantly, no scientific law would result from such statistical analysis.

Useful knowledge does arise from empiricist connection-- from the accumulation of observations and generalizations to form the propositions of Empiricist Theory. This, as already emphasized was the basis of all "technology" until recent centuries, is certainly the case in ordinary life, and may be the case with some of the empiricist knowledge which has been produced in empiricist research on teaching. *Performing empiricist research, while believing one is performing scientific research does not mean that there have been no useful findings*. Indeed I believe that some of the research has been helpful in a pedestrian sort of way: it has reminded the pedagogical community of some very general, commonsensical principles of instruction-- principles which have been ignored by poorly-trained teachers, or have not been grasped by those educationists and researchers who have themselves spent only minimal, or no time teaching primary and high school children.

In normal circumstances we do not require scientific precision, or a knowledge of scientific propositions, or formulae, or laws, to be able to act in the world, such as to use a pool pole in a reasonably effective fashion. Moreover, most people are quite content with empiricist generalizations such as: poles, like other rigid objects, appear to be bent in pools of water. They construe this as a general fact or truth about the world, which does not require any further explanation. Generalized empiricist knowledge, which in its connected relationships forms Empiricist Theory, is sufficient and useful in most everyday circumstances for most people.

Perhaps the most interesting technical example of such Empiricist Theory, which is indeed useful, is meterology, even though weather forecasting is still notorious for being unreliable. Nevertheless its observational connections and generalizations, linked to the technology produced by modern scientific research, such as computers and global surveillance by satellite have resulted in a relatively powerful body of Empiricist Theory, which is continually improving and which can be used to make predictions, which, though not scientific are nevertheless usable in a general way. Indeed, because of the innovations of the recently-developed mathematical theory of chaos (Gleick,1987; Stewart, 1989) meterology may be at the point where it is developing Scientific Theory which relates abstract concepts. But we should note carefully that the computers and the satellites, which provide the advanced and complex data for this useful empiricist knowledge are themselves technology which could only be developed on the basis of Scientific Theory.

In short, for many of the purposes of everyday life, empiricist knowledge, empiricist connections, and more or less developed Empiricist Theory are sufficent, and in such cases the application of scientific research would be otiose. But that is not so where we *do* need scientific understanding, as in constructing suspension bridges, in producing pharmaceuticals, in building radio telescopes, or in sending satellites "Mariner" and "Galileo" to the far corners of the solar system. There, only Scientific Theory[19] will be good enough. It is not that one kind of theory and its knowledge are better *per se*. Indeed, that has been the ironic mistake made by too many empiricist researchers, who have believed that a scientific approach to pedagogy must be superior, and will solve all. *The point is that both Empiricist Theory and Scientific Theory have their legitimate place.*

Examples of the Confusion between Different Kinds of Theories

Distinctions between different Clusters of theory cannot just be ignored. Consistent failure to make the differences explicit leads amongst other things to the following interrelated problems: to systematic ambiguity in pedagogical discourse and confused thinking and writing, and to faulty pedagogical research. The second issue will be thoroughly discussed in the body of the book. Here I shall offer some quick consideration of the first.

Recall the example mentioned at the beginning of this chapter, where my colleague said the conference paper was highly theoretical, and another delegate said it entirely lacked theory. Clearly, they were referring to different Clusters of theory. But no progress could be made in their discussions until this confusion was removed.

. An excellent example of systematic ambiguity is provided by the quotation from Kerlinger near the beginning of this chapter. Kerlinger is using the word in

---

19. What I mean by Scientific Theory will emerge in the following chapters. Why the Empiricist Theory produced by present-day empiricist research is inadequate will also emerge. For one thing, Empiricist Theory in everyday life is conceptually exceedingly sophisticated (Austin, 1962, 1970) and takes a multifarious variety of subtle intentional action, refinement of language, and accumulated personal experience into account. In contrast, the Empiricist Theory produced by present-day empiricist pedagogical research is exceedingly limited, is conceptually naive, and has largely ignored our sophisticated knowledge of intention in teaching, and the complex ordinary language knowledge of experienced and perceptive practitioners.

numerous different ways:

Thus, "theories" on his line 2 alludes to Cluster 1, Type 1 (Theory cf. Fact);

"theorists" on line 3 demarcates scientists' work as Cluster 2, Type 2 (Opposite of Practice) (here used in the derogatory sense);

on line 4, "theory" is either Cluster 2, Type 3 (as Instructions to be Applied), or Cluster 3, Type 4 (Evolving Explanation);

on line 7 Kerlinger is clearly referring to some features of Scientific Theory (Cluster 9);

on line 9 he means Type 8 (Hypothesis);

on line 10 he is using "theory" as Scientific Theory (Cluster 9) again, though given his earlier uses above, and what he says on the following two lines, we may well wonder what *his* understanding of this is;

on line 11 and line 12, "theories" and "theory" may be Cluster 8, or they may be Cluster 3, Type 4 ;

with line 13, "theory" seems to be back with Cluster 9;

on line 15, "theory" seems a mixture of Cluster 9 and Cluster 3, Type 4.

Kerlinger acknowledges none of these differences, so what *he* means is hard to say. His account is pseudo-explanation.

This confusion is disquieting in respect to research on teaching, because the approach Kerlinger advocates is typical of the kind of research which has been the most influential of all forms of research in education. It is the approach which not only underpins the prominent, empiricist, Process-Product research on teaching, whose findings have been the basis for legislation of pedagogical policy in many American states and school districts, but in one way or another his approach has been imitated in a diverse range of educational fields from educational administration, to special education, to comparative education. But as the remainder of this book will argue, such theoretically-inadequate (many meanings) empiricist research cannot perform the task it sets itself.

It is now time to move to a consideration of the nature of scientific research and Scientific Theory. Only through having some conception of what has gone on in scientific research, can the inadequacies of present-day empiricist research on teaching be exposed, and can some suggestions for its amelioration be made.

CHAPTER II

SCIENTIFIC RESEARCH

To support the claim that empiricist research on teaching is not scientific, it is necessary to provide some succinct examples from research which is indeed scientific. I shall consider the astronomy of Copernicus and Kepler, the physics of Galileo and Newton, the chemistry of Lavoisier, Crick and Watson's discovery of the structure of DNA, and Wegener and the plate tectonic theorists in the earth sciences. It is by contrast with these accounts that inadequacies can be indicated. The scientific accounts are also of use in providing some indications of what would have to be done to change empiricist research on teaching, into scientific research. At appropriate places I shall also draw upon some fundamental work of the last thirty years in philosophy of science.

But prior to the accounts themselves, there needs to be a brief rejection of a possible meta-challenge. For it may be asked why my critique of the scientific pretensions of empiricist research uses only examples from the natural sciences and none from the social sciences. The answers are fourfold.

First, "science" has become an honorific and hortatory term. Not only are physics, chemistry, biology and geology called sciences and not only is it agreed that their practitioners make use of scientific methods, but universities also teach subjects, or areas, with such titles as agricultural science, computer science, criminological science, and library science— to mention but a few areas claiming the kudos and significance of science. The situation is confusing, the last mentioned for instance being merely a set of systematic techniques for recording, storing, and retrieving information. Again, one of the more prosperous businesses on Main Street says that it practices mortuary science. Marxists argue that their ideology is science – even though they always reject by redefinition and redescription, the many obvious real world empirical falsifications of their claims. There is a religion called Christian Science. And some Fundamentalist Christians promote a bizarre hybrid they call Creation Science. There is also a trend in some academic writing to refer to studies of the nature of science as "science of science", although these are really literary, or philosophical, or historical and/or empiricist analyses. This usage is unfortunate because it blurs epistemic categories.

The point is thus, that unless there is some deliberate restriction of the category, the term, "science" becomes so vague as to be useless for philosophical discussion. The word becomes as wide as all knowledge and belief. Its stipulated meaning is therefore here confined to what are often called the natural and biological and earth sciences.

Second, it is to such disciplines as physics, chemistry, biology and the earth sciences that both scholars and the man-in-the-street all agree the great successes have accrued, both in the cogency of their Scientific Theory, and in their ever-more-spectacular technological applications. Third, natural scientists *never* refer to the so-called social sciences, or pedagogical research, for ideas or inspiration, or to demonstrate the fundamental insights or effectiveness of their research. Fourth, in laying claim to the kudos of scientific status, social scientists and pedagogical researchers certainly *do* make claims that their methods are like, or at least on the way to becoming like those of the natural scientists. And they copy the titles: for example, pedagogical researchers at such institutions as the Center for Research on Learning and Teaching at Ann Arbor, Michigan call themselves "Senior Research Scientist", "Research Scientist" and "Assistant Research Scientist". In more moderate moments social scientists and pedagogical researchers may say that they are still searching for their Galileo or their Newton. In short the approach and status to which *social scientists and pedagogical researchers themselves aspire* is that of natural science.

For all these reasons it seems sensible to confine the examples to astronomy, physics, chemistry, biology, and the earth sciences.

Copernicus and *De Revolutionibus Orbium Coelestium*

After thirty long years considering the problem, Copernicus eventually reached the conclusion that the explanation of the universe should be reversed, the idea of a stationary earth should be abandoned, and the earth be seen as one amongst other planets orbiting the sun. Copernicus did not observe the earth's moving just like any other planet. The notion of the motion was theoretical (several Clusters). As Bronowski has pointed out, a fruitful way of construing scientific research is as a *search for hidden liknesses* in nature, liknesses which no one has *conceived of* previously. As he says,"The symbol and the metaphor are as necessary to science as to poetry" (1965, p.36)[1]. Suppose, just let us suppose, said Copernicus to himself, that the earth is moving around the Sun. What then would the universe look like? The Aristotelian-Ptolemaic account of the universe with the sun at the center had been accepted for 1500 years. In challenging this conception, Copernicus was beginning a revolution which was to overturn Aristotle and Ptolemy. He was, as Kuhn (1962/1970) would put it, developing a new "paradigm"[2].

As is now generally well-known in academia, Kuhn has suggested that sciences pass through stages. Though I believe, like Toulmin (1970;1972), Bohm and Peat (1987), and others, that Kuhn's overall view of scientific change is incorrect, aspects of his ideas have their usefulness. In the pre-paradigm state

---

1. Indeed, R.S.Jones has written a whole book called *Physics as Metaphor* (1982).
2. Kuhn did not originate this term. Philosophers had used it for decades to designate a central case of a concept, or for a type of argument which claimed that particular things must exist because particular expressions had a standard correct use in our language.

of a science, says Kuhn, there are several competing "schools", or else a theoretically-unconscious fact gathering– a natural history sort of Baconian garnering of observations and drawing of inductive generalizations[3]. Eventually one particular theory (Cluster 8? Cluster 9 Type 14?) begins to gather adherents and the discipline's first "paradigm" is established. Though Kuhn uses the term "paradigm" in a grossly ambiguous manner, the intuitive idea is obvious enough. A paradigm interrelates key concepts, methods of solving problems, mores of behavior by scientists, ways of introducing students to the discipline through laboratory sessions and text-books, and so on. For instance, the Ptolemaic paradigm which Copernicus inherited and challenged, claimed that the movements of heavenly bodies could be reduced to a complex of circles running on circles. Budding Ptolemaic astronomers had to master the techniques for the geometrical representation of movements, and ways of collecting germane observable data; they also had to master the astrological ideas of divination which at that time were still intertwined with astronomy, methods of working on the calendar, and navigation. With the establishment of a paradigm, says Kuhn, there occurs a period of "normal science", in which the potential of the the paradigm is being rendered actual. In normal science the paradigm is assumed to be the appropriate way of going about all the questions and explorations within that branch of science. Aspects of nature which do not fit are ignored, or explained away, or absorbed with the use of *ad hoc* hypotheses. Normal science consists of what Kuhn calls "puzzle solving"– working out the implications of regarding nature in the manner of the paradigm. On this view, what counts as an acceptable solution to a puzzle is determined more by what the paradigm allows, than by any independent relationship between theories (several Clusters) and observations. A scientist's failure to find a solution is thus construed as a failure of his own skill, rather than as an inadequacy in the paradigm. Normal science is successful for a time, but eventually the number of still unsolved problems, anomalies and *ad hoc* hypotheses build to overwhelming proportions, the paradigm is no longer satisfactory and a period of "crisis" occurs. Alternative paradigms are suggested, metaphysical and presuppositional underpinnings are hotly debated, thought-experiments become common. Eventually from the whelter of controversy a revolution occurs and another paradigm takes over, followed by another period of normal science and puzzle solving. Thus Kuhn sees scientific change as anything but an orderly accumulation and evolution. It is here that his claims are most controversial. On Kuhn's view, successive paradigms are "incommensurable", in that they cannot be compared on rational grounds– the same term, say, "planet" *means* different things in Ptolemy's paradigm and in Copernicus's paradigm. So, says Kuhn, the manner in which scientists change paradigms cannot be a rational matter, but has to be more like a religious conversion, or a *Gestalt* switch.

This was at least Kuhn's views in his earlier work. But in response to his critics, Kuhn somewhat reduced the nature of his "revolutions". His paradigms

---

3. I would see this as empiricist activity.

became much less monolithic and all-encompassing.

I believe that Kuhn was wrong in his details. Changes certainly occur, but not in the radical and incommmensurable way he originally suggested[4]. Kuhn nevertheless began some important reorientation of thinking about scientific research. The notion of paradigm, though ambiguous, is useful in the sense that it does point both to the significance of the phenomenon of similar ways of approaching some scientific problems, and to the importance of developing eventual new forms of research as Scientific Theory evolves.

All of this is relevant to the consideration of Copernicus. For prior to Copernicus, Ptolemy had indeed standardized the ideas of his Greek predecessors, and it had become normal for astronomers to do and see things Ptolemy's way. On the Ptolemaic view, the sun, moon and planets all circled the earth. To predict future planetary positions and to explain such things as the observed retrograde motion of Mars, Ptolemy made use of small circles (epicycles) running on large circles[5] (deferents), and where necessary he placed the point around which the deferent moved, somewhat to one side or another of its true center (eccentrics); moreover, the center of any planet's deferent also moved around an "equant"– a point on the diameter of the deferent but at a position opposite to that of the earth from the center of the deferent.

In the dedicatory letter to his great book, *De Revolutionibus* (1543), Copernicus states that it was the unsatisfactory state of astronomy[6], which led him to look for an improved account and that accordingly he examined the writings of the Greeks. There he found the teachings of Aristarchus of Samos, Heraclides of Pontus and the Pythagoreans on the motions of the earth. By "unsatisfactory state of astronomy" Copernicus meant: (1) the mounting inaccuracies in the Ptolemaic system (the differences between what was observed and what should have been the case as predicted by the system); (2) the vast and increasing complexity of the system; (3) the *ad hoc* moves which, in order to "save the appearances", allowed astronomers to shrink or expand a deferent, or add an eccentric, and so on; and (4) the fact that *the system no longer described uniform, circular motion* : equants were for Copernicus a glaring example.

Qualitatively, Copernicus explained the retrograde movements of the planets better: such bodies *appeared* to move backwards in their paths because they were seen by an observer on the earth. Because the earth had a smaller orbit than Mars, Jupiter and Saturn, those planets as viewed against the background of the fixed stars seemed to have retrograde movements at the particular stages of their

---

4. See final section of this chapter.
5. Movement in circles was Presuppositional Theory, Cluster 6.
6. Scholars such as Nickles (1978), Wrightsman (1980), and Boorstin (1984) argue that there are also deep theological and metaphysical arguments underpinning Copernicus's work. This would indeed seem to be the case, and in the sense that it shows how scientific development often results from ideas from other domains, is germane to the points being made in this book about the nature of scientific research. Such matters are not considered here, but are addressed in Chapter VII.

orbits when they were being overtaken by the earth. And Venus and Mercury appeared to retrogress when the earth was being overtaken by them in their own smaller orbits.

It was also to the scientific credit and to the consistency of the abstract conceptions of Copernicus that, unlike in Ptolemy, it was not possible in his system to make *ad hoc* shrinkings or expansions in one planetary orbit without causing disastrous changes in others. Furthermore, like all sound scientific developments, Copernicus's theory carried with it explicit and implied predictions. For instance, if Mercury and Venus really were circling the sun, then phases similar to that of the moon should be observable. Though such phases were invisible to the naked eye, this prediction was confirmed by Galileo with his telescope within fifty years. *De Revolutionibus* was also on a different level of mathematical sophistication from Ptolemy.

To overturn Aristotle-Ptolemy, Copernicus had initially to accept all his own theoretical presuppositions (Cluster 6) about circular motion, epicycles, etc. For neither Copernicus nor we ourselves could make sense of what it would be to question all theories (various Clusters) at the same time. This is a logical could not. But contra Kuhn's ideas of *Gestalt* switches and incommensurabilities, this does not mean that these theories could not be removed and replaced piece by piece– and indeed they gradually were, and it was Copernicus who began the removal. Nevertheless, there were still many problems in the Copernican physics, which awaited explanation. For instance, plain observation seemed to contradict Copernicus's claims that the earth rotates– it is the sun which we observe to move, not the earth. If Copernicus were correct, and the earth really were rotating, then loose objects such as stones and philosophers should be thrown from its surface, as from a spinning top. The fact that this fails to occur supports Ptolemy and Aristotelian physics, not Copernicus.

Such observational "facts" were key reasons for the rejection by the Aristotelians. Why, they thought, give up the traditional system for a new one, when it was so contrary to the then commonsense idea that the earth does not move, and when it contradicted the only system of physics (Aristotle's) that then existed?

Kepler and His Three Laws of Planetary Motion

It is true that Kepler was in possession of Tycho Brahe's detailed observations of planetary positions. But it is not true, as is commonly claimed, that Kepler began with these. Kepler did not, and logically could not so begin. On the contrary, just like Copernicus, *he began with numerous and various presuppositions and theories* (Clusters 5, 6, 8 and 9 and perhaps even 7). At various times he believed in Ptolemy's ideas of perfect celestial spheres and the music they made in their moving, the magic Pythagorean properties of numbers, as well as the powerful, modern idea of the mathematical basis of all nature. What is more, Tycho's own observations were theoretical in the Cluster 6 presuppositional sense, for Tycho assumed the correctness of descriptions of the various planetary movements, he looked for these planets to be traveling along the ecliptic and within the Zodiac, and he took for granted the reliability of his huge measuring instrument.

Kepler's presuppositions initially led him into a dead end. After having tried and failed to make the planetary orbits fit musical intervals, Kepler developed the idea that there was a relationship between the orbits of the planets, and the five regular geometrical solids. They seemed to fit just like Chinese boxes. It appeared to Kepler that this fit had the stamp of divine providence. Here Kepler had produced a beautiful theory (Cluster 5, Type 6) of abstract concepts. But in his attempts to achieve interpretation to the observable world, he saw his error. The regular geometrical figures had no theoretical power other than to explain an approximate spacing of the planets. Nothing else followed. So Kepler began in an immense labour all over again, to try to work out the explanation for the differences between observed positions and what Copernican theory (Clusters 6 or 9) predicted. He made use of his presuppositions, the mathematical knowledge of conic sections, which he had learned from Tycho, his own continuing development of these, and the observations of Tycho, and of Longomontanus. It took ten years.

Kepler tried one theory (Cluster 5, Type 6), then another. For instance, if the earth travelled in uniform circular motion, it should describe equal angles about the sun in equal times. Alternatively, this could be construed as covering equal areas in equal times. This did not happen. He then demonstrated the same thing for Mars. After calculating the distance of Mars from the sun, discrepancies occurred. In order for the orbit of Mars to be circular, its eccentricity[7] from the sun had to be unacceptably great, ". . so great in fact, that the resulting equations concerning the orbit's elements were either false or inconsistent" (Hanson,1972, p.74).

The stable element in all this was Kepler's reliance on the mathematical fact that uniform circular motion required the covering of equal areas in equal times. He therefore continually directed the modus tollens at features of the account other than the circular orbit. He continued to argue that: the axiom of equal areas in equal times is to be maintained; the orbit of Mars is circular; with the accurate data of Tycho, errors of measurement should be no greater than 1"(one minute) in excess or defect. But he saw that errors of 8" in excess and defect were occurring. His initial reaction was to suspect that he was making an error in his finding of areas. Gradually and reluctantly, Kepler speculated that the axiom of uniform circular motion might be incorrect. *This was a magnificent, revolutionary thought* .

So the orbit was not a circle, but what was it? With his normal, meticulous care, Kepler computed twenty Mars-Sun distances at various points in the planet's orbit. These indicated that the orbit must be some kind of oval. After immense labors, he at one point actually held the secret of the shape (viz. an ellipse), but failed to recognize it as such. For with the mathematical tools then

---

7. The imaginary point around which Mars would have to orbit to save the calculations.

available to him, he computed the orbit given by the formula, but erroneously derived an ovoid. This ovoid he rejected as not being a possible orbital shape. And on the same basis he also rejected the correct formula. For he now wanted to test a new hypothesis– that the orbit was an ellipse! Soon after, he stumbled upon the answer that the original formula was actually a formula for an ellipse[8]. He realized his ironic mistake. That planets follow elliptical orbits is known as Kepler's First Law of Planetary Motion.

Let nothing detract from the epochal nature of the reluctant move to an ellipse. With the use of a single *abstract mathematical concept*, the convolutions and complexities of Copernicus and circular motion were swept away. But Kepler did not just decide, "I'll use Tycho's observations and see what I can do with them." Rather, he already entertained a nagging problem, and recognized the significance of the data for it.

After further immense intellectual and mathematical labor, Kepler produced a geometrical relationship: an imaginary line drawn from the sun to a planet, in equal intervals of time, traces out equal areas of space within the elliptical orbit. This is his Second Law of Planetary Motion. Despite Copernicus's revolutionary conceptual leap, nothing in his work can compare with Kepler's laws. Here is indeed a powerful *relationship between the propositions linking abstract concepts and the observable world of the planetary universe.* Kuhn says, ". . for the first

---

8. A recent book by W.H.Donahue (1990) has claimed that Kepler fabricated his data. Such a claim reveals a profound ignorance of the nature of scientific research– which develops Scientific Theory involving propositions which link abstract concepts (see below), and relates consistently to the empirical world. Such consistency and precision are quite lacking in the social sciences and empiricist research. If Kepler had wished to fabricate or "fudge his data" as social scientists say, it would have been at this point. For he could then have fudged to make things fit a circular orbit ! As the Columbia University astronomer, Motz says, "[Kepler]. .had tried to squeeze the orbit of Mars into a Copernican circle. Though he had almost succeeded in doing so after lengthy calculations and could have fudged the data then if he had been given to cheating, he rejected the circle and turned to the ellipse as the correct shape" (1990, p.35).

The irony here is that fabrication of data in the social sciences and in other areas which use the statistical approach, such as medical research, is unfortunately becoming all too common. It is no accident that social scientists invented this phrase,"fudge the data". "Fudging" occurs so much, because unlike in scientific research, most social science is statistical and empiricist, and as a result usually impossible to check or replicate precisely. There is now so much illicit medical research that in 1989 the U.S. National Institute of Health set up the Office of Scientific Integrity (OSI). In 1990 it found 5 cases of plagiarism, 6 cases of fabrication of data, 7 cases of "deviant practices" (referees using privileged data as their own; fabrication of entries in bibliographies; selective reporting of primary data), and 6 cases of misconduct (*Science* , Vol.253, p.1084). Given that many cases will never be detected and that OSI errs on the side of leniency, these are appalling figures. But Kepler was doing scientific research, not social science, or statistical-health, or empiricist statistical research, and he did not "fudge".

time the predictions are as accurate as the observations" (1957, p.212). Long years later, Kepler conceptualized his third law: that the squares of the periods of revolution of the planets are proportional to the cubes of their mean distances from the sun, or, $t^2 = k\, r^3$.

But why did the laws take this form? What was now needed was a new physics, which would provide a more coherent explanation of the motion of all bodies, terrestrial or celestial. The basis of such a new physics was laid in a spectacular way first by Galileo, then by Newton. Their work is of particular interest, because it has sometimes been suggested that pedagogical research has yet to find its Galileo or Newton.

It has often been suggested[9] that Kepler's fitting of an orbital shape to Tycho's observed positional data was an early and scientifically paradigmatic example of a generalization derived from observations. *This is a mischievously misleading and distorting misunderstanding of Kepler's powerful insight. For without theoretical presuppositions (Cluster 6) about orbital shape and/or about orbital speed, the observational data would necessarily have remained barren, and nothing would have been discovered.* As Kuhn correctly points out, "Unless the planetary orbits are assumed to be precisely reentrant (as they were after Kepler's work but not before), a speed law is required to compute orbital shape from naked eye data" (1957, p.216).

Galileo's Dynamics

If ordinary mortals, physical scientists, or philosophers were forced to rely on what they could observe, and to generalize from this, then they would have to agree in general with an Aristotelian physics. For all the objects which we observe on this earth do require forces to move them, and when such forces cease to act, the moving object eventually stops. To grasp what was needed for understanding of such matters, imagination was required, not observation. Losee emphasizes Galileo's creative imagination, when he says that propositions involving abstract concepts can be obtained, ". . neither by induction by simple enumeration nor by methods of agreement and difference [see Chapter III]. It is necessary for the scientist to intuit which properties of phenomena are the proper basis for idealization, and which properties may be ignored"(1980, p.55). The scientists, Ripley and Whitten say that, "Galileo's great contribution to science, in addition to the many particular discoveries was to establish *a method that combined hypothetico-deductive procedures involving idealized and abstract [concepts] with empirical verification through experiment* "(italics added)(1969, p.123).

Although Galileo when necessary was a careful observer, he achieved his success by the initial conceptual leaps of thought experiments, not by observation (Koyre,1939; Kuhn,1950). Galileo used observation to help find the relationships between the propositions which contained the abstract concepts

---

9. Most notably by J.S.Mill (1843/1967) and Reichenbach (1938, p. 371).

(velocity, acceleration, time, distance), and their interpretations in the empirical world. As Butterfield says in respect to Galileo's conceptualization, "The modern law of inertia is not the thing you would discover by mere photographic methods of observation— it required a different kind of thinking cap, a transposition in the mind of the scientist himself" (1957, p.17). Taking his clue from the Oxford and Paris Scholastic critics of Aristotle, Galileo's final conclusion about horizontal motion was that, ". . any particle projected along a horizontal plane without friction . . .will move along that same plane with a motion which is uniform and perpetual, provided the plane has no limits"(1950, p.144). This claim is not a generalization from observations. In the empirical world we cannot eliminate friction and there is no infinite plane, and even if these existed Galileo could scarcely have observed them in perpetuity.

Consider his thinking in more detail. Galileo produced two different arguments. His first argument, *a pure thought experiment* involves the conceptual assumption of a frictionless plane. He writes:

> .. we may remark that any velocity once imparted to a moving body will be rigidly maintained as long as the external causes of an acceleration or retardation are removed, a condition which is found only in horizontal planes, for in the case of planes which slope downward there is already present a cause of acceleration, while on planes sloping upward there is retardation; from this it follows that motion along a horizontal plane is perpetual (Galilei,1950, p.215).

The second argument is *partly a thought experiment, partly the result of observation*. (And even the observational aspects involve considerable insight.) Galileo observed that if the bob of a pendulum is released, it will swing through a descent and into an ascent, to reach approximately the same height at which it began its descent. He then showed that if a nail is placed directly below the point of support so that the descending pendulum strikes the nail part way along the pendulum, the bob will rise to the original height, though naturally, over a shorter arc.

Historically, the next move may have been by way of insights gained through experimentation, but it reached its crucial conclusion only through its *extension into a thought experiment*. Galileo's remarkable perception was to grasp that what happens with the bob of the pendulum, would be duplicated when a ball rolls down one inclined plane and up another, were it not for the loss of momentum when the ball hits the angle between the planes. Without such a loss of momentum, and if friction could be disregarded, the ball would reach the original height up the second plane. Now if the slope of the second plane is progressively reduced the ball will travel increasingly farther along the second plane. So when the angle of slope has reached zero, the plane will be horizontal, and the original height from which the ball was rolled will be an infinite distance away. Thus the motion of the ball must then be perpetual. So, contra Aristotle, contra most children in elementary schools, and contra a majority of the world's present adults as well, Galileo had thus demonstrated that a force does not cause a velocity, but rather a change in velocity, either accelerating or retarding.

In establishing his Law of Falling Bodies, Galileo's initial conceptual move was to propose that, again contra Aristotle, in the absence of air all objects should fall to the ground with the same speed, and that it is the resistance of the air which has a greater effect on the descent of very light bodies. He reached this conclusion by two routes. The first was a thought experiment which showed that the Aristotleian view was self-contradictory, the second involved an empirical experiment. I shall consider the second.

In developing his empirical experiment, he argued that air resistance must increase as the speed of the falling bodies increases. If this is the case, then any test of falling bodies will be more accurate, i.e. less affected by the air, when conducted at low velocities. Galileo realized that a lower velocity would result if the tests, instead of being made upon falling bodies, could be made upon bodies "falling" along an inclined plain, or along the arc of a pendulum. To be noticed is his *theoretical assumption* (Clusters 5 and/or 6) that acceleration along an inclined plane is the same as that of vertical acceleration in relevant respects. He tested his pendulum hypothesis against experience and found that lead and cork bobs traversed the same length in the same time. He had thus shown indirectly that objects of different weights fall at the same rate. But he still had to establish in what manner the velocity of falling bodies changes.

It is logically possible that bodies might gain velocity (v) in any number of ways. Logically, v could equal ks or kt, or ks^2, or kt^2, etc. (where v is velocity, k is a constant to be found by experiment, s is the distance travelled, and t is the time taken). With an economy which characterizes the best scientific research, Galileo considered the simplest possibilities first. Initially he thought that velocity might increase in proportion to distance, but his *examination of the mathematics* showed an inconsistency, so he then concentrated on $v = kt$. As he writes, "A motion is said to be uniformly accelerated if, when starting from rest it acquires during equal time intervals, equal increments of speed" (1950, p.162). But how to find the form of this motion in the primitive circumstances of early seventeenth century measurement? He tested it by testing the hypothesis indirectly, through its particular consequences. The particular consequences were that if $v=kt$, then $s=Kt^2$ (K is another constant), i.e. distance is proportional to the square of the time. This situation is easier to measure, because instead of velocity and time, distance and time are now measured.

Galileo then wished to observationally test this mathematically-derived relation. To increase the accuracy of measurement by increasing the time of fall, as already mentioned, he rolled balls down a groove in an inclined plane, rather than dropping them vertically. Galileo varied the conditions of the experiment, by varying the angle and by allowing the ball to roll a quarter, then a half, then two-thirds, and so on, of the length of the groove, and measuring the time taken in each case. After making allowance for the slope of the plane, viz. the result multiplied by the sine of the angle of the plane, he found, for instance, that to reach the quarter mark, the ball did in fact take half the time for a full descent, i.e. distance is proportional to the square of the time.

With such abstract concepts and experiments, Galileo established the acceptability of his view that all objects, independently of their weight, fall to the earth with an equal and constant acceleration, assuming the abstract condition of no air resistance. His experiments were not Baconian[10], but were tests of the direct *consequences of propositions involving his abstract theoretical (Cluster 9) concepts (velocity, distance, time)*.

Equally significant was Galileo's extension of his Law of Acceleration into his mathematical account of projectiles. Prior to his mathematical description, firing cannon balls was empiricist. It depended on experience and trial and error, together with the approximations of charts of combinations of straight lines and parts of circles generated empiricistically, used to describe a trajectory (Ripley and Whitten,1969, pp.121-122; Mendelssohn,1977, p.80). Galileo showed that the trajectory of projectiles does not involve parts of circles, but is parabolic.

Rejecting a possible empiricist approach of seeking regularities, and the misunderstandings derived from observation of supposed regularities, Galileo moved to the level of theory (Clusters 5 and 9) and abstract concepts. With conceptual brilliance he divided the motion of a projectile into two abstract components, one vertical, one horizontal. (And where would one observe *them* ?) In consistency with his law of inertia, he conceptualized that the horizontal motion is constant; in consistency with his law of falling bodies he conceptualized that the vertical motion is accelerating. Thus in consecutive equal periods of time, while the horizontal distances traversed will remain equal, the vertical distances will increase proportionately. The resulting path will be a parabola.

With Galileo's mathematics, trajectories of cannon balls could be accurately computed, and artillery began to be a true applied science.

Newton's Laws of Motion and Gravitation

Newton's great work, the *Philosophia Naturalis Principia Mathematica* , normally known as the *Principia* , was published in 1687. It was however the result of spectacular work which began at least as early as 1665 (Dampier,1949, p.151). Ripley and Whitten provide a good resume of its contents :

> The book sets forth both the framework for later analysis and specific suggestions, which kept the finest minds of mathematicians and scientists busy for 200 years. . .the three laws of motion and their consequences are established. The law of gravitation is set forth as a mathematical hypothesis from which the three laws of Kepler can be deduced as consequences. . . The motions of all the planets, the satellites of Jupiter, and the orbits of the comets are studied in detail and shown to fit the law of gravitation. The complicated motions of the moon, the precession of the equinoxes, and the tides are all examined in detail and systematically explained (1969, p.128).

---

10. i.e. random: Let us see what happens, if we . .

As stated in the *Principia*, Newton's three laws of motion and his law of universal gravitation read thus:

Law 1. Every body continues in its state of rest or of uniform motion in a right line, unless it is compelled to change that state by forces impressed upon it.

or, if $F = 0$, then $a = 0$ (where F is the net impressed force and a is the acceleration),

Law 2. The change of motion is proportional to the force impressed; and is made in the direction of the right line in which that force is impressed. (In modern terms this would be stated as: change in momentum is proportional to the impressed force.)

or, $F = ma$ (where F is the net force acting on a body of mass m, and where a is the acceleration in the direction of the force F).

Law 3. To every action there is always opposed an equal reaction; or the mutual actions of two bodies upon each other are always equal, and directed to contrary parts.

(In modern terms we should state this as: to every force there is an equal and opposite force.)

or, $F_1 = -F_2$

or, $m_1 a_1 = -m_2 a_2$

The Law of Universal Gravitation.

(i) That there is a power of gravity pertaining to all bodies, proportional to the several quantities of matter which they contain.

(ii) The force of gravity towards the several equal particles of any body is inversely as the square of the distance of places from the particles.

or, $F = G \dfrac{m_1 m_2}{d^2}$

(where m_1 and m_2 are any two masses, d is the distance between them, F is the force of gravity between them, G is the universal gravitational constant.)

Newton's first law of motion will be seen to be similar to the Law of Inertia of Galileo, though Galileo still thought in terms of great circle inertia. Newton's second law was set forth in a round about way by Galileo. But Newton's third law was entirely his own. Newton moreover stated all three laws in a manner in which interrelated conclusions could be drawn in a precise, mathematical form. What is more, by combining these laws with his law of gravitation, Newton produced a single, systematic account which could explain both terrestrial and celestial motion. "Here, indeed, is a 'system of the world', so broad in its range, so powerful in its methods of analysis, so elegant and simple in conception, that its dominance over the minds of men for many succeeding generations was inevitable" (Ripley and Whitten, 1969, p.130).

Let us consider the example of the Law of Gravity, or The Inverse Square Law. Newton had to make a speculative leap. Bronowski's idea of searching for hidden liknesses applies here, for Newton grasped that the same abstraction, viz. the "gravitational force" which caused an apple to fall to the earth might also be that which deflected the moon from flying off at a tangent to its orbit. Taking his hint from Kepler's Third Law, Newton, by mathematical methods, established that the centripetal force (that force which prevents the body from following its straight-line path) which acts towards the centre of motion of an orbiting body, must follow the inverse square law. His next move was to argue that if it were correct that the force responsible for objects falling to the earth were the same as that which kept planets in their orbits, then it should follow that the value of g (acceleration due to gravity on the earth) could be worked out by calculating the acceleration of the moon (the orbit of the moon being close to a perfect circle). Performing the calculations, Newton found they were very close to his expectations.

Newton's own research, and its further developments and applications by later scientists have been described by Lakatos (1970) as a splendid example of a "research program". A research program, less restrictive and more sagacious than Kuhn's paradigm may be construed as an evolving Scientific Theory together with various procedures which guide research with it and on it in both a positive and negative way, so that it continually evolves and the result is the rational progress of that branch of science.

Newton's program involved penetrating *conceptual moves*. The Law of Gravitation relates to attraction between particles. Fortunately for science, Newton was able mathematically to show that spheres made up of particles can be considered *as if* they were point masses. To analyse gravitational forces, therefore, he regarded any planet, including the earth, as instantiating *the abstract concept of a point mass*. In working out the details for *Principia*, Newton initially calculated for a fixed point-mass sun and one point-mass planet. From this he derived his inverse square law for Kepler's ellipse. Then he improved on Kepler by developing an account in which both sun and planet revolved around their own *centre of gravity* . "This change was not motivated by any observation. . .but by a theoretical difficulty in developing the programme" (Lakatos,1970, p.135). He then proceeded with the other planets, for the sake of simplicity assuming that there were *no interplanetary attractions* but only heliocentric ones. This being completed, he worked out the situation as though the heavenly bodies were not mass-points, but *mass-balls* . After all, because mass in the empirical world was assumed to occupy space, mass-points were an impossibility in Newton's own (unarticulated) presuppositional theory (Cluster 6), being used merely as a conceptual and mathematical abstraction, which allowed certain calculations. This development involved immense mathematical difficulties, taking something like ten years. Having achieved success, Newton then went to work on *spinning* mass-balls and their wobbles. He next introduced *interplanetary attractions* so as to allow for observed perturbations from Kepler's perfectly elliptical orbits.

On Lakatos's account, a research program has both a "hard core", a "protective belt" of auxiliary hypotheses, and positive and negative "heuristics". The hard core is the set of laws, propositions, and assumptions within the Scientific Theory, which are beyond the challenge of any *modus tollens*. To change the hard core would be on Lakatos's view, to change the research program itself. To change the protective belt is to work within it, deflecting challenges from the core. Calling this approach, "epistemological conservatism", Garrison says that it is a good policy, for it makes sense not to abandon too quickly principles which, ". . have worked well in the past, just because they don't happen to work well just this once. . Epistemological conservatism is pragmatic; it follows the commonsense expedient that declares that it is *reasonable* to preserve what has worked well. ." (1986, p.13).

In Lakatos's conception, the heuristics guide the scientist in achieving the necessary modifications in the protective belt. A research program may at different stages in its development be either "progressive" or "degenerating". If a program is leading to new discoveries which can be comprehended within the terminology of the Theory by adjustments and additions to the protective belt, then there is a "progressive problem shift" within the program, for the developments are then increasing that which the program can explain. But in other circumstances there is a "degenerating problem shift", when additions are made, but also accompanied at places by *ad hoc* assumptions or hypotheses, which have no function other than to save the Theory from apparent refutation.

Newton's work provides a beautiful example in full flourishing, of what Lakatos has in mind, developing for some 250 years by way of a continual progressive problem shift, as the Scientific Theory developed. For most physicists and astronomers there was certainly a Newtonian hard core of assumptions which remained sacrosanct, so that for example, they blamed observed anomalies upon the disturbing effects, either mechanical, or gravitational, of masses not hitherto considered, i.e. they diverted them to the protective belt. This was the case with certain anomalies in the motion of the moon, which were explained, not by rejecting Newton's laws, but by developing the new abstract concept of "angular momentum". But this was not just an *ad hoc* move, because such notions were able to be incorporated into the Scientific Theory proper, thus making it more powerful in both explanation, application and prediction in other places.

Keeping to the "hard core" also made possible the discovery of Neptune. Only if the Sun and a single planet were the sole bodies in the system, would the orbit of a planet strictly follow Kepler's First Law. In accordance with Newton's Law of Gravitation, the planets pull one another from their abstractly conceptualized elliptical orbits in proportion to their masses and proximity. By the early 1800s, astronomers were familiar with the perturbative effects which Mercury, Venus, Mars, Jupiter and Saturn[11], the Earth, and Uranus[12], had one upon another.

An orbit for Uranus had been determined by 1790. But by 1820, it had become clear that the assumed orbit did not quite match all the observations of its position now available. This was still the case after allowance had been made for the crucial perturbational effects of Jupiter and Saturn. An unexplained "error" of about 0.01 of a degree remained.

---

11. The five planets known since antiquity.
12. Discovered in 1781.

Some astronomers began to wonder about the accuracy of Newton's Law of gravity. Others said that an unknown planet farther out in the solar system would explain such effects. But the problems of finding it were immense. It is difficult enough to compute the effect of a known planet upon another. Here was the reverse: the planet was unknown and its mass, distance, direction, motion, all had to be postulated and tried on the basis of a tiny deviation of Uranus from a pure Keplerian orbit[13].

The construal of the pertubations of Uranus as due not to some inadequacy in Newtonian theory, but as most likely being the result of the gravitational attraction of a previously unsuspected planet is a good example of the diversion of the challenge from the hard core of Newtonian Theory to the protective belt. Adams in England and Le Verrier in France set out to calculate the position of the postulated planet. After two years on the problem, Adams wrote to Challis, the then Astronomer Royal, telling him where to look– Adams was actually within two degrees. Just as occurs today, leading scientists were annoyed by lots of letters from cranks, and Challis took no notice. Meanwhile, equally hard work allowed Le Verrier also to predict its position– near to that of Adams. He also wrote to Challis, who only then arranged for a (very unhurried) search. But Le Verrier also wrote to Galle, in Berlin, who organized an immediate search and within a few nights had sighted the planet. The discovery electrified astronomers, and all the world's major observatories soon confirmed the sighting.

Who discovered Neptune? Was it Galle, or was it Adams and Le Verrier contemporaneously, but independently? Neptune provides a good example of how scientists work within and pursue the questions of evolving research programs. The fact that we can ask who actually made the discovery shows the interrelatedness of the activities.

A similar situation appeared to exist in the later nineteenth century when, as mentioned in Chapter I, it was observed that the orbit of Mercury precessed at perihelion. Once again the two alternatives were raised: perhaps a hitherto unknown immensely fast-moving planet (this time between Mercury and the Sun) might be the cause, or perhaps Newton's Theory might need revision. As an example of the first view, some astronomers went searching for "Vulcan", as the postulated planet was called. Though Lescarbault claimed he had observed it as a small round black spot against the Sun, despite careful and repeated searches, no other plausible evidence emerged. As an example of the second, one scientist suggested that gravity might fall off not to the power of 2, but to the power of 2.0000001612 ! [14]. This phenomenon is now explained by Einstein's General

---

13. In our age of hand-calculators and computers of ever-increasing power, we little realize the immensity of the mathematical labours required for calculations in earlier times.

14. Thus, Poynting writes, "..the enigmatical motion of the perihelion of Mercury has not yet found any plausible explanation except on the hypothesis that the gravitation of the sun diminishes at a rate slightly greater than that of the inverse square– the most simple modification being to suppose that instead of the exponent of the distance being exactly -2, it is -2.0000001612" (1910, p.384).

Theory of Relativity.

For Lakatos, a program is an improvement on an earlier or rival one when it explains everything explained by the other, when it explains phenomena unexplained by its rival, and also when it predicts novel facts unexplainable by, or not conceivable by its rival. Once it was a going concern, the program developed by Newton and his successors and exemplified in the discovery of Neptune and later of Pluto (and also in a host of other applications, such as ballistics) certainly did this in relation to Aristotle-Ptolemy. Like his mentor, Popper, Lakatos believes in the rationality of science, that scientific research can be demarcated from other endeavors, and that all claims to knowledge are fallible. Lakatos's account is useful in providing a rational explanation of a Scientific Theory in its early stages of development. For with many, perhaps most, theories in their early days (e.g. with that of Copernicus), some observations will seemingly falsify them, but these observations will be construed in a different way when the Theory has developed more fully. In the meantime, the proselytes of the program ignore the seeming falsifying observations of the supporters of the earlier Scientific Theory, concentrating upon the new applications and the burgeoning features which their program does explain.

The Scientific Achievement of Galileo and Newton

Let us be clear what the methods of Galileo and Newton involved: ideas of empiricist observational concepts and of generalization from numerous empirically observed cases are grotesquely inadequate. Both savants demonstrated and insisted upon the importance of abstraction from the world. They extrapolated conceptually and then trialled their conceptualizations, either by thought experiment, by mathematics, or by empirical experiment.

Galileo continually made use of abstract concepts such as frictionless planes, free fall in a vacuum, vertical components. These are not entities exemplified directly in empirical phenomena. These are not references to any supposed regularity in the empirical world. A frictionless plane is a concept arrived at by abstraction from, not generalization from observed behavior of bodies on real surfaces. Free fall in a vacuum[15] is an idealization suggested by the observed behavior of bodies dropped in a series of fluids of decreasing density. Newton made use of abstract concepts such as invisible centripetal forces, point masses, interplanetary attractions, centers of gravity. His First Law specifies the behavior of bodies which are free from impressed forces. But empirically there are no such bodies. And if by some celestial sleight-of-hand such a body could exist, no Newton could have any empirical knowledge of it, because such knowledge would require somewhere in the universe the presence of an observer, or his apparatus. This would mean the presence of impressed forces. In any case, the

---

15. This analysis of acceleration in a vacuum was at a time in history when such a physical state was assumed by most of the scientific community to be an impossibility.

empirical existence of such a body conflicts with Newton's own Law of Universal Gravitation. Consequently, as Losee says (1980, p.84), ". . the [Newtonian] law of inertia is not a generalization about the observed motions of particular bodies. It is, rather, an abstraction from such motions." Similar points can be made about his other laws.

The superb laws of Galileo and Newton are better science than Aristotle's ancient claims, or the intelligent claims of the medieval impetus theorists, *not because they represent more careful observations, but because they go behind the superficial regularity disclosed by empiricist observation* to something more fundamental.

Lavoisier and the Discovery of Oxygen

Lavoisier was aware of the work of Galileo and Newton, with their emphasis upon mathematical interpretation, and of the underpinning of an atomistic or corpuscular metaphysical theory (Cluster 6: Presupposition), which viewed matter as composed of unobservable but massy, hard and indivisible, moving particles.

Chemistry at that time was based on a sort of mixture of Empiricist Theory and Scientific Theory, and possessed a large number of observed, supposed facts and relationships, in respect to metallurgy and combustion (e.g. the *Pirotechnia* of Biringuccio (1540/1990))– derived from the lore of miners and metal workers, and from the natural historical and conceptually-naive Baconian experimental work of the alchemists. Such chemical views were however plagued by masses of conceptual confusion and alchemical misconstruals[16]. Begun by Cavendish, a vague notion of "gas" was gradually being conceptualized.

Lavoisier initially worked within the then-accepted theory of Phlogiston (Clusters 5 or 6 : Hypothesis or Presupposition; and (putatively) 9: Scientific Theory), and it was this theory which he applied in his laboratory. Phlogiston theory had been developed after the fact by the German physician-chemist, Stahl to explain the empiricist, craft details of smelting, of combustion, and of what was then called "calcination" (actually oxidation). It had become gradually accepted that in calcination the residues were actually heavier than the original substances. And despite the radical developments in physics, this was still accepted in Lavoisier's time, on the assumption that substances *became heavier*

---

16. Consider the following account from Biringuccio, still in use at the time Lavoisier first took an interest in chemistry. Mercury is described as, ". . a body that consists of flowing and liquid materials, almost like water . . it is composed by Nature of a viscous, subtle substance with an overflowing abundance of moistness and coldness together. . [gold] it embraces and instead of holding it in suspension, it draws it into itself to the very bottom. It does not join nor approach any other thing without the aid of art, even though that thing be liquid and moist, but it does not leave gold because its humidity is well mixed with viscosity. .All the mountains or places where it is engendered have abundant water and trees. . (1540/1990, pp.79-82).

by losing their "volatile" or "spiritous" or "airy" parts. In respect to lead, Biringuccio writes that, "Since those watery and airy parts are removed by the fire"(1540/1990, pp. 58-59) "calcined" lead naturally becomes heavier. Phlogiston theory was a more sophisticated version of this view. Details of smelting had been known since ancient times, and discussed, "such and such an amount of charcoal" to be heated to, "some particular heat", this heat being described in observable, empiricist terms of the appropriate color of the fire. The presence of air was also central to the theory, because it had been observed that in a vacuum, neither combination nor "calcination" occurred.

The phlogiston theory is an interesting departure from the rest of the then current claims of chemistry. It *postulates the existence of an abstraction*– phlogiston, which could not be seen. As Gale suggests, any attempt to pass beyond observable relationships into some explanatory scientific scheme is risky. "It requires a certain amount of bravery to postulate unobservable objects in order that they might explain observable events" (Gale,1979, p.118). And as the history of scientific research consistently shows, it requires skill and intelligence, and a deep knowledge of one's subject, to postulate effectively.

Phlogiston theory also grew out of the Aristotelian theory of earth, air, fire and water. Associated with each of these basic substances were others which were classified as having similar qualities, e.g. metallic ores, rocks, and clays were classified with "earths". Iron ore for instance, was seen as earthy and called a "calx", i.e. a substance produced by calcination. It is not hard to see why alchemists came to construe chemistry as a matter of qualitative explanation and the transformation of particular substances into other substances. Phlogiston theory was consistent with the ancient, observation-based idea, that when a substance burns, something streams out of it, with the original substance thereby being reduced to more elementary ingredients.

But paralleling such ideas and embedded in them, was that of the atomistic theory (Cluster 6 : Presupposition) already mentioned. On such a theory, chemical reactions were supposed to involve some kind of rearrangement of the smaller, unobservable particles of which matter was constituted. As a result, reactions were construed as of two types: movement of particles toward each other, and movement away, i.e. combination and dissociation. All explanations were to be given in such terms.

Phlogiston was supposed to have various properties which helped to explain the observed behavior of substances of which it was part. Thus, its most important property was its apparent inflammability. Substances rich in phlogiston burned readily, and left little residue. Thus charcoal was seen to burn readily, leave a small ashy residue, and liberate water. In phlogiston theory, combustion is a matter of decomposition. Thus charcoal's combination of phlogiston, earth, and water breaks up during burning, resulting in its three now-separated constituents. But there was no attempt to make these claims quantitative.

It should be noticed how this view, ". .was based upon one of those fundamental conclusions of *commonsense observation* which may set the whole of men's thinking on the wrong track" (Butterfield, 1957, p.206). Butterfield has in mind the Aristotelian view of motion as something requiring an impetus, the Ptolemaic view of the heavens with the earth immovable at the center, and the supposed ebb and flow of the blood pre-William Harvey (1958, p.206); to which we may add, the idea of fixed species prior to Darwin and Wallace and their precursors, and of an immobile crust of the earth prior to Wegener. These all involved revolutionary scientific thinking involving abstract concepts. Rejection of phlogiston and development of the oxygen theory of combustion required the same kind of intellectual revolution. None of these scientific discoveries, it should be stressed, arose from empiricist endeavor— from generalizations from observables. Generalizations from observables supported the opposite views.

The first time Lavoisier made use of his preconceptions about the importance of quantitative dimensions was in 1764, when he showed by way of extremely close attention to the weights involved, that plaster of paris hardened because of the actual chemical combination between the plaster powder and the water added to it at the time of mixing. He was also well aware of Priestley's experiments, and like Priestley, he became deeply interested in the chemical reactions of combustion and smelting. In 1772 he began to weigh the amounts of the substances before and after burning.

His results violated his phlogiston-based expectations, and surprised him in a crucial way. " 'About eight days ago,' he wrote,' I discovered that sulfur in burning, far from losing weight, on the contrary, gains it; it is the same with phosphorus'" (quoted in Conant, 1950, p.16). Here was the kind of anomaly which Kuhn rightly points to as being important in posing a challenge to the existing views and indicating the need for something new. Indeed, as Lavoisier continued, his work was instrumental in setting up a new research program involving a completely new Scientific Theory.

The matter now became complex, because this unexpected result could indeed be fitted into phlogiston theory, if it is argued that weight gained during combustion results from the loss of phlogiston, so phlogiston must have negative weight. This idea, however presposterous and merely *ad hoc* to the modern mind, seemed to have some empirical support: for instance the account could be used to explain the success of what we nowadays call "hot-air" balloons (Gale, 1979, p.131), which were in the public eye because of the activities of the Montgofier brothers. Moreover, it seemed simple enough to posit the existence of a physical property of "levity", the opposite of gravity, as Aristotle had done. This is indeed what some of Lavoisier's contemporaries did. This issue of levity provides a good example of the difficulty in science of *developing not merely abstract concepts, but abstract concepts which can be made to work* (cf. Kepler's regular solids). For many contemporaries it appeared that with negative weight phlogiston theory had taken a step ahead.

But this was not how Lavoisier saw matters. As a result, he took the powerful and revolutionary step of rejecting phlogiston theory itself. In 1772 he writes,

> This increase of weight arises from a prodigious quantity of air that is fixed during combustion and combines with the vapors. This discovery, which I have established by experiments that I regard as decisive, has led me to think that what is observed in the combustion of sulfur and phosphorus may well

take place in the case of all substances that gain in weight by combustion and calcination (quoted in Conant,1950, p.17).

So although he has not yet arrived at a full Cluster 9 Scientific Theory of oxygen combustion, he is asserting that combustion is a matter of combination of substances rather than a dissociation. His explanation is *the exact opposite of that of the then current observationally-based one*, and has massive implications.

Lavoisier eventually came to understand that it was not air as such but another X which was involved in combustion. This, together with the rejection of phlogiston theory (and thus the need for an entirely new explanatory theory) was the true, revolutionary move. The real discovery occurred in the conceptualization of an X, a substance which was later found in the empirical world. This search for the X took him the next five years, on the way continuing to use his quantitative methods and to produce the first true chemical equation.

And the search was not easy. For he was encountering what might be called the paradox of chemical discovery, viz. that, "Chemical analysis depends intimately upon the purity and identity of the substances under investigation. On the other hand, the purity and identity of the substances depends necessarily upon the correctness of the chemical analysis" (Gale,1979, p.238). The substances which Lavoisier had to make use of were never pure. The abstract concepts of "element", "mixture" and "compound" had not yet been conceptualized. For they could not be, such conceptions having to wait upon the very conceptual and quantitative work Lavoisier was himself slowly sorting out. What is more, he was working with gases— which are mainly invisible, odorless and tastless as well as being difficult to handle. To understand what was happening we must remember, ". .the incredible complexity of the situation faced by chemists. Everywhere they turned they found messy, complicated, impure, real-life situations which they had to try to sort out" (Gale,1979, p.241). The example also throws doubt upon the apology too often made for empiricist pedagogical research and the social scences: that the reason for their lack of potency is the incredible complexity of the situation with which they have to deal.

At this point Lavoisier thought that the X was exceedingly pure air. Within six months Priestley had shown it could not be merely air, but one of several gases of which air was made up. But, Priestley continued to remain entirely within the phlogiston research program! He even made the seeming facts consistent with the phlogiston theory by his very naming of the substance, which he called, "dephlogisticated air". In contrast, Lavoisier took the new substance and made it the central element in his revolutionary theory (Clusters 5 or 9). Initially, he tentatively named it "eminently respirable air", because of its efficacy in animal respiration, and finally, "oxygen". As he said in a memoir to the Academy in 1783, ". .Stahl's phlogiston is imaginary. . all the facts of combustion and calcination are explained in a much simpler and much easier way without phlogiston than with it" (quoted in McKie,1962, pp.110-112).

Thus, between 1777 and 1783 Lavoisier showed by experiments of meticulous collection and weighing of the masses, that with no losses of mass a heated metal + eminently respirable air produced a metal calx and also that a heated metal calx produced a metal plus eminently respirable air. Or, as it is so well put by Bronowski (1973, p.150),

> The idea was simple and radical; run the experiment in both directions, and measure the quantities that are exchanged exactly . . . suddenly the process is revealed for what it is, a material one of coupling and uncoupling fixed quantities of two substances. Essences. . .phlogiston, have disappeared. Two concrete elements, mercury and oxygen, have really and demonstrably been put together and taken apart .

The kind of experiment performed by Lavoisier was guided by the abstract concepts of his Scientific Theory. Its success was judged by the relation which could be established between the rationally connected propositions of abstract concepts and the derivations to the observable world. In such an experiment, hypotheses are drawn from and reflexively develop the Scientific Theory involved, and are performed to discover their scope.

The successes of Lavoisier ought not to blind us to the immense complexities he faced. He had to evolve the Scientific Theory at the same time he was doing the experimentation. But, with the aid of insights produced by Priestley and others, he gradually fought his way through the confusion to develop the necessary abstractions and their interrelationships. For instance he laid it down explicitly that an element was the most basic stage of chemical decomposition and that in all experiments an equal quantity of matter exists both before and after the experiment, and nothing takes place beyond these changes and modifications of the combination of elements. In his climb out of phlogiston and into the oxygen theory of combustion, he had come to understand that the fundamental kinds of matter such as oxygen, combined in precise proportions. These conceptions have become chemical presuppositions which are obvious to us, but were not obvious to the eighteenth century.

In performing his experiments, Lavoisier was able also to predict precisely the amounts of substances produced in his reactions. This again was in immense contrast to the merely qualitative descriptions of phlogiston. In other words, Lavoisier had now connected his abstract concept of oxygen, the Scientific Theory in which it was embedded, and the various effects of oxygen in the observable world— such as its quantifiable proportions, and its combination with mercury to form mercuric oxide. Although in the year 1992, we know that oxygen is of course an element, a real part of the empirical world, which can be made to exist independently and can be enclosed in test-tubes and gas cylinders and in that sense is not today an abstract entity, it is still an abstract entity in the sense that thanks to the work of and the Scientific Theory slowly produced by such people as Priestley and Lavoisier, the concept of oxygen can be rationally connected and manipulated in chemical equations. Moreover, it was indeed an abstract concept in another sense, when Lavoisier grasped its need for existence. It was an abstraction, a theoretical notion, an X which was construed as existing somehow, but whose precise nature and properties remained to be discovered.

The clash of theories (Clusters 5 and 9) ought also to be noticed— the clash between phlogiston theory and the evolving oxygen theory of combustion, for without such a competing, evolving, oxygen theory, the challenge to phlogiston would have remained mute[17].

Equally important in the evolution of the oxygen theory are Lavoisier's statements about the nature of chemical elements, in his ground-breaking book of 1789, *Traite Elementaire de Chimie* (known as, *Elements of Chemistry*). Here, he not only suggested a tentative table of the then claimed elements, including oxygen, but provided the above account of the abstract concept of an "element"-- a working conception of "element" into which future discoveries could be made consonant.

So Lavoisier had begun the development of a research program in oxygen chemistry, and at the same time a new paradigm for chemistry as a whole, viz. that the fundamental building substances of chemistry are elements and that scientists ought to try to distinguish them one from another; that chemistry consists of reactions, and that to do chemistry one should try to find which substances or elements are the reagents, and which the products of the reaction; that by applying the theory of conservation of mass (Cluster 6 Type 9) one can observe whether any of the substances go missing, and if so, the chemist goes looking for them.

In 1989, in a university seminar series on the varieties of science, I heard a chemist who was so trapped in his subject as it is presently constituted, that he espoused a view that chemistry provided *the facts*. He seemed completely unaware that chemistry had a vast and controversial history of theoretical change (several Clusters), as in this example of Lavoisier, which relates to what can count as a chemical fact. It is because there are such scientists, and also such empiricist researchers, who need to be shaken out of their dogmatic slumbers, that Kuhn's account of paradigms has been important despite its faults.

The Discovery by Crick and Watson of the Structure of DNA

How inherited characteristics pass from one generation to another was discovered in 1953. The code rests in the structure of deoxyribonucleic acid (DNA), and its discovery provides an excellent and intriguing example of hypothetico-deductive scientific research, as it is carried on in a continuing and cyclical flux:

> Hypothesis and inference, feedback and modified hypothesis, the rapid alternation of imaginative and critical episodes of thought-- here it can all be seen in motion, and every scientist will recognize the same intellectual structure in the research he does himself. It is characteristic of science at every level (Medawar,1968).

DNA is a nucleic acid, i.e. an acid which occupies the nucleus of cells. It had

---

17. More will be said below about the clash of theories.

been found during the preceding decade that nucleic acids somehow mysteriously carry the codes of inheritance across generations. But how? Thanks to the earlier work of Griffith, Dawson, Sia, Alloway, Levine, Avery and others, it was clear that the chemical building blocks of DNA were sugars and phosphates, together with four small molecules or bases. Two of these are small molecules, of thymine (T) and cytosine (C), in both of which there are atoms of carbon, nitrogen, oxygen and hydrogen, arranged in a hexagon; two, guanine (G) and adenine (A), are somewhat larger, in each of which the atoms are arranged in a connected hexagon and a pentagon. But what was the "architecture" of these bases as expressed in the DNA molecule?

It was also clear by 1951 that the DNA molecule was a long but rigid chain-- a sort of "organic crystal", and highly likely that this chain was a helix. "Helices were in the air, and you would have to be either obtuse or very obstinate not to think along helical lines" (Crick,1988,p.60). But how many helices?

When Watson reached the Cavendish Laboratory in Cambridge, England, in the autumn of 1951, he was exceedingly fortunate to become part of a small and amazingly talented group of leading scientists: Bragg, the Director, himself a Nobel Prize laureate and a great molecular theorist; Huxley researching into the chemistry of muscles (also a later Nobel winner); Perutz, specialist in X-ray diffraction methods and a keen experimenter, and Kendrew who was studying the muscle protein, myoglobin (these two scientists were to share the 1962 Nobel Prize in chemistry); Crick, a physicist, whose interests were between Bragg the theorist and Perutz the experimentalist; nearby was the laboratory of Todd, an organic chemist concerned with how the the atoms of DNA were linked (Nobel Prize, 1957). Although early scientists such as Kepler and Newton worked alone, in modern times scientists have available the expertise and progressive results of colleagues and/or competitors.

Watson teamed with Crick within two days of arrival. Copying Pauling's approach in discovering the structure of the alpha helix, they decided to build models of the DNA helix, using templates like children's "Leggo". "As long as we could be sure it was a helix, the assignment of the positions for only a couple of nucleotides automatically generated the arrangement of all the other components"(Watson, 1980, p.34). So, after quickly dismissing one chain helices as incompatible with Wilkins's X-ray photographic evidence of the diameter of the molecule, they assembled several different models, several nucleotides in length. (Wilkins, and Rosalind Franklin were also working on DNA at King's College, London, using the method of X-ray diffraction patterns, which had been pioneered by Bragg. Their X-ray diffraction photographs showed that the diameter of the DNA molecule was wider than would be the case, were only a one chain helix involved; and Wilkins reported this to Crick.) But there were problems with the inorganic ions, for which there were then no simple rules to tell at what angles they formed their respective chemical bonds. There were also questions about the chemical nature and the shape of the forces which held the chains together. But after some months of trial and error adjustment, a three-chain model began to look very promising. It complied with an earlier idea which they had called the Cochran-Crick theory. So well did it appear to meet all the requirements, that they believed they really had the solution. In their enthusiasm they asked Wilkins and Franklin to come to look at it[18].

Wilkins was not impressed, saying that his colleague, Stokes had made much the same moves in a much simpler manner. Franklin pointed out that her own X-ray data showed clearly that any correct depiction must contain at least ten times as much water as this one. Although this fact did not necessarily show the model to be wrong, for the water could conceivably be placed elsewhere, this would be a terribly *ad hoc* move. And with this amount of water, ". . the number of *potential* DNA models increased alarmingly" (Watson,1980, p.59). The day was an humiliating disaster. Bragg even ordered them to forget DNA. This rejection of Crick and Watson's first theory (Cluster 5, Types 6 or 8) is an appropriate place to discuss some pertinent ideas from Popper (1968;1969). Neither of Kuhn's notions of "normal science" nor "revolution" seems to fit what Crick and Watson were doing— they were not just embellishing, or filling in gaps, and though they were doing something that would prove revolutionary, they were not developing a new kind of science. Their work seems to fit into Lakatos's notion, as a part of a developing research program— they were extending our understanding of biology and biochemistry. But it is the ideas of Popper (1968, 1969) which bring out the most significant features.

It had long been agreed by philosophers, that although there was no *logical* way of inductively establishing the truth of a proposition from a limited number of instances, nevertheless a continuation of positive results of tests of an hypothesis somehow supported it. Popper rejected this suggestion entirely. His radical proposal for solving the centuries-old problem of induction was to argue that scientists and philosophers had been looking at matters the wrong way around. For although there is no *logically* valid form[19] which can prove a claim from specific instances, there is a *logically* valid form which can disprove a claim. This form is the *modus tollens* , or the denial of the consequent, viz.

If H then C
Not C
Therefore not H

In ordinary language this means that if we have an hypothesis or claim H, and C is a consequence which follows logically from it, and it transpires that C is false, then we know that logically the hypothesis H must also be false. From this elementary, but profound logical point, Popper has proposed a revolutionary general approach to the understanding of scientific research. The scientific com-

---

18. As mentioned in my earlier discussion of theory, in the research of Crick and Watson, theory as model was often literal-- and the hypotheses (Cluster 5, Type 8) which they evolved were often able to take a literal three-dimensional shape.
19. Induction is, I believe, as Popper claims, faulty *logically* . But this does not mean that as a *process* within the totality of scientific endeavor, it cannot be preliminary to, or suggestive of, the imaginative conceptual leaps required for the formation of abstract concepts and the propositions which link them (see below).

munity ought to propose radical hypotheses and theories and then use all their efforts *to try to falsify* them[20]. As he says, ". .there is no more rational procedure than the method of trial and error-- of conjecture and refutation: of boldly proposing theories [Cluster 5, Type 6 or 8]; of trying our best to show that these are erroneous; and of tentatively accepting them if our critical efforts are unsuccessful"(1968, p. 51). Popper grasped that the gift which renders us both human beings and scientific researchers is the gift of imagination– of conceiving what *might* be. Moreover, ". . it is always advantageous to try the simplest theories first [Cluster 5, Type 8][21]. These offer us the best chance to submit them to severe tests: the simpler theory has always a higher degree of testability than the more complicated one" (Popper,1969, p.61). If, despite all the attempts to falsify it, the theory still survives, then it has been as Popper calls it, "corroborated". (It has not been proved, for this is logically impossible.) Popper's view is really a kind of corollary of Hume's centuries-old argument that there can be no logical basis for inductive reasoning. Popper's second revolutionary book was called, *Conjectures and Refutations* (1969), and the title is an aphoristic version of his philosophy of science.

So Popper separates science from non and pseudo-science by his criterion of "falsifiability". Scientific propositions or sets of propositions, theories, hypotheses, laws, are all *in principle falsifiable*. Non-scientific and pseudo-scientific propositions are not falsifiable even in principle. This is why non-science, pseudo-science, dogmas, and ideologies can explain everything– or so their proponents claim, for they just "explain away" the counter evidence. Here Popper was paralleling, drawing upon, and developing for his own purposes, the insights of mathematicians such as Hilbert, Godel, Turing and Taski, who demonstrated that attempts to uncover an ultimate and comprehensive deductive system of the phenomena of the world were impossible because: (a) every axiomatic system is limited, and such limitations cannot be forseen or circumvented, and (b) any axiomatic system cannot be guaranteed to be consistent. In short, the explanation of nature cannot be formulated in a deductive, formal and unambiguous system, which is also complete. Hence all genuine Scientific Theory (Cluster 9) is potentially falsifiable.

However, in the real world of scientific research, actual falsification may be an immensely convoluted matter. It was Quine who made this point in the most striking way: ". . our statements about the external world face the tribunal of sense experience not individually, but as a corporate body" (1953/1961, p.41). Quine says that we must reject the view that we can use experience to confirm *or*

---

20. Though he did not follow it in his own experiments, in his *Novum Organon* (1620) Francis Bacon was an early advocate of what we now call "falsification" or "falsifiability". Bacon writes perceptively, ". . it is the peculiar and perpetual error of the human intellect to be more moved and excited by affirmatives than by negatives. . .Indeed in the stablishment of any true axiom, the negative instance is the more forcible of the two (1960, pp. 50-51).
21. cf. Galileo's work on free fall.

to falsify hypotheses. He agrees that when observation apparently falsifies, then we may need to modify our science— somewhere. Yet this can be done in a variety of ways. The idea is succinctly put by Garrison: ". .no scientific proposition, including explanatory hypotheses can be tested in isolation" (1986,p.14). Thus, theory (various Clusters) is underdetermined by evidence[22]. Indeed the situation is even more complex. For, whereas the notion of underdetermination by observational evidence at least acknowledges the significance of that evidence, because of the arguments introduced by Hanson (1972), observations themselves become equivocal in meaning, because they are themselves theory-laden with (at least) Cluster 6 presuppositional theory. Yet it may be just such presuppositional theory which is the problem. It is also germane to point out, as Siegel does that,". .the history of science is full of examples of scientists who stubbornly maintained their theories in the face of negative evidence, only to be vindicated in the end"(1988, p.146). The early Copernicans, the Newtonian planetary astronomers, and the early continental drifters are notable cases.

However the problems introduced by Quine-Duhem[23], and by the fact of theory-ladenness have been greatly exaggerated. For one thing, a scientist cannot change just any arbitrary part of the "network" or "corporate body" to suit his pet hypothesis. For the network or corporate body works both ways: it certainly makes possible the diversion of a challenge to some other part; but it also places reciprocating constraints on where this challenge can be sensibly placed. For another thing, as Giere points out in a recent book (1988), the instruments and technologies of scientific research provide a particularly stable, powerful, and reliable context upon which the scientific community in a particular discipline clearly and properly places great store. For a third thing, a number of philosophers, particularly Scheffler (1967), Hollis (1982), Bernstein (1983), F.C.White (1983), D.C.Phillips (1983, 1987), and Siegel in a whole string of probing articles and books (1980, 1985, 1986, 1988, 1991) have pointed out the multiple errors in the assumptions that are supposed to follow from these theses[24]. For a fourth, the history of philosophy is replete with arguments, which pragmatic, down-to-earth action in the real world has rendered null. Observational evidence has never been quite the hopelessly subjective or permanently equivocal thing these post-positivistic arguments suggest; pragmatic decisions are taken by real scientists; falsification clearly occurs. The philosophical-pretension-pricking power of this pragmatic action is most effective in technological applications of Scientific Theory. Technological applications just do not work in the empirical world unless the Theory is apt.

Of course, Popper is too sophisticated and too sensitive to and familiar with

---

22. More will be said on this point, towards the end of this chapter.
23. This view that observational evidence may not itself determine our judgment of an hypothesis is normally called the "Quine-Duhem thesis", because the French physicist, Duhem advanced related ideas in 1906 (Duhem, 1906/1954).)
24. More will be said in the final section of this chapter.

actual scientific practice to fail to realize that scientific behavior is wonderfully complex, and that, for instance, scientists normally continue to use the currently most successful theories (Cluster 9), even though these may have been faced with apparently contradictory facts. The important point is not that the theory must be abandoned immediately, but rather that such contradictions and anomalies must be recognized and carefully noted, until such contradictions become overwhelming, or until new evidence or conceptualization can explain the anomaly. Contradictions and anomalies must not be just explained away, or merely redefined so that they fit the theory once again— as occurs with theories of Cluster 7, Type 11 (Doctrine and Dogma).

Though Popper's account may not describe all moves and episodes in the history of scientific research, it matches many, a good example being our present concern, the research of Crick and Watson. Moreover, Popper's account is popular with many practising scientists, perhaps the most vocal of whom have been the Nobel Prize winners, English biologist Sir Peter Medawar and Australian physiologist Sir John Eccles. Indeed, Crick is Popperian in reporting on his collaboration with Watson that:

> If either of us suggested a new idea the other, while taking it seriously, would attempt to demolish it in a candid but nonhostile manner. This turned out to be quite crucial. In solving scientific problems of this type, it is almost impossible to avoid falling into error. . Now to obtain the correct solution of a problem, unless it is transparently easy, usually requires a sequence of logical steps. If one of these is a mistake, the answer is often hidden, since the error usually puts one on completely the wrong track. It is therefore extremely important not to be trapped by one's own mistaken ideas (1988, p.70).

It should be noticed that in talking about conjectures and refutations, Popper's account also introduces the issue of how science changes, and thus implicitly the question of whether science is cumulative, or at various stages advances in discrete leaps (as Kuhn argues), or whether there is some *via media* .

Under the cover of alternative research, Crick and Watson went back to the "drawing board" for a year, to study the biochemistry, and to do some Popperian "tinkering" and debating, when likely ideas arose. During that year, Franklin's X-ray photographs were continually improving, though according to Watson she was still opposed to helices. However, she thought there was evidence that the sugar-phosphate backbone was on the outside of the molecule. This was indeed what was finally established by Crick and Watson, and seems to have been a major insight by Franklin (Klug, 1968). But Watson claims that there was no easy way to judge at that time whether this idea was well-based.

They still had far to go: ". . we still remained much at the same place we were twelve months before" (Watson, 1980, p.86). Every now and then they would become enthusiastic and make what looked like promising changes. But sooner or later the creative genius of Crick would grasp, ". . that the reasoning which had momentarily given us hope led nowhere" (p.87). So Crick and Watson were doing what Popper says is so characteristic of research. Each of their successive versions of the DNA helix was checkable against the constraints of contemporaneous theory (Clusters 6 and 9) and/or observable evidence, and was falsified there and then (despite Duhem-Quine, and Hanson)– until their final version.

During the first week of February 1953, a manuscript copy of a paper by Pauling was provided by his son Peter. Like that of Crick and Watson of a year previously, it was a three helix model with the sugar-phosphate backbone inside. They immediately realised that something was amiss. Then they saw that the phosphate groups in Pauling's model were not ionized, but on the contrary that each group contained a bound hydrogen atom and so had no net charge. Thus its nucleic acid in a sense was not an acid at all. Moreover, the uncharged phosphate groups were significant features of the model. Again, because the hydrogens were part of the hydrogen bonds which held the three chains together, then without the hydrogen atoms the chains would immediately fly apart and the structure vanish. Watson's evolving knowledge of nucleic acid told him that phosphate groups never contained bound hydrogen atoms. Yet Pauling, the world's greatest biochemist, had drawn the opposite conclusion. This view of Watson and Crick was supported by other scientists at the Cavendish.

Crick and Watson realised that by mid-March Pauling's paper would have been published and he would then be apprised of his error. They had about six weeks to produce something better.

They examined new photographs provided by Wilkins, taken with a second form of DNA, the "B" form, originated by Franklin. This form gave clearer pictures, and Watson became even more convinced of their helical structure. "With the "B" form. . mere inspection of its X-ray pictures gave several of the vital helical parameters" (1980, p.98), for example,". . that DNA was a helix which repeated its pattern every 3.4 A [Angstrom Units] along the helical axis" (p.99). Watson tried to convince Wilkins of the urgency of the situation, if they were to beat Pauling. Wilkins remained unmoved making the point that,". . *if we could all agree where science was going, everything would be solved and we would have no recourse but to be engineers or doctors* (italics added)"(p.99)– itself an acute and astute comment about the nature of science.

Wilkins seems to have preferred three helices, because of the value of the water content(Watson, 1980, p.99). Crick and Watson decided to stay with two chain models. This was a momentous decision. Especially important was Watson's observation that the meridional reflection at 3.4 A was in fact much clearer than any other reflection. Further consideration made him realize that the 3.4 A thick purine and pyrimidine bases must be stacked on top of each other in a direction perpendicular to the helical axis. In addition, they felt sure, from both electron-microscope, and X-ray evidence, that the helix diameter was about 20 A.

In alluding to the importance of earlier knowledge in the discovery of new knowledge in research, Crick has recently written,

The main difference of approach was that Jim and I had an intimate knowledge of the way [Pauling's] alpha helix was discovered. We appreciated what a strong set of constraints the known interatomic distances and angles provided and how postulating that the structure was a regular helix reduced the number of free parameters drastically. The King's [College] workers were reluctant to be converted to such an approach (1988, p. 68).

A delay of some days intervened before the machine shop could turn out the further components required for model building. For one and a half days, Watson tried strenuously to produce a two-chain model with the sugar-phosphate backbone inside. He says that all the possible models compatible with the 'B' form X-ray data, however, looked stereochemically even more unsatisfactory than their earlier three-chain model. The fundamental problem was how to accommodate the bases if they were inside the helix. "If they were pushed inside, the frightful problem existed of how to pack together two or more chains with irregular sequences of bases"(1980, p.102). But they did try some versions with the bases inside. Shapes compatible with Franklin's X-ray pictures and also "stereochemically reasonable" (p.104) were easy enough to produce. Crick and Watson felt that the most satisfactory angle of rotation between adjacent bases was between 30 degrees and 40 degrees. By contrast, angles twice or half this size, ". . looked incompatible with the relevant bond angles" (p.102). This meant that if the sugar-phosphate backbone were on the outside, then the crystallographic repeat of 34 A had to represent the distance along the helical axis required for a complete rotation.

Further, because of a happening, the exact form of which is disputed in the historical accounts, Crick and Watson gained access to Franklin's precise experimental data. They then realized that only minor modifications were necessary in their backbone configuration. But what modifications? Watson says that every time he attempted to produce a solution, he encountered the fact that all four bases had quite different shapes, so that,". . unless some very special trick existed, randomly twisting two polynucleotide chains around one another should result in a mess (1980, p.106). Though for more than a year they,

> . . had dismissed the possibility that bases formed regular hydrogen bonds, it was now obvious to me that we had done so incorrectly . . . a recent rereading of J.M.Gulland's and D.O.Jordan's papers on the acid and base titrations of DNA made me finally appreciate the strength of their conclusion that a large fraction, if not all, of the bases formed hydrogen bonds to other bases. Even more important, these hydrogen bonds were present at very low DNA concentrations, strongly hinting that the bonds linked together bases in the same molecule (p.106).

By the middle of the following week, ideas began to flow. A DNA structure might have an adenine residue whose hydrogen bonds were similar to those of pure adenine. Watson began to wonder if each DNA molecule might consist of two chains with identical base sequences, held together by hydrogen bonds between pairs of identical bases. This was a notion of like-with-like base pairs, i.e. of A (adenine) with A and T (thymine) with T, and so forth. However, "The resulting backbone would have to show minor in-and-out buckles depending upon whether pairs of purines or pyrimidines were in the centre" (Watson, 1980, p.106). Watson writes that despite the messy backbone, his pulse began to race, because the existence of two intertwined chains with identical base sequences could not be a matter of chance. Rather, it strongly suggested that one chain in each molecule had at an earlier stage served as the "template" for the synthesis of the other. In such a situation the replication of the gene would begin with the separation of its two identical chains. Two new daughter strands would be made on the two parental templates, thereby forming two DNA molecules identical to the original. So the essential mechanism of gene replication would come from the feature that each base in the newly synthesized chain always hydrogen-bonded to an identical base (p.108).

But though it was a good idea, and gene replication does indeed begin with the separation of two identical chains, Watson was wrong about the like with like base pairs— but not so wrong. Their continuing attempts would finally be able to get things right with an even better, simple idea.

The singular good fortune of Crick and Watson in being surrounded by men of genius at the cutting-edge of their fields, once again showed itself. Watson outlined his scheme, but the crystallographer, Jerry Donohue, himself second only to Pauling in knowledge of hydrogen bonds, said that the new idea was unworkable (Watson,1980, p.149). He pointed out the awkward fact that they had chosen the wrong tautomeric (the same chemistry but different shape) forms of guanine and thymine, even though Watson had copied these from the leading book in the field! Donohue suggested that for years organic chemists had been favoring certain forms on flimsy grounds, and that most textbooks had diagrams of unlikely forms. Donohue argued that it was the keto tautomeric forms which must apply rather than the enol forms, which they had required to make the model fit. This meant a significant change in the position of the hydrogen atoms in the molecules. Crick saw two further problems. A like with like structure would give a rotation of only 18 degrees between successive bases, which sat badly with the recent adjustments to the model. Secondly, the structure did not fit the discovery of Chargaff, which showed that A always paired with T and G with C.

The following day, further play with the model showed that an adenine-thymine pair held together by two hydrogen bonds was identical in shape to a guanine-cytosine pair held together by at least two hydrogen bonds. Moreover, the hydrogen bonds all formed naturally. No changes were required to make the shape of the two types of base pairs identical. Donohue had no objection to the new version. It was clear that two irregular sequences of bases could be packed in a regular way in the centre of a helix if a purine always hydrogen-bonded to a pyrimidine. Moreover, the need for hydrogen-bonding indicated that adenine must pair with thymine, while guanine must pair with cytosine. At this point, "Chargaff's rules then suddenly stood out as a consequence of a double-helical structure for DNA. . .Always pairing adenine with thymine and guanine with cytosine meant that the base sequences of the two intertwined chains were complementary" (p.114).

Crick also noticed that the bonds which joined each pair of bases to the sugars were systematically related by an axis which was perpendicular to the structure. The contemplated structure was such that replication in one of the chains would act as a template or a mould for synthesizing the other chain: the sequence of the bases in one chain would determine the complementary sequence in the other. Over several days their further fastidious and precisely measured work with the model corroborated their views. At last they had achieved it– *not by creating new abstract concepts, but by ingeniously rearranging relationships amongst established abstract concepts.*

Wilkins and Franklin supported the Crick-Watson interpretation. Todd, the organic chemist was equally accepting of the structure. On April 23rd, 1953, the 900 carefully chosen words of what was to become one of the most famous papers in the history of science was sent off to *Nature*. The paper explains that DNA molecules are long chains of repeating groups of sugar and phosphate organized into two right-handed helical chains, with the bases on the inside of the helices and the phosphates on the outside. A base is attached to each sugar, with the sugar-phosphate chains forming "backbones" from which the bases radiate inwards and make connections with one another. There are four different bases: adenine and guanine, and, cytosine and thymine. There is also,". . a residue on each chain every 3.4 A in the z-direction. We have assumed an angle of 36 degrees between adjacent residues on the same chain, so that the structure repeats after every 10 residues on each chain, that is, after 34 A . ." The heart of the paper comes in two statements. The first is,". . it is found that only specific pairs of bases can bond together. These pairs are: adenine (purine) with thymine (pyrimidine), and guanine (purine) with cytosine (pyrimidine)." This pairing was seen to be necessary because it had been shown that two pyrimidines could not reach far enough to cover the space between the sugar-phosphate helices, while two purines would be too crowded. Furthermore, an adenine would always be opposite a thymine and a guanine always opposite a cytosine. The four different bases were sufficient to account for combinations to produce the twenty different kinds of amino acids which make up the proteins. The second statement is, "It has not escaped our notice that the specific pairing we have postulated immediately suggests a possible copying mechanism for the genetic material." This reads as a splendid understatement[25], for, given their final model, the way in which the replication occurs at least looks obvious: how DNA splits apart so

25. Crick's comment is revealing both of the complex motives of scientists and of their awareness of the complexity of that with which they are dealing:
This [statement] has been described as "coy", a word that few would normally associate with either of the authors. . In fact it was a compromise, reflecting a difference of opinion. I was keen that the paper should discuss the genetic implications. Jim was against it. He suffered from periodic fears that the structure might be wrong and that he had made an ass of himself. I yielded to his point of view but insisted that something be put in the paper, otherwise someone else would certainly write to make the suggestion, assuming we had been too blind to see it. In short, it was a claim to priority (1988, p. 66).

that one chain becomes a sort of mold for the other, to make more DNA exactly like itself.

These discoveries were subsequently corroborated in experimentation. And, dominated by Crick and Watson independently, a vast new research program in biochemistry and molecular biology resulted, one of the latest developments of which is the immense, mind-stretching project to map the human genome.

Wegener, Continental Drift, and Plate Tectonics to 1991

In profound contrast to present-day views, official geophysical tradition in the early twentieth century ignored anomalies, emphasized the properties of a solid earth and denied the possibility of any lateral movement of continents, basing explanation mainly on a contracting crust and its local effects. Thus geology was in no mere natural history, or "pre-paradigmatic", or empiricist stage, but had theoretical orientations (from various Clusters).

There had however been earlier suggestions that continents had been linked in various ways, and in the late nineteenth century, the Austrian, Suess argued that most of the continents had once been concentrated in a single, giant land mass, which he called, "Gondwanaland"[26]. But it was the American, F.B.Taylor, who was the first to develop an account involving large-scale conceptual leaps, akin to ideas of continental drift and plate tectonics. He argued for, ". . a mighty creeping movement" of the earth's crustal "sheets" from the north towards the periphery of Asia. Some of his views were partly-correct precursors of more recent ideas. Prophetically, he wrote, "It is probably much nearer the truth to suppose that the mid-Atlantic ridge has remained unmoved while the two continents on opposite sides of it have crept away in nearly parallel and opposite directions" (1910, p.218). However, the first scholar to deal in a fully systematic manner with the idea of "continental drift", as movement of the continents came to be known, was the German meteorologist, Alfred Wegener (1880-1930)[27].

Wegener says that his first intimation that the continents might have moved laterally came in 1910, when he, like others before him, became intrigued by the continental fit on each side of the Atlantic. Initially sceptical, in autumn of 1911

---

26. After Gondwana, a key geological province in India.
27. During the 1920s, geologists called the idea of drift, "The Taylor-Wegener Hypothesis". Taylor delivered his paper to The American Geological Society in 1908, but it was published only in June 1910. Wegener says he developed his ideas independently of Taylor. Taylor disagreed (see Totten, 1981). Some geologists, such as my former colleague, Mr.David Hannan, are convinced that Wegener purloined Taylor's idea. Wegener makes quite explicit the following point about precursors: "I myself only became acquainted with these works-- including Taylor's-- at a time when I had already worked out the main framework of drift theory, and some of them I encountered much later on" (1966, p.4). Whatever the explanation, it is clear that Wegener's case is typical of discoverers of science. He immersed himself in geological ideas, and his theory sprang out of that general *Zeitgeist* .

he came "quite by accident", upon a "synoptic report" on paleontological similarities between Brazil and West Africa. The similarities were seen as evidence for a former land bridge. "Land bridges" were the skeleton in the cupboard of orthodox geology, because they were *ad hoc* and defied other then-accepted principles. Wegener realized that they could equally be evidence for drift. Having the abstract concept, Wegener sought empirical evidence. By January 1912, he was sufficiently convinced to write two papers, in which he also briefly adverts to Taylor, in particular to Taylor's ideas on the separation of Greenland. His great book, *Die Entstehung der Kontinente*, with the English title, *The Origin of Continents and Oceans*, came out as a slim volume in 1915[28]. Wegener begins by saying that scientists still do not understand sufficiently that, ". . . all earth sciences must contribute evidence towards unveiling the state of our planet in earlier times, and that the truth of the matter can only be reached by combining all this evidence" (1966, p.xxix). These words were prophetic of post-World War II developments.

He visited Greenland four times. Giere argues that the experience provided Wegener, ". . with a valuable 'cognitive resource' " and that the glaciers and icebergs furnished him with, ". . a rich source of metaphors, for large masses moving imperceptibly across the surface of the earth" (1988, p.230). Certainly Wegener seizes upon measurements of the longitude of Greenland from 1823,1870,1907 and 1927 and concludes that it is moving away from Europe at the rate of about thirty-six metres a year[29] (1966, p.29). We should note the reasoning: *if the abstract concept of continental drift is apt, then there must be empirical interpretation* .

Wegener's visionary suggestion is that at one time in the remote past, all the continents were united in one vast mass of land in one world-girdling ocean. He calls his supercontinent,"Pangaea" from the Greek word meaning "all Earth". Pangaea fragmented and the various sections drifted apart. Following his own advice, he provides geodetic, geophysical, geological, paleontological and biological arguments.

His *geodetic arguments* challenge the view of the slowly contracting earth assumed by orthodoxy. He argues persuasively that a fairly uniform contraction of the earth's surface would have to provide a much more even coverage of fold mountains spreading across continents, rather than the lineal belts of mountains actually found.

Amongst various *geophysical arguments*, Wegener points out that statistical analysis of the topography of the earth indicates two predominant levels. This he shows is inconsistent with the then-accepted idea of random subsistence and up-

---

28. There were longer, revised editions based upon additional studies, in 1920, 1922 and 1929. The 1922 edition claimed most attention, being translated into English, French, Spanish and Russian. References used in the present book are from the 1966 translation of the 1929 edition.
29. This suggested movement has not been supported by modern measurements. But that is not important, for there is nowadays ample corroboration of *plate* movements.

lift. To support this idea of lateral movement, Wegener argues that the upper earth lives a sort of dual-life, as do glass and pitch, behaving as a solid, elastic body when acted upon by short-period forces such as seismic waves, but as a plastic fluid under forces applied over geological time scales (Wegener,1966, p.55). Forty years later, Runcorn was to support a reasonably similar, but more geologically sophisticated view (1962a, p.311), as do modern earth scientists.

He marshalls convincing *geological* arguments, e.g. the similarities between the two sides of the Atlantic. Thus the Buenos Aires range finds a continuation in the Cape of Good Hope fold belt. His argument then is something like: "If a mobilist approach is correct, the existence of strong geological similarities at corresponding locations in Africa and South America is very probable [but on a static view their existence] is highly improbable" (Giere, 1988, pp.232-233). Wegener explains the complex geology of the Moluccas, by the impact of a northward moving Australia-New Guinea. Such accounts are fascinatingly close to those of modern plate-tectonic theory.

He gives intriguing *paleontological and biological arguments*. A link between Australia and South America is indicated by the marsupial opossums of the latter. "Even the parasites of the Australian and South American marsupials are the same" (Wegener, 1966, pp.108-109). Given sufficient time, such resemblances might be explained by the theory of land bridges (Cluster 5). But it also suits drift equally satisfactorily. Here we have an instance of the not uncommon situation in the history of science, of the underdetermination of theory by data, where two different theories (Clusters 5 or 9) (land-bridges, and continental drift) can seemingly explain the same observations, and we have to await further developments before we can adjudicate.

He also cites paleoclimatic evidence, which indicates similar ancient climates for presently widely-separated regions.

Wegener's theory (Clusters 5 or 9) was dramatic, bold, and all-embracing. No one else had mounted such a unified challenge to geologists, geodesists, geophysicists, paleontologists, botanists, zoologists, and climatologists. The fierce antagonism to Wegener in the U.S.A. did not develop until the mid-1920s. One event has been argued as seminal, viz. the first international symposium in New York in 1926, of the American Association of Petroleum Geologists. Those who spoke most persuasively were very decidedly against him, ". . and they used geological evidence together with logic, wit, and sarcasm with telling effect" (Marvin, 1974, p.88). The fit of the Atlantic continents was challenged, as was the purported geological and glacial evidence. R.T.Chamberlin thought that to match moraines, ". . to prove continental linkage in the Quaternary was positively ludicrous" (Hallam,1973, p.23). "Can we call geology a science when there exists such difference of opinion on fundamental matters as to make it possible for a theory [Cluster?] such as this to run wild?" (Marvin,1974, p.1). This response is deeply ironic when we remember that Chamberlin's father, T.C.Chamberlin was known for a paper famous in the philosophy of science (1897), in which he advocated a "method of multiple working hypotheses". By this he meant that scientific research works better when it is undogmatic and makes use of several alternative interpretations at the same time.

More recently, Feyerabend (1968, 1976) has said similar things. There is, says Feyerabend, no method of scientific research: that idea is an unacceptable conservatism. Moreover, he believes scientists ought not necessarily to wait until a theory (Cluster 9) has been falsified in the normal course of events. Scientists ought to consider, even deliberately create alternative theories (Cluster 5), hoping to produce observations which might falsify current ideas, because some of the limitations can be exposed only by developing a rival. Even more, scientists should consider what can still be learned from theories (Clusters 5 and 9) which have been falsified. Thus, says Feyerabend, there is no normal science as Kuhn would have it: indeed, were this form of scientific research to exist, it would drastically restrict progress. Not only does Feyerabend reject Kuhn's idea of normal science, he argues diametrically that there should be methodological anarchy, so that "anything goes" (1970c, p.22) and it is this which has allowed science to progress. In later writing he describes this as methodological "dadaism". Taking Galileo as his prime example, Feyerabend claims that every supposed scientific principle has been violated at one time or another by a great scientist. He argues that Galileo succeeded only by sabotaging Aristotelian physics by propaganda and circular arguments, and by making Aristotelians look like obscurantist fools. Moreover, he argues, metaphysical beliefs strongly influence theoretical development (Cluster 5), and whether scientists accept a new theory (Cluster 5 becoming Cluster 9) depends as much upon aesthetics and their nationality (1981a,p.60-62) as upon any rational considerations.

While Feyerabend's ideas of the deep significance of new and competing theories, and of the profound influence of metaphysical beliefs are important for scientists to bear in mind (and also for those who do research on teaching), in other ways he has to be rejected. For instance, it should be pointed out that although the Aristotelians may well have been defeated by propaganda and circular arguments, *it was not these which made Galileo's science good science* – as Feyerabend acknowledges. Galileo has been amply vindicated by the fruitfulness, force and technical applicability of his science. Methods of persuasion are irrelevant to the power and truth of Galileo's propositions. However, propaganda and persuasion and the implied threat to one's scientific career may be temporarily powerful, and it was these I believe, more than rational considerations, which destroyed Wegener's ideas in the United States. As a meterologist turned earth scientist, he did not quite fit any regular scientific mold. Though some acknowledged Wegener's evidence, which was strengthened by the work of the South African, Du Toit[30], in the late 1920s and 1930s, thirty-five years were to pass, before American geologists would meet again for the purpose of seriously discussing drift.

However, there may be a little more rationality in the rejection than I am suggesting. For Wegener's critics pointed to what was on then-accepted geophys-

---

30. Du Toit's fieldwork so powerfully supported Wegener, that he included Du Toit's map of Paleozoic glaciation in the 1929 edition of *Die Entstehung der Kontinente*.

ical theory (Cluster 9) a massive problem: how could lighter continents plough their way through more dense sea-floor? To this Wegener had no convincing response. Giere says (1988, p.235) that from the point of view of decision theory (Cluster 3? Cluster 9?), such a geophysical argument *ought* to have been the most significant (physics being the most prestigious and and well-established science of all), and for many it was.

In continental Europe there was never the depth of antipathy to Wegener's ideas which occurred in the United States. "Sympathetic interest" would have been an appropriate description of many Europeans (Hallam,1983, p.130). In an encouraging letter to Wegener, the German, Milankovitch also expressed an important philosophical point, germane to the argument of this book,

> I am still totally under the influence of your brilliant lecture. . .I am not in the least bit disturbed that not all geologic details will fit neatly into your picture. The same applies to those who occupy themselves with the investigation of the mechanism behind complex natural phenomena. Those who on the contrary *have spent their whole lives collecting and recording facts , are incapable of penetrating beyond these facts* ; their view is fixed on the surface. You should not be discouraged *if you find it more difficult to persuade empiricists than students of the exact sciences*, quite the contrary in fact (italics added) (quoted in Hallam,1983, p.130).

In the British Empire supporters and critics were more evenly divided— with more supporters in the Southern Hemisphere, where the positive geological and palentological evidence is more immediately compelling. Holmes, of Edinburgh University had even proposed a mechanism— convection currents in the mantle (1929; 1944). This problem of mechanism is an example of what Laudan (1977, pp.50-54) in his account of scientific development based on the idea of "problem-solving effectiveness"[31] has called a "conceptual problem" in the development of Scientific Theory: yet as Giere points out, Holmes's conceptual solution had little impact. Giere attributes this to, ". . a simple realism among scientists. They are, in general, not going to decide that a model is right, or even worth pursuing, without some basis for that decision beyond its mere physical possibility" (1988,p.250). True. And several related points need to be made. First, "solving a conceptual problem" is itself an ambiguous notion[32]. It may mean proposing a putative solution, as in the present instance; or it may mean proposing an actual solution, as in Lavoisier's conceptualizing of oxygen (which quickly fitted into a whole new research program). Secondly, Giere's "mere physical possibility" (often ambiguous) is actually one of the features of models – they often remain suggestive, not substantive. That is part of their psychic attraction. It is also both their strength and their weakness. Moreover, to

---

31. See also Frankel,1980.
32. I am not referring to Laudan's distinction between inconsistencies within or between theories, and inconsistent ontological assumptions.

become powerful, the model has to be developed so that it is no longer just a model, but has evolved through one or more sets of propositions of Cluster 5, Type 8, to form Type 14 (First-order Scientific Theory), as in the oxygen case, and as in that of Maxwell's equations of electromagnetic radiation. Holmes's conception was at the putative stage: perhaps the conception still is, for even today there is a huge amount of debate about the actual physical form of such convection currents.

Post World War II American criticism continued. Bailey Willis who had contributed to the 1926 symposium, continued to be sarcastically contra, saying, "The theory of continental drift is a fairy tale, *ein Marchen* . It is a fascinating fancy" (quoted in Hallam,1983, p.136). And in 1967, a year or so after the large-scale change in beliefs in 1966 (see below), Harvard University Press published the *Source Book in Geology: 1900-1950* (Mather,1967). A compilation of the important geological ideas, of the first half of the twentieth century, the book contains *no* mention of Wegener, Taylor, or continental drift!

The situation was changed by an explosive outburst of new abstract concepts, observations and techniques from the expanding scientific community in the 1950s and 1960s. For a time after the later sixties, it was often suggested that the change from ideas of stabilism to drift was a fine example of Kuhn's revolution and paradigm change. But scientific change is indeed immensely complex, and the more details one notes, the less does Kuhn seem to apply. Except for the work of King (1983) in South Africa, and Carey in Tasmania (1958, 1974, 1976) (and they believed in an *expanding* earth which caused the movements) one does not see the steady development of the rival mobilism as a research program, or research tradition (Laudan, 1977). Moreover, most of these developments occurred quite independently of any committed interest in stabilism or drift.

With hindsight, we can see that one key reason for the lack of acceptance of Wegener, had been the *lack of evidence about the nature of the ocean bottom* : as oceans cover some 70% of the earth's surface, knowledge of their nature is crucial. Even though conceptual abstraction is essential to the development of Scientific Theory, theory must still provide interpretation to the empirical world, and to do that, sufficient observational evidence must be available and checkable. From the 1950s, new devices made for increasingly subtle mapping and testing of ocean topography. Menard writes,

> In 1950 the first generation of the new marine tools was assembled: echo sounders, magnetometers, temperature probes, explosion seismometers, piston corers, and dredges. Within five years they would be adequate to discover the magnetic anomalies that take the measure of the world, the great fracture zones that offset them, and the median rift where the sea floor is born (1986, p.3)

Oceanic observation and theory (several Clusters) began to work hand-in-hand. One crucial theory (Cluster 5, Type 8) was Hess's sea-floor spreading hypothesis (SSH) from the mid-ocean ridges, which he "grafted" onto the theory of continental drift, thus leading to more powerful problem-solving, at the same time (like Holmes) explaining the mechanism. He argued: "The continents do not plough through oceanic crust impelled by unknown forces; rather they ride passively on mantle material as it comes to the surface at the crest of the ridge and then moves laterally away" (1962, p. 608)[33]. As Hess also says in good Popperian style, "Without hypotheses to test and prove or disprove, exploration tends to be haphazard and ill-directed. Even completely incorrect hypotheses may be useful in directing investigation towards critical details" (quoted in Frankel, 1980, p.349).

But the most significant development, unsuspected pre-war, was the theory of rock magnetism or paleomagnetism (moving from Cluster 5 to Cluster 9, but also involving Cluster 9). During the 1950s and early 60s through such people as Runcorn and Blackett in England (Runcorn, 1962; Blackett and others 1965), and Cox, Doell and Dalrymple in California (Cox, Doell and Dalrymple,1963; Doell and Dalrymple, 1966), it became known by studying the magnetism of surface volcanic rock, and deep-sea sediments, that the earth's magnetic field had reversed many times during the past. In the oceanic crust for example, this had produced anomalies of alternate magnetization, discovered initially in the Pacific, but later located in every ocean. In 1963 Vine and Matthews published a crucial paper on such anomalies, which detailed the patterns of magnetic anomalies on the Carlsberg Ridge in the Indian Ocean. Blending Hess's SSH and the magnetic anomalies, they realized that if Hess were largely correct, then each side of the mid-ocean ridges should show symmetrical anomaly patterns of bands of 'normal' (present) and 'reversed' polarity. This was the case. Fossils entombed in ocean sediments raised by the oceanic-drilling ship *Glomar Challenger* also produced good evidence for the uniform drifting apart of the Americas from Europe and Africa. Vine then put together all this evidence for sea-floor spreading in a powerful paper in December 1966 (Vine,1966), which stunned the earth science community. In respect to the changing position of Australia, for instance, estimates from paleomagnetism agreed closely with those of paleoclimatology. This all began to look like dramatic corroboration of Wegener's ideas.

There were also several cooperative efforts between M.I.T. and the University of Sao Paulo, to search for confirmatory evidence. Using a map by Bullard as one tool, they predicted that a continuation of a sharp boundary between Precambrian rock provinces in Ghana would be found in the vicinity of San Luis in north-eastern Brazil. The predictions were splendidly borne out.

Plate-tectonics is the final brilliant and revolutionary theory (Cluster 9) which has resulted from Wegener's original conceptual abstraction of continental drift. Through a synthesis of the wide range of conceptions such as strike-slip faults, patterns of magnetic anomalies, a multitude of further data, and the arguments of Vine and others, it was conceived independently by the young scientists, Morgan and McKenzie and Parker (Menard, 1986). Early in 1967 they separately thought of applying their conceptions to the surface of a sphere, by

---

33. Rates of spreading vary from about one centimeter per year per ridge flank near Iceland, to six per ridge flank in the equatorial Pacific (Vine, 1985, p.877).

way of of Euler's Theorem[34] (Menard,1986, p.233; p.284). Their theory of plate tectonics suggests that the earth's surface is made up of a number of large so-called "plates" and several smaller ones which, riding on currents in the mantle, move and jostle one another, to produce the multifarious features of the earth's crust. It is hard to think of notions more abstract and less easy to be discovered by some sort of direct observation than those of crustal plates riding on global arrays of convection cells deep in the mantle beneath them. *For there is no generalization or precision of empirical observation in the conceptualization* : the *concept* of plate is the important matter– the precise, empirical nature, particularly the "bottom" of plates, remains obscure. But that does not matter; it does not prevent incisive extrapolation, unified Scientific Theory, explanatory power, and an important variety of quantitative prediction. Indeed the point should be put the other way around: the abstractness of the concept of plate makes the scientific usefulness possible. As Menard says ". .as the geological world tried to accommodate the shock of Vine's paper in December [1966], Jason Morgan was already developing the *quantitative theory* of plate tectonics that would subsume the qualitative miracle of sea-floor spreading" (italics added) (1986, p. 285).

Another key development was the concept of transform fault. A conception developed independently (Menard,1986) by Coode (1965) and Tuzo Wilson (1965), they are faults on which the displacement suddenly stops or changes direction, when it is transformed into a zone of extension or compression. Wilson provides a simplified theory (Cluster 5: Hypothesis) of the opening of the Atlantic, showing how transform faults develop. He shows how the off-sets or displacements within the Mid-Atlantic Ridge are unrelated to the distance through which the continents have travelled, merely reflecting the shape of the initial break between continents. Marvin says they are, ". .a structural necessity in ocean-floor spreading" (1974, p.160). To be noticed again is that *the idea of transform faults was the result of conceptual extrapolation*. Of course, once conceived, such a concept could be seen to be a refinement of earlier depictions, and additional examples of transform faults could *then* be sought for and used in explanation and prediction of further observational evidence.

In 1968, seismologists Sykes, Oliver and Isacks drew attention to how much better the theory of global plate tectonics is as an explanatory and predictive system for earthquakes. For earthquakes occur largely in the circum-Pacific island arcs, the Alpine-Himalayan mountain belts, and the mid-ocean ridges– all of which involve edges of plates.

Thus by 1970, the theory of plate tectonics was almost universally accepted

---

34. Euler's Theorem demonstates that if a block can move on a sphere, then a single constant rotation about a given axis can define its motion. "The velocity of relative motion of any two blocks is proportional to the angular velocity about the axis of rotation and to the angular distance from the axis" (Hallam,1973,p.68). Thus we can expect differential rates of spreading.

by the Western earth science community. What we should not fail to notice is the magnificient simplicity and the seeming obviousness once evolved, of the abstract concept of "plate". Since that date, there has been a continually-improving technology, and a continuing theoretical-observational cycle. Embedded in theory (Clusters 6 and 9), observational evidence is continually being produced by the new technology. Such theory-based observation produces hints as to the nature of the phenomena. Hypotheses involving new abstract concepts ("propagating rift", "overlapping spreading center", "DEVAL", etc.) are suggested and tested against the theoretical background[35].

But it is the earth scientists who take the hints and they who create the concepts. Untheoretical observers would not see the hints to take. For in the earth sciences the observations are theory-laden at two separate levels: there is the theory of Clusters 5, 6 and 9 which the earth scientists *bring to* their making of observations; and there is the Scientific Theory *which informs* the technology itself. Thus, in a recent article by Macdonald and Fox (1990), there are several strikingly colorful maps of the East Pacific Rise and the Mid-Atlantic Ridge. These are the result of a complex sonar system known as SeaBeam, and the theoretical (Cluster 9) contributions of Macdonald and Fox. Observations by way of sonar depend upon electro-magnetic theory and the theory of sound. Macdonald and Fox interpreted the sonar "observational evidence", and with the help of graphic artists turned it into two-dimensional maps, which represent the three-dimensional details. Without the abstract concepts and the theory which links them, such maps provide no information, summarize no data, have no meaning, but are merely colorful patterns.

So Wegener's early theoretical moves (Clusters 1, 5 and 9) have resulted in the marvellous and complex theory (Cluster 9) of plate-tectonics, which has now become the hard core of an ever-expanding research program in the earth sciences. Popper and Feyerabend should be very pleased with Wegener.

Cluster 9: Scientific Theory

It is now time to say something more specific about the nature of Scientific Theory-- which results from the scientific research of such people as Copernicus, Kepler, Galileo, Newton, Lavoisier, Crick and Watson, Wegener, Wilson, and all the others, in their respective fields.

With Kepler's production of his three planetary laws, Galileo's development of his laws of inertia and of the acceleration of falling bodies, and Newton's creation of his three laws of motion and his law of gravity, something new came into the universe: such laws and their interrelationships were the embodiment of

---

35. Not that things are proving easy. The vast complexity of Scientific Theory and evidence promotes continual controversy in its development. Thus Dalziel and Moores, to the astonishment of colleagues, recently proposed that western North America abuted Australia and Antarctica in the late Precambrian (Monastery, 1991).

a new epistemological status. That which they embodied was Scientific Theory. I am not arguing that, say, Galileo or Newton were themselves fully aware of this status. Indeed it is noteworthy that Newton occasionally made some empiricist remarks about his own work (Bronowski, 1973, p.234; Oldroyd,1986, p.83). However, I am arguing that they and the other scientists great and small from that day forth were at work developing an evolving type of knowledge, which accounted for the workings of the empiricial world, but which epistemologically involved perhaps more than they explicitly realized.

To get to the nub of these matters we need to consider some specifics. For example, textbooks of physics point out that Snell's Laws of refraction were important discoveries in the early development of the Theory of Light , viz. :

1. the normal (the straight line at right angles to the surface where the ray strikes) together with the angles of incidence and of refraction of the rays of light all lie in the same plane.

2. for a given pair of transparent media the ratio of the sine of the angle of incidence to the sine of the angle of refraction remains constant regardless of the angle of incidence of the ray of light,

or,

for a given pair of transparent media, $\sin i / \sin r = \mu$. (Where i is the angle of incidence, r is the angle of refraction, μ is a constant for the particular pair of media.)

As a typical section of Scientific Theory more generally, the theory of light refers to the whole set of facts, laws, claims and presuppositions of that particular branch of physical science. Such theory is interpretive, descriptive, explanatory and predictive. For instance, Snell's Laws explain why a pool of water appears to be shallower than it actually is, the effect being due to refraction of the light on passing from water into air. Moreover, Snell's Laws, together with simple mathematical extrapolations can be applied to an amazing range of lenses, prisms and reflectors. Thus this kind of Scientific Theory explains why light behaves as it does, can be used to predict that behavior, and can be applied as a powerful and precise technology.

It has often been suggested that there are two broad levels of Scientific Theory: a first-order level which is active and which encompasses observational, factual, descriptive, and predictive matters, and which involves such entities as Snell's Laws, and a second, more meta-theoretical level which in its turn both makes possible and explains the facts, laws and predictions of the observational level— provides the underpinnings for and the reasons why first-order theory works, and perhaps hints as to where further discoveries might be made. Thus Bergmann argues, "Statements of individual fact are explained by laws; laws are explained by theories"(1957, p.78). And Nagel writes,"No matter how far the question 'why' is pressed— and it may be pressed indefinitely— it must terminate in theory" (1954, p.29). Perhaps the earliest example comes from Newton, who distinguished carefully between the scientific knowledge of Snell's Laws that light is refracted in various ways, and the more speculative theories of waves and corpuscles, which, it was argued were responsible for such refractive phenomena.

That which Toulmin calls "ideals of natural order"(1953) are examples of this more meta-level, such as his example of the principle of rectilinear propagation of light-- without which there could be no Snell's Laws.

It should be stressed that this distinction though sometimes useful, *is by no means absolute*. When I wish to draw it, I shall refer to the the lower-order level as First-Level Scientific Theory (here called Type 14) and the more meta level as, Second-Level Scientific Theory (here called Type 15). On such a division, the Wave-Particle Theory of Light, and the Wave-Particle Theory of Electromagnetism more generally (Type 15: Second-Level Scientific Theory) explain why we might *expect* laws such as Snell's to apply (Type 14: First-Level Scientific Theory). Moreover, what is sometimes being referred to are ideas of Presuppositional Theory (Cluster 6, Type 9), and philosophers certainly seem to be more interested in such a separation than scientists themselves. Nevertheless, it is the case that within Scientific Theory, higher-level propositions can be used to support more specific propositions: there are certainly at times differences of degree. There are layers of scientific observations, claims and propositions.

It is necessary to demonstrate the crucial differences between on the one hand an enterprize such as empiricist research on teaching and the Empiricist Theory which has resulted, and on the other, scientific research and the Scientific Theory which this has evolved. In order to do so one must discuss the contrast between the philosophical status of scientific and empiricist concepts and of their relationships. The argument is that scientific concepts are abstract-- abstract concepts which relate in complex ways to form the propositions of the Scientific Theory in particular, but also relating in one way or another, or at one time or another, to theory in the sense of Clusters 5 (Hypothesis) and 6 (Presupposition). In contrast, empiricist concepts are general notions which interrelate to form the generalizations of Empiricist Theory[36].

In being abstract, scientific concepts can enter into precise relationships with one another by way of conceptual definition and mathematics. That so much Scientific Theory can be mathematicized, with the use for instance of non-observables such as equals signs (not approximations, not probabilities, not positive or negative correlations) shows that scientific research does not involve generalizations from experience, but rather universal claims from conceptualization: for there are rarely or never such equalities in observed nature.

Some philosophy which has so far been only implicit now requires making explicit[37]. Scientific Theory makes use of three different kinds of connection between concepts. Empiricist connection is empiricist or "sensible" and relates empiricist concept of observable to empiricist concept of observable. Rational connection is purely conceptual and relates abstract concept to abstract concept. Abstractive connection straddles the abstract and the empirical and relates abstract concept to empiricist concept of observable.

Empiricist connection relates empiricist concept of observable, to empiricist

---

36. These matters are considered in some detail in Chapters VI and VII.
37. The inspiration for the following distinctions and discussion is Willer (1971) and Willer and Willer (1972).

piricist concept of observable— the falling rain is making the ground muddy; the teachers expressed strong support for the report forms; the bob of Galileo's pendulum rose to its original height; the pointer lies between the 31º and the 31º 30" marks. Rational connection relates abstract concept to abstract concept, e.g., $F = m\ a$; adenine bonds with thymine, and guanine with cytocine; $2NaCl + H_2SO_4 \rightarrow Na_2SO_4 + 2HCl$; $Sin\ i / Sin\ r = \mu$ (for a given pair of transparent media). Abstractive connection relates abstract concept to empiricist concept of observable, e.g. $F = m\ a$, *as applied in calculating* actual wind stresses; $2NaCl + H_2SO_4 \rightarrow Na_2SO_4 + 2HCl$, *as used in the production of* hydrochloric acid; using $Sin\ i / Sin\ r = \mu$ *in calculating the refractive index of a plastic* to be used in spectacles, with the use of a Pulfrich Refractometer.

To make the point about abstractive connection clearer, let me provide an example: the use of Snell's Laws and the Pulfrich Refractometer.

Snell's Laws show that a ray of light passing from a more dense to a less dense medium is refracted away from the normal, i.e. the angle of refraction is in such cases always greater than the corresponding angle of incidence. If the angle of incidence is increased sufficiently, there can be no refracted ray and all the light is reflected back into the original medium. Refraction has become reflection. This phenomenon is called "total internal reflection". It is this phenomenon of the theory of light which Don Campbell is referring to when he says (see p. 204, below) that in the film he shows the students, the feet of the swimmer who is sitting on the edge of the pool appear to separate from the rest of the person.

The Pulfrich Refractometer employs total internal reflection to measure refractive indices of transparent media. We place a symmetrical block of the plastic to be tested, adjacent to a cubical block of glass with a high refractive index, and mount a telescope at the far side of the block in order to encounter the rays of light which pass through both the plastic and the glass. Once a certain angle of refraction "i" is exceeded by the rays of light passing from the glass into the air, total internal reflection occurs. So there can be no light leaving the block at angles less than "i". Hence if the telescope is placed to receive the ray at that angle, there will be a sharp edge to the field of view, half the field being illuminated, and half being completely black. The telescope is mounted in such a way that this angle can be read from a scale. We know the high refractive index of the glass block as reported by its manufacturers, and by a suitable use of simple trigonometry and algebra we can calculate the refractive index of the plastic. Using such methods the index can be measured to an accuracy of some 0.0003% (Daish, 1981, p.140).

But notice what has happened here. We have to make use of observable interpretations of the abstract concept of "light ray", and we theoretically assume that such a ray exists at the interface between illumination and blackness as detected by a human eye through the telescope. "Light ray" is interpreted in that manner in the observational world. Again, we say that the ray enters the telescope at an angle between the 31º and the 31º 30" marks. But our knowledge of this angle depends upon our fallible human judgment in reading a scale. The abstract concepts of "angle of refraction", "angle of incidence", "normal", "light ray", and the propositions in which they occur, have been interpreted by way of observable marks and performable operations in the empirical world. The proposition, The angle of refraction was between 31º and the 31º 30", *means* , The angle (abstract concept) of the light ray (abstract concept) as received through the telescope as measured from the normal (abstract concept) appeared on the scale (empiricist concept of observable) to lie between the 31º and the 31º 30" marks (abstract mathematical/scientific concepts, interpreted through empiricist concepts of observables). This then is abstractive connection.

So Scientific Theory consists of interrelated sets of propositions, of rational connections between the rational abstract concepts, of empiricist connections between concepts involving observables, and of abstractive connections between abstract concepts and the corresponding empiricist features, the observables of the empirical world. We should also carefully notice that *there has been no reference to any supposed regularity of the empirical world.* From the point of view of Scientific Theory with its rational and abstractive connections, the issue of whether the world is or is not regular is irrelevant. This is in contrast to Empiricist Theory, which is based upon the Presuppositional Theory (Cluster 6, Type 9) that the empirical world is informed by regularities-- and so empiricist researchers go looking for such regularities (generalizations).

Another fundamental and often misunderstood difference between Scientific and Empiricist Theory concerns measurement. *For two reasons, first because they use mathematics rather than just statistics, and second because they use abstract concepts embedded in the propositions of Scientific Theory, scientific researchers are enabled to use ratio scales in measurement; in contrast, because they use statistics and do not use the abstract concepts of Theory, but merely propositions relating general empiricist concepts, empiricist researchers on teaching just cannot use the ratio scales of measurement used by scientists* [38].

Thus, one of the features which help make Scientific Theory and scientific research *Scientific* Theory and *scientific* research is the use of ratio measures such as meters, grams, seconds, seconds per second, Newtons, Volts, Angstrom units, moles, calories, and so on. In scientific research, this use of ratio measures occurs in the rational connections within and amongst the propositions of the Scientific Theory and in the abstractive relationships between the propositions of Theory and the observable world, i.e in the propositions of Theory and in the interpretations to and applications of that Theory in the observable, empirical world.

The general point is that ratio measures as such cannot be made use of in theory or research until there is conceptual abstraction, as discussed in the above examples from the history of scientific research. For it is not the ratio scales of measurement which have made the Scientific Theory possible, but the Scientific Theory which has made the ratio measurements possible. Measurement in scientific research is *one way in which* abstract concepts are related (the other is conceptual connection) and are abstractively interpreted to the observable world. Thus length and mass in Newtonian Theory are not what meter rules or balances

---

38. See also, discussion in Chapter VII.

measure, rather the meter rules and the balances are merely tools which make possible the interpretations of the abstract concepts, "length and mass". And because the meaning of the concept derives from its interrelationships within the Scientific Theory, there may be *numerous such tools* for any particular concept, which will involve measurements which apply the Scientific Theory.

In contrast, Empiricist Theory and empiricist research on teaching use so-called nominal, ordinal (and, it is claimed) interval scales. These are supposed to be becoming increasingly powerful and precise. The hope is that when "the measurement problem in education" is solved, they will somehow turn into ratio scales. This is an illusion. For these are not scales of numbers, but merely *ranked categories masquerading as numbers*. Certainly the categories are assigned numerals, but this does not turn them into numbers. Such empiricist scales are the result of a kind of routinized procedure for sequencing or averaging views about some empiricist general concept in a given population. With the use of some face-saving rules (e.g. Chi Square test with one or more groups and with nominal or ordinal scales, t test with one or two groups and with interval scales, and so forth) the numerals of the scales are used in various statistical procedures and tests. But empiricist scaling is not scientific measurement.

Empiricist Theory is the result of observation of assumed causal connections, or empiricist relations and generalizations in a presumably ordered world; Scientific Theory makes use of abstract concepts embedded in scientific laws and principles, which impose their own Theoretical order on the world. Moreover, because of the order imposed by the Theory, its propositions can be expressed mathematically, can be measured using ratio scales, and can be interpreted within degrees of error to observable events, prior to the fact.

Science became powerful both in theory and in technological application when people such as Galileo, Newton and others established such relationships. Newton's work for example concerned abstract concepts and rational connection and was theoretical and mathematical rather than observational, or experimental. Certainly it was not empiricist. And so he proceeded until his grand synthesis had been achieved. Consider also the lesson about the nature of Scientific Theory to be learned from Newton's relations with Flamsteed, the first Astronomer Royal. When Newton was at work on refinements to his account of lunar motion, he visited Flamsteed to point out that some of Flamsteed's observations must be in error, *because they contradicted Newton's theory*. Flamsteed checked and had to agree. Indeed, over the years Newton constantly criticized and corrected Flamsteed's observational results, to the latter's increasing chagrin and frustration.

Recall the empiricist account of using a pool-pole to clean the bottom of a swimming pool. The account made use of propositions linking empiricist concepts of observables. Scientific explanation, in contrast, does not make reference to generalizations of supposed causes, but to scientific laws and principles, which are determinative and universal. *Logically*, scientific laws must apply in all times and places, else they cannot be laws. In contrast, empiricist generalizations can be true or probable at one time and/or place, yet not at others. Of course, the *scope* of application of any law is limited: Snell's

Laws do not apply to all crystalline substances, Newton's First Law does not apply to electromagnetic radiation, and so on. Whether a law does or does not apply to additional phenomena is decided by further conceptualization, experimentation, and observation. But this does not alter the general point about the logical universality of laws.

Thus with respect to the increase in the seeming bending of the pool-pole, I might begin my scientific explanation with a statement of Snell's second law of refraction as described at the beginning of this section, viz. $\sin i/\sin r = \mu$ (where i is the angle of incidence, r is the angle of refraction, μ is a constant for the particular pair of media, in this case for water and air and is thus 1/1.33 or 0.751 approx.) Moreover, in pushing the pole farther from me, I am thereby increasing the angle of incidence of the ray of light coming from water into air, and into my eye, and concomitantly increasing the angle of refraction in the precise proportion of 1: 1.33. Thus when I push in my pole at, say, an angle of 45 degrees from the surface of the pool, the angle of incidence will be 45 degrees, and the angle of refraction approximately 60 degrees, but as I push my pole farther, say to an angle of 30 degrees to the surface of the pool, the angle of incidence will be 60 degrees and the angle of refraction approximately 80 degrees. In contrast to empiricist research and experiment, such measurements being integral to the Scientific Theory may be made as precise as the need requires. So the angle of refraction gets closer and closer to 90 degrees, the appearance of the pole gets increasingly different from its material reality, and thus the manipulation of the pole on the bottom becomes increasingly difficult.

Scientific Theory is not a summation, but a new creation. To be scientific, Theory must be capable of producing something new; it must be able to solve problems which could not previously be solved. Toulmin points to an essential of much (not all) Scientific Theory when he writes that through science,"We are led to look at familiar phenomena in a new way, not at new phenomena in a familiar way" (1953, p.20).

Consider for instance what physics has to say about the phenomena of shadows and of the apparent bending of objects which we place under water [39]: the phenomena which are in part explained by the above-listed Snell's Laws. For a million years human beings have recognized a sort of regularity in the way shadows are cast by the sun and the manner in which sticks appear to be bent in water. It is worth noticing that although human beings did not know their precise form, all these phenomena were indeed already-recognized empiricist regularities, which were in various ways puzzling. What science has achieved, by the work of people such as Snell in the last few centuries is to provide an intellectually-satisfying and practical account of why it is that we can observe such regularities, of what underpins these two regularities. In so doing, scientists have had to create *new* ways of conceptualizing. Thus, in relation to shadows, physical theory suggests that we shall construe light both as travelling, and travelling in straight lines. Scientists then push this abstraction of the rectilinear

---

39. An example drawn from Toulmin, 1953.

propagation of light to its limits to see what is its scope and range, what it can and cannot explain. The point is that if we construe light as travelling in straight lines, we have a very fruitful way of considering the various phenomena of light as they occur in the observable, empirical world. Similar features apply to more complex abstractions, such as chemical elements, DNA helices, or tectonic plates.

Snell found that this ideal had to be modified in respect to transparent bodies, e.g. the bending of a ray of light when it passes from the air into another transparent medium, and that such deviation could be expressed by way of the laws stated above. That is, he discovered that provided we still construe light as travelling and within a particular transparent medium as travelling in a straight line, then for any such medium the number produced by dividing the sine of the angle of incidence by the sine of the angle of refraction remains constant no matter at what angle a ray of light may be projected. This number or this ratio of sin i over sin r is called the refractive index. The deviations from Snell's Laws (such as diffusion of light) are explained in further modifications. But there are of course limits to such principles: for example these regularities and inferencing devices do not help us to calculate or even to consider the speed of light, for which we need to have recognized different regularities and for which we require additional, or alternative creations, or ideals, or abstractions of natural order.

Scientific concepts and explanations, unlike empiricist concepts and explanations, are not defined by reference to observation; they do not consist of descriptions of observable facts; on the contrary, they enable us to interpret the empirical world anew, and so *observation in science is defined by reference to the abstract scientific concepts, theories and explanations* . To put it another way, the statements of Scientific Theory are statements involving symbols and the connections between them, from which statements about observables can be derived.

Thus whereas there has been empiricist knowledge in all societies at all times (the Australian aborigines, the Polynesians, the Bushmen), and Empiricist Theory in all more developed societies (Rome, Sung China, Mediaeval Europe, Abassid Baghdad), Scientific Theory is an European invention, especially during the last 400 years. This fact is revealing: if Scientific Theory really were a matter of observables and generalizations from them, why would it be so temporally and geographically restricted? But if we construe it as based upon striking, imaginative, conceptual leaps to produce abstract concepts which are linked by rational connection, then, such mental moves being far from simple or observable, strict limitations on where science has evolved look much more explicable. Until such moves were made, the knowledge upon which technology was based was necessarily empiricist, not scientific.

In an attempt to explain away the comparative lack of results in empiricist pedagogical research, many writers have argued and continue to argue that social science, including research on teaching, suffers from having to deal with immensely complex phenomena— phenomena which are much more complex than that with which established science deals. This may be true, but in my view is not the heart of the matter. The important reason why natural science has been successful has not been *because* the phenomena are any simpler, but because it has worked with abstractions from empirical phenomena, rather than with generalizations about it. And such abstract concepts, unlike such general ones, can then be related to seemingly quite diverse empirical phenomena. Falling rain, the moon, Halley's Comet, baseballs, bullets, and space shuttles do not look alike, but all can be construed under Newton's three laws of motion and his law of gravity. Rainbows, magnetic compasses, generators and television sets ostensibly do not appear to have anything in common, but they can all be understood through electro-magnetic theory. The behavior of submarines, Olympic swimmers, the car-hoist in Bill's Corner Garage, the hydrometer in the local swimming baths do not seem alike, but can all be explained on the principles of hydrostatics. The explanation of success is not *that* scientific phenomena are any simpler than social or pedagogical phenomena, but that brilliant conceptual insight has been able to abstract concepts and relationships, *which when used make the phenomena seem* more simple. Central to this book is the claim that Scientific Theory is not the same as the generalizations of Empiricist Theory. Whereas the generalization of empiricist research gradually builds up as more and more approximately similar cases are observed, abstraction, such as in Galileo's or Newton's mechanics, or Lavoisier's theory of combustion, immediately defines a potentially infinite universe of conception which provides the framework for interrelationships, observations, measurements, and understanding. As already pointed out, laws such as those of Galileo and Newton are important improvements on earlier Aristotelian and mediaeval impetus conceptions, not because such savants observed more precisely, but because they penetrated behind the superficial regularity disclosed by the senses. As Kuhn says, "To verify Galileo's law by observation demands special equipment . . . Galileo himself got the law not from observation. . . but by a chain of logical arguments," (1957, p.95). Such points are general to scientific research.

Furthermore, in using abstract concepts and their interrelationships, and thus making possible the development of Scientific Theory, Copernicus, Galileo, Newton, Lavoisier, Crick and Watson, the Plate-Tectonic theorists, and all the others enabled their concepts to remain *stable in meaning* and also (though conceptually related by their rational connection within the Theory) *conceptually distinct*. Moreover such abstract concepts, because of their abstraction and because of their conceptual distinctness were able to be rationally connected by way of mathematics and measurement, and thus that precision for which most of established science has become so noted became possible. Stability of meaning, conceptual distinctness and mathematical and measurable connectedness of their concepts also made possible the abstractive interpretation to the observable world, and thus gave power to their Scientific Theory when used in that world.

Natural history does not a science make. Indeed, scientific research gets going when, as Galileo originally pointed out, and Dearden more recently put it, there is a "rape of the senses" (1968, p.119). Thus, as will be argued in detail in later chapters, one of the most unfortunate misunderstandings of empiricist pedagogical research is the belief that Scientific Theory is a matter of observational generalizations. Indeed, a good argument can be made that, paradoxically, it is because it is not a matter of generalization, that its theories, laws and principles do apply with accuracy in the empirical world. It is because Empiricist Theory and empiricist pedagogical research are matters of generalization, that their findings apply in the empirical world only "more or less".

There are three main reasons for the preceeding discussion of Scientific Theory, and the earlier discussion of other kinds of theory. The first is to indicate the exceeding complexity of the theoretical underpinnings of scientific research, and by implication, of any future scientific research on teaching. The second is to provide intimations of the form any such future fully scientific pedagogical theory would have to take. The third reason is to emphasize the instrumental value of such awareness *for an improved empiricist research* on teaching: lack of awareness of the subtle distinctions between different types of theory stops empiricist researchers from understanding how a wide range of theory equally fundamentally affects how they construe and perform their tasks of research. Unless researchers are aware of the differences between these theoretical types, they will be unlikely to avoid the kind of confusion which can invalidate even the best intended empirical research, scientific or empiricist, for they will be failing to address the appropriate questions.

Scientific Theory developed when thinkers began to address the appropriate questions: when they added action to contemplation; when they tested in the empirical world, the imaginative and abstract conceptions and explanations they had produced in their heads; when they realized as Francis Bacon wrote in 1620, at the very beginning of the scientific revolution, that knowledge is power–KNOWLEDGE is power.

Through Scientific Research, Scientific Theory Improves.

In the section which discussed DNA, it was pointed out that the role of falsification is often more complex than it initially would seem. The Duhem-Quine thesis and the theory-ladenness or Presuppositional Theory discussed by Hanson raise the issue of subjectivity and objectivity. As Bechtel describes the fear, "If we lack an objective, theory-neutral reference point, it is claimed, scientists who hold competing theories will simply see what they are prepared to see by their theory and there will be no theory-free reference point to which we can refer to settle disputes"(1988, p.47). The implication is that although we know that Scientific Theory changes, do we know that it improves?

Prior to the challenges by such philosophers as Hanson, Kuhn and Feyerabend, there existed a standard view widely-held by scientists, philosophers, and laymen, that Scientific Theory was an interrelated body of objective fact and laws, that scientific research was a systematic enterprise, pursued by way of that Scientific Theory, observation and experiment, and that such theory and research formulated an evolving truth about the empirical world. On this view, as Scientific Theory changes over the years, scientific knowledge accretes and accumulates. As change takes place, an increasingly more sophisticated, accurate and refined (but not necessarily more complex) account of the nature of the empirical world emerges: Scientific Theory changes and gets better. There may be debate amongst scientists over what should be done and the conclusions to be drawn as research develops, but it is still assumed that scientific research and Scientific Theory remain objective because there exist accepted scientific practices, replications, traditions, logic and methods of activity and criticism. These include the presupposition that the findings of scientific research must be intersubjectively agreeable and its activities must be publicly replicable.

The burden of this section is to show that much of this standard version of scientific research is acceptable. A concomitant task is to show that the controls of the scientific community are themselves applied in a manner which allows science not merely to change, but also to improve. In short, the task is to reject the more extreme claims of Kuhn and Feyerabend, and the problematic issues raised by Hanson's arguments about Presuppositional Theory. The challenges they posed for the objectivity of scientific knowledge have been grossly overestimated. D.C.Phillips, in a recent perceptive and witty book has usefully called these overestimations "rampant Hansonism" and "rampant Kuhnism" (1987).

I am not suggesting that scientific research has produced the *best* sort of knowledge that we have, for that would be counter to my views of the deep significance of an improved variety of empiricist research on teaching, which I discuss in the final chapters. It is nevertheless true that scientific research has produced an exceedingly well established kind of knowledge. It has been a model for 400 years. It has made an institution of the activities of justification under its public controls. Even if we can suggest with some cogency that scientific claims at times distort; even if an occasional scientist plagiarizes, and cheats; even if scientific research is at times misused and misapplied to subjects for which it is unsuitable; even if empiricist activity is often mistaken for scientific research; and even if we ought not to view scientific research as *the* single way to truth and light; nevertheless, if the very possibility of the objectivity of scientific research becomes problematic, then epistemically we are in exceedingly deep trouble.

Whatever the precise form of the criticism of objectivity, its core always relates in some manner to the *questionable nature of observation* in scientific research. When it is remembered just what an important place observation plays in research, directly and indirectly, in the context of justification and perhaps as some sort of preliminary in the context of discovery, and when it is remembered that it is observational implications which tie the propositions of Scientific Theory with their abstract concepts by way of abstractive connection to the empiricist propositions which relate observables, then the significance of any criticism of the objectivity of observation becomes obvious.

It is a normally-agreed fact of psychology that observation (by scientists, or by laymen) is not merely passive, but rather involves an active even constructive role by the observer. Such a role for the observer is also reasonably well known in ordinary experience outside psychology. We sometimes see what is not there (as every magician and teacher knows) and we sometimes do not see what is there (as every proof-reader and teacher knows). Such evidence is sometimes assumed to demonstrate that there are questions to be asked about the objectivity of such observation. This however is a mistake. It may be granted that such evidence weighs against a fixity of observation. But it does not in the long run weigh against objectivity, because it is also agreed by laymen and psychologists that such problematic observations can be explained and that where they have gone wrong they can be corrected. (Unless this were the case, there would be a paradox of observation here.)

Objections of a more philosophical kind to any pristine observation are however another matter. Although it was probably Popper who introduced the notion of the theoretical basis of all observation to current generations of thinkers, it has been the writings of Kuhn, Feyerabend, and Hanson which have had the deepest effect on current views about the nature of observation in scientific research. As Kuhn dogmatically says in a key passage, "Like the choice between competing political institutions, that between competing paradigms proves to be a choice between incompatible modes. . . When paradigms enter, as they must, into a debate about paradigm choice, their role is necessarily circular. Each group uses its own paradigm to argue in that paradigm's defence" (1970, p.94).

When sophisticated academics and intellectuals propound seemingly consistent theses, which nevertheless do run so counter to pragmatic action, *and reject the beliefs and actions of most scientists themselves*, there is a need for extremely careful analysis of the content of such theses. It has to be suspected that there are unjustified assumptions and ambiguities, which, when examined and exposed, make the case much less forbidding. It has to be suspected that the intellectuals are involved in a theoretical conundrum of their own devising. That indeed is my own view. An analytical challenge will now be attempted, based upon the three key issues of *observation, meaning,* and *change and progress*. The result is an account rid of the excesses of the sceptical critique.

One of the less modest ways of putting the idea that our theories, concepts, etc. determine what we observe, so that there is no possibility of common observation, comes from the sociologist, A.Blum who says, ". . it is not an objectively discernible and purely existing external world which accounts for sociology; it is the methods and procedures of sociology which create and sustain that world"(1971,p.131). Fortunately we do not need to take the delusions of people like Blum at face value.

For in the discussion of what we bring to our observations, there are considerable ambiguities. Sometimes writers are referring to the concepts which are linked together in the propositions, sometimes to the claims or propositions themselves. As Scheffler says,"Conceptualization relates both to the idea of categories for the sorting of items and to the idea of expectation, belief, or hypothesis as to how the items will actually fit available categories" (1967, pp.37-38); it relates to both category and hypothesis. Scheffler's point also provides a reason for keeping conceptually separate the issue of discovery in scientific research and the issue of verification.

The point is that we can use the same concept or category system, but make use of it with different hypotheses; the converse is also true, that the same hypothesis may be formed compatibly with different category systems. This last point is one of considerable importance in ordinary life, and in pedagogical and scientific research. People continually mistake *an* hypothesis for *the* hypothesis, *an* explanation for *the* explanation. For example, the same claims to facts or the same actual facts may be implied by different theories. Alluding to the underdetermination of theory (several Clusters) by observation, O'Neill has stressed that, "It is a point of very great significance (for the understanding of research), that different theories may imply the same observed facts" (1969, p.67). For instance, once observed, the phases of Venus could be explained both by Copernicus and by Tycho; the existence of related flora and fauna on different continents can be explained by continental drift, but it was also explained for a century or more by land-bridges; the observed magnetic symmetry on each side of the Juan de Fuca Ridge was seen by Vine and Matthews in England as evidence of sea-floor spreading, but by scientists at the Lamont Observatory in New York, as merely the result of some control of the magnetic pattern by geological formations, in or underneath the central ridge; by Walter Alvarez, the extinction of the dinosaurs is seen as the result of a global winter caused by effects in the atmosphere from a crashing asteroid at the end of the Cretaceous Period, but Robert Bakker sees the extinction as the result of exotic diseases contracted by dinosaurs as they wandered over land-bridges brought about by shrinking oceans between the Jurassic and Cretaceous.

What is important in these examples is that, just because we have conceptions by which we categorize, this in itself does not prejudge what will fit into such conceptions, or how that which does fit will interrelate. This point is considerable. It argues that although observation will be thoroughly shaped by conceptualization, categorization, language, and theory (several Clusters), in an important sense, any decision about just what is the case will still be independent of these.

There is a further Kantian-type of logical point which is of fundamental importance. It should be stressed that *there must be some principles of organization of what there is in the given world,* which will affect how human beings mentally grasp that world. There must be some basic organizational "principle" of experience for human beings to have experiences at all: some sort of interdependence between what is and what is experienced. We cannot state the nature of this interdependence in words, for if we could we would be describing the ineffable, which is logically impossible. Nevertheless we can *point to* the interdependence in words. The argument must go something like this. We cannot seriously conceive of a world which contains no features or principles of organization which are independent of how the observer grasps them: there logically must be some features independent of the observer. For if the world of experience were such that there were nothing in the objects of observation which *forced* the observer to organize his descriptions in one way rather than another, *there would be no possibility of grasping things observed and no possibility even of having a conception of a thing, of oneself, or of anything else.* Indeed, "Most philosophers who accept the concept-ladenness of experimental research acknowledge that experience is not totally structured by theory, but rather is at least partially pre-structured independently of the researcher's conceptual framework" (Garrison, 1986, p.15). I believe this understates the case. On the basis of his own experience, Einstein puts it that, "Nobody who has really gone into the matter will deny that in practice the world of phenomena uniquely determines the theoretical system, in spite of the fact that there is no theoretical bridge between phenomena and their theoretical principles" (quoted in Pirsig, 1976, p.107). Or as O'Hare says in good Popperian spirit, "We impose our concepts and theories on nature, but nature kicks back"(1981, p.22).

Furthermore, although observation is deeply influenced by theory (Clusters 5, 6 and 9), as White has shown (1983, p.11), intelligent encounter with *any* phenomenon of experience must involve some common elements. These include such matters as assumptions of a cause and effect sort, similarity of spatial properties and relationships, assumptions about time, some sort of inductive reasoning and learning from experience, assumptions about the perdurance of material objects, belief in the general trustworthiness of the senses, and reliance upon what they indicate in order to engage in pragmatic action, and even many of the common features of languages themselves.

So although, say, Marxists or Behaviorists may be strictly and deliberately limiting the categorial and conceptual possibilities in their accounts of what is the case, they are not limiting in quite the same way, what can be observed as the case. An important possibility for testing any conceptual system still exists. For, "Observation may be considered as shot through with categorization, while yet supporting a particular assignment which conflicts with our most cherished current hypothesis" (Scheffler,1967, p.39). It is often forgotten that the English-speaking Marxist and Behaviorist continue to use a common language and conceptualization of commonsense, which supports and is in addition to any esoteric technical terminology or conceptualization embedded in their theory (Clusters 3, 5 and 7).

As part of the challenge to objectivity, the Kuhnian/Hansonian-type suggestion is made that because falsifying observations must themselves be theory-laden, they are only as good as this presupposed theory. The two following passages are typical. Eisner writes, ". . evidence is determined by the theoretical system within which one operates. . conclusions about reality cannot be dissected from the theoretical and methodological procedures used to generate those conclusions" (1984, p.14). And in the fifth edition of Borg and Gall's popular manual on how to perform pedagogical research, we find the following criticism of qualitative research. The authors call it "a major problem" of qualitative research, at the same time failing to realize that the same point, if true, would apply equally to their own, preferred, quantitative research:

> If all observation is theory-laden . . how can we hope to find a common ground of neutral observational data by which to judge the validity of competing knowledge claims derived from different theories? For example, how can we judge whether the theory of evolution is more valid than the theory of creationism ? Researchers from each theoretical camp can define constructs and select observational data that support their theory and refute the other (1989, p.21).

But there are at least three kinds of confusion in these quotations from Eisner and Borg and Gall. The first is the ambiguous use of the word, "theory"; the second is the confusion between conceptualization and the evidence which makes use of that conceptualization; the third is a specific variety of the first confusion, viz. the confusion of ideology (Cluster 7, Type 11) with Clusters 3, 5, and 9.

First, because the distinctions made in Chapter I have not been drawn, the word, "theory" is normally used ambiguously. Thus Eisner's "theoretical system" might refer to any of the Clusters, except Cluster 2, with different implications following from different meanings. And it is immediately obvious how Borg and Gall have confused matters by juxtaposing the word, "theory" in their question, "How can we judge whether the theory of evolution is more valid than the theory of creationism?" so that Cluster 9 theory looks to be the same thing as Cluster 3? or Cluster 7? (Moreover, what "observational data" could possibly support creationism? Actually, the authors would have improved their case by using a more cogent comparison.)

Secondly, the quotations massively exaggerate the difficulty caused by the conceptualization of the "theoretical system". That much of what person A and person B observe is itself theory-laden (several Clusters) is not the problem assumed. For one thing, such observation always uses ordinary language with its common, commonsense conceptualization[40]. Furthermore, it is a necessary truth that any test based on observation must assume and make use of *some* categories and concepts. The theory being tested must be couched in some categories and concepts, and the most obvious tests of Behaviorism or Marxism will make use of Behaviorist or Marxist terminology. It is true of course, that such theories as Behaviorism and Marxism (Clusters 3, 5, 6, 7 or 8) will both indicate and restrict the sort of observational data which can be allowed to count, but this does not stop testing. Intellectually-honest users of such conceptualizations can devise astringent tests using Behaviorism's or Marxism's own conceptual base. And although any single check or test of a procedure or conclusion may rarely be sufficient, nevertheless, failures, falsifications, falsified predictions, anomalies, insufficiencies, inadequacies in explanatory power, and inconsistencies, can gradually amount to overwhelming proportions, as has been the case in many historical instances such as in procedures and conclusions using Ptolemaic-Aristotelian astronomy, or phlogiston chemistry, or the stablist view of continents, or indeed, in Behaviorism and Marxism. That there are no hard and fast rules for rejection of any procedure, proposition, explanation, account, or theory, is not an argument for claiming that we never know when these should be rejected, or that we can never tell when one theory is better than another. Even though for the Quine-Duhem reasons, matters are normally complex, in the long run propositions involving such conceptualizations either survive these tests or do not. In ordinary life, in professional life, and in scientific research we

---

40. Feyerabend has argued (1970) that even commonsense is theory-laden. He may be correct. But the point is irrelevant for the present argument, because I am referring to the shared commonsense language and conceptualization, i.e. the *common* commonsense.

continually make such tests and comparisons. While theory-ladenness or conceptualization may orient the evidence, *they do not make it thus and so* .

Thirdly the above point is cogent for scholars, scientists and theorists who understand the logic of anomaly etc. However, if believers in a theory (several possible Clusters, but particularly Cluster 7, Type 11) use such a theory as an ideology, then there is much truth in Eisner's claim that conclusions about reality cannot be dissected from the theoretical procedures used to generate them. In an ideology the conceptualization functions both to orient the evidence and to say that the evidence *must be thus and so prior to and after the fact* . Such ideologues will (for their political, or theological, or moral reasons) introduce continual ad hoc hypotheses, or changes in the meanings of key terms, so that falsification is no longer even potential. Ideologues want things both ways. They will make their ideology immune to falsification by making it non-empirical as far as falsification is involved, but empirical in so far as they still say it is true and does apply in the observable world. Such is the common response of too many student leaders, theologians, social scientists, politicians, and trade union secretaries. Such persons may acknowledge the general point about falsification, but will not allow it to apply to their favored ideology. Equally, the cogency of anomaly, falsification etc. cannot be acknowledged by anyone who lacks the sophistication to grasp the logic of different kinds of statement, or when the counter-arguments apply: people such as Evans-Prichard's (1937) Azande believers in the chicken oracle, or that gullible quarter of the present American population which unfortunately believes in astrology.

There are many Clusters and Types, and also levels within Types of theory. Observations conceptualized in some particular theory or level of a theory can be free from the use of conceptualization of another sort or level of theory; such observations can then provide evidence for the views expressed in the other sort or level of theory, without making use of *its* conceptualization, presuppositions, or beliefs. Although it is a logical truth that all conceptualization involves theory of *some* sort (we cannot construe the world without construing it in some way) it in no way follows that our observations cannot be independent of some particular theory which these same observations may later come to support (White, 1983, p.28). This is why Eisner, and Borg and Gall are begging the question. For instance, when the lay tourist in the limestone caves observes that long pointed stones hang from the roof, that there are bats in the caves, and that water seeps through the rocks, his observations are affected by his commonsense theory (Clusters 1 or 8), but they are not affected by any complex scientific theory (Cluster 9). If, later in his life the tourist learns some chemistry, zoology and geology, his observations in such caves will begin to be modified by the concepts and theories (Cluster 9) of those disciplines. He will now be able to observe by way of his scientific theories *or* by way of his commonsense conceptualization. And in this particular case, his scientific theorizing will not contradict or be incommensurable with his commonsense observations, so that, ". . a sort of hierarchy of theory-ladenness is built up" (White, p.28). This indicates that any set of observations,". . can be theory-laden with respect to one theory (relatively low in the hierarchy) but quite free with respect to another, (higher up)" (White, p.28). It is incorrect to claim that observations being theory-laden can never be separated from the theories they are used to support. "For the plain truth is *that while all observations are theory-laden, observations are not laden with all theories* " (italics added) (White, p.28). Indeed, because of the vagueness, ambiguity, or suppressed nature, of their theories (Clusters 3, 6, 7 or 8) and/or the difficulties of checking and replicating their statistical approaches, the kind of point made by Eisner and Borg and Gall applies more to the studies of empiricist pedagogical researchers, and to social scientists, than to the work of scientists.

Thus the theory-laden nature of all observation does not pose the problem it is often assumed to pose for scientific research.

Equally, it is not impossible for people to share common observations even though they may bring different theoretical backgrounds (several Clusters) to what they observe. For we have seen both that falsification can occur using a particular theory and also that there exists a hierarchy of theories such that the lower-level theory does not beg the question regarding observations which use or have implications for any higher-level theory. And any account of scientific research which suggests that common observation is impossible is just ignoring plain human experience. Ptolemy, and Tycho, and Copernicus, despite their different theories, all agreed that the planets followed seemingly erratic paths; Lavoisier, and Priestley both agreed that materials can burn; Wegener with his views about continental drift, and his opponents like Chamberlin, all agreed on the facts of plant and animal distribution; Newtonians and Einsteinians can both observe that light bends near large masses; Darwin and Lamarck both agreed that evolution occurs; Marxists and Behaviorists can point to the occurrence of revolutions in Russia in 1917 and 1991, and to changes in behavior brought about by what adults say and do to children. Such examples could be multiplied endlessly.

It is of extreme importance to note that opposing theorists are not necessarily forced into opposing categorizations of what occurs. White makes the incisive and fundamental point, that if observations which are to count as supporting or falsifying could not be described both by using the conceptualization of a particular theory and by not using such conceptualization,

> *. . men would rarely if ever be able to explain what they set out to explain .* If astronomers, for example, could speak only of bodies with Newtonian mass, acceleration, and so on, they would not be able to explain what first they set out to explain: namely, why it is that in the sky certain luminous bodies wander about (italics added) (1983, p.34).

It is also worth mentioning that there is no limit to the development by scientists, ". .of overlapping categorizations where none existed before; were this not so, we should be faced by a problem greater than the one posed by our original paradox, that is, to *explain how theoretical agreement could ever arise"* (italics added) (Scheffler, 1967, p.40).

The negative point is however, that athough it is true that it is possible for there to be observation of what happens, which is independent of categorization in the very important sense distinguished above (i.e. in the sense that it can be tested despite the particular categorization), nevertheless once a particular claim about what is the case has been accepted, once a particular theory has been seen to apply in many examples, the categories and concepts of the theory and the claims made within it do tend to limit people's perceptions of alternatives. What is more, as Bacon indicated so long ago, they also disincline people from seeking to look elsewhere. Expecting something to fall into the category and its typical relationships, we do tend to see it there. Expecting the child to be dyslexic, the guidance counsellor sees her as such. Believing the pupil to be of low I.Q., the teacher treats him accordingly.

Nevertheless, such categories, concepts, hypotheses, expectations, theories, certainly do not wholly restrict what is observed to be the case, as is well attested by common experiences of both laymen and scientists that they are surprised by seeing something unexpected. Indeed it is one of the well-established empiricist generalizations *about* scientific discovery, alluded to by greats such as Pasteur, Fontenelle and Crick, that discovery favors the well-prepared mind— well-prepared in the conventional way of observing. Discovery is then consequent upon recognizing something as *not* fitting what one is prepared to expect, and learning from this experience.

There is a second, somewhat different challenge which must also be rejected.

Under this challenge, meaning, it is said, is not an intrinsic quality of the term or concept itself. Meaning is not an automatic concomitant of sound or printed shape. Meaning resides in both the words used and the context in which they are used. Change the particular words surrounding the term, change the order, change the context, and the meaning will change. Meaning is the result of the total situation. Thus, it is said, different concepts, and different words, embedded in different contexts and forms of language, and with different theoretical underpinnings will certainly involve different meanings.

On this view, sometimes called "semantic holism", even if two scholars or scientists use the same observational concepts, but surround them with different contexts or presuppose different theories (various Clusters), they will not be allocating the same meaning, despite their common categorization. Thus, Tycho and Copernicus both used the same category,"planet", but their meanings were different; Priestley and Lavoisier both used the category, "mercury", but their meanings for this term differed. On such a view, communication from within one such system to another is impossible. The scientist is trapped within his own meanings.

Views such as those of the Logical Positivists, earlier in this century, tried to avoid this problem by strictly separating an observational level from a theoretical level (some aspect of Cluster 9), with the observational level making it possible to state facts supposedly neutral of any theoretical clash. There are some advantages in holding to such a view. It allows laws to be saved, and it provides a way in which conflicting theories can be adjudicated between, at the observational level. But such an account runs afoul of the Quine-Duhem thesis. If meaning resides in categorizations, contexts, forms of language used, and the particular theoretical (various Clusters) underpinnings, then it is not possible to sustain the separation of meaning as between observational level and theoretical level in the required manner. Writers such as Quine, Kuhn, and Feyerabend and the many scholars who have followed their lead have subscribed to the account of meaning just described: meaning inheres in the context, in the theoretical background, and so on.

Unless this second account of how meaning is acquired can be re-formulated, once again it seems, there can be no common observation, no objective observation, no permanence of laws, because meanings of terms change with change of theory, no accumulation of scientific knowledge-- only *replacement* of one total, integrated, theoretical point of view by another. In such a situation, there is no justification of scientific views, but merely the personal proclamation of them.

Indeed, unless the above account of meaning is rejected, we can scarcely have such an entity as *scientific* research. For it has normally been assumed that one of the defining characteristics of scientific research is its objectivity and its replicability: we have no science unless not merely American and British and Australian, but also Russian and Chinese and Esquimo scientists can in large measure agree about the questions being asked and the evidence which is relevant. Even more, on the above solipsistic account of meaning there is the immense problem of how language is supposed to acquire meaning in the first place.

An improved account of meaning is however possible. For the word,"meaning" as used in the sceptical account begs several questions.

I think it is generally agreed that in some complex manner the meanings of terms do indeed derive from context, form, order, role, etc. of and in the language. But the word, "meaning" may itself refer to different things. For instance, I.A.Richards (1924;1929) distinguishes in a prose work the several distinct features of "sense","feeling","tone" and "intention", all of which in their own subtle manner contribute to the meaning of a passage. For present purposes it is not necessary to make a four-fold distinction, but merely to call the last three aspects,"connotation" and the first, "denotation". So on the one hand may be distinguished denotation: reference, extension, application; that which denotes and provides the basic meaning. On the other hand may be distinguished connotation: intension, attribute; that which connotes and provides the particular penumbra of associations. Clearly enough, connotation will depend upon context, form and so on. But as Scheffler points out, denotation also depends upon these. As he says, both are,". . elements not of nature, but convention" (1967, p.55). And as long as we interpret the word, "convention" broadly, Scheffler has a point. For instance,"Venus", "the Evening Star" and "the Morning Star" have the same denotation, but different connotations. The same point applies to, "the President of the United States", "Mr.Bush", and "the most important man in the Republican Party". Consider, for example, "Mr.Bush". To a true-blue Republican and a blood-red Marxist, the two words denote the same. But the connotations are decidedly different. Conventional beliefs about the political good and the metaphysical nature of human beings and their history can color and change the connotations of the words, "Mr.Bush" for the two different party members. The key point for present purposes is that it is the sameness of denotation or reference which is important if we are to keep the same meaning for scientific discussion and comparison. Connotation may be ignored. What is more, "As for deduction within scientific systems, it should be especially noted that it requires stability of meaning only in the sense of stability of reference in order to proceed without mishap" (Scheffler, 1967, p.58).

The examples of Venus and the President of the United States are now instructive. For what caused their different connotations to be carried by the same denotations was either scientific discovery, or else actions taken by people in the empirical-social world– not linguistic synonymy. The sameness of reference of the three connotations in each case was not the result of linguistic convention or linguistic decision, but of scientific astronomical research in the one case, and social conventions such as actions of voting by human beings in the other case. The point is instructive because it shows how shifts in belief and knowledge and in human situations do *not* have to affect the denotations of words.

To argue as Kuhn and Feyerabend do that changes in theories produce changes in meaning is to suppose that meaning so depends upon the language used that it encompasses language in both senses: language not merely as a consideration of synonymy and the ways in which claims may be put into words, but also as including beliefs, hypotheses and claims supposed to be true. Only on such an assumption can any argument proceed to the conclusion that shared meanings are only possible on the basis of shared beliefs and the whole connotational paraphernalia of a particular theory (of whatever Cluster). But surely, although the number of things believed to be true may be different from one theory to another, this fact is hardly significant in relation to meaning. Indeed, that we can say that the theoretical beliefs *are different* assumes that we can agree on meanings as being the same in propositions which describe those different beliefs. Change in connotation does not mean change in denotation. Identity of reference may survive through widespread changes of belief and knowledge.

Meaning is thus seen to be quite different from the self-referring entity which makes it appear necessary for each scientist to be isolated within his own theoretical world. Thus the key subjectivist argument about meaning is countered. It can now be seen that it is possible for scientists to share meanings even though they differ in belief and in theoretical presuppositions. It is also shown that it is possible to have real theoretical disagreement and debate (in the sense of explicit contradictions) rather than mere equivocation, talking past each other, changing the subject, advocacy, proclamation, proselytizing, ideological browbeating, tablethumping, etc.,which are all that is possible on the subjective view of meaning offered initially by Kuhn.

Because theory reduction, derivation of laws, sameness of observation, comparison of accounts, and so on require only sameness of denotation, but allow for differences of connotation, we can now see how it is possible to argue that Kepler, Newton, Lavoisier, and Morgan, McKenzie and Parker are not merely changes from Ptolemy, Aristotle, Stahl, and R.T.Chamberlin, but are improvements upon them. "Planet", or "force", or "combustion", or "oceanic trench" are those entities responsible for particular specified effects, expressed in ordinary language.

To paraphrase Newton-Smith (1981,pp.164-165), that which unites scientists across different ages is their intention to use a specific scientific term to indicate whatever magnitude it is which is construed at the time, to be responsible for certain effects. Despite the fact that earlier and contemporary scientists held or hold quite different views of its scientific nature, they are still referring to the same entity. If we place too much emphasis upon the limitations to their use of the term, there will seem to exist variances of meaning, and as a result we shall begin to draw invidious and misleading contrasts between their attempts to refer to the entity, and those of a contemporary scientist. However, if it is stressed that all that a present-day scientist means by a particular term is merely that magnitude responsible for certain stated effects, and if the scientist agrees upon the effects in question or agrees that the effects in question are those that were produced by that magnitude which the first users of the term had in mind, we, because we believe that there exists a physical magnitude which produces the sequence of effects in question, will be perfectly well able to understand that to which the original and contemporary use of the term refers.

But although there is a stability here, and although we can see that connotational and theoretical differences may be accommodated to the same referents and denotations, it ought not to be therefore deduced that there must be a sort of fundamental descriptive language, *fixed for all time* , which scientists must use to fit their explanations of the empirical world. What has been said above in no way rules out evolution of the descriptive language of scientific research. New theories (various Clusters) and new language do develop as research proceeds. It is for precisely this reason that I believe that Shapere (1983) is beside the point when he suggests that the history of science shows that this kind of denotation-connotation distinction is untenable.

There is a third challenge which must be rejected also: that different Scientific Theories are incommensurable, and thus that scientific research makes possible merely change in Scientific Theory, rather than progress.

Much of this task of rebuttal has been achieved in the previous discussion of observation and meaning. But a somewhat different rebuttal is still required.

To achieve this rebuttal it will be important to emphasize the distinction introduced by F.C.S.Schiller (1917), revived by Reichenbach (1938) and stressed by some modern philosophers of education (e.g. Siegel, 1980), between on the one hand, the context of generation, genesis, discovery; and on the other hand that of evaluation, acceptability, justification. The distinction is important because it is the features of justification within the open and replicable practices of science to which we must turn, in order to show the objectivity of scientific change.

As demonstrated in the case studies of this chapter, in their actual work, scientists are continually engaged in *both* the generation of ideas and the logical justification of them. The process of critical appraisal is a continuing feature of scientific research at all levels, operating by way of institutionalized controls-- the traditions, replications, debates, conferences, independent checks, discussions in learned journals, and so on— by which scientific research as *an institution* itself evaluates novel ideas.

Kuhn is the doyen of those whose work has implied change rather than improvement in science, and the other subjectivists provide variations on his theme. So I shall institute a quick rejection of some of his typical key arguments[41]. The faults in his account can be exposed by keeping clear the distinction between genesis and justification; at the same time, the basic irrelevance of psychological and sociological considerations as aspects of *justification* becomes clearer. It will be argued that scientific change is a matter of objectivity, rather than persuasion, psychological change, etc. *Reality is not merely belief reified*. Some quotations from Kuhn will be used as a basis for rejection, and for showing the possibility of progress.

Kuhn writes that once there is a paradigm, ". . a scientific theory is declared invalid only if an alternate candidate is available to take its place. No process yet disclosed by the historical study of scientific development at all resembles the methodological stereotype of falsification by direct comparison with nature" (1970, p.77). He then argues (all the following italics are added), ". . the choice between competing paradigms regularly raises questions that cannot be resolved by the criteria of normal science . . . In the partially circular arguments that regularly result, each paradigm will be shown to satisfy more or less the criteria that it dictates for itself and to fall short of a few of those dictated by its opponent"(p.110). And, ". . the transition between competing paradigms cannot be made a step at a time, forced by logic and neutral experience" (p.150). Later, he makes the even more extraordinary claim that, "Scientists change for many reasons, few of them rational, such reasons as metaphysics, personality, nationality. ."(p.152). Scientists who adopt a new paradigm at an early stage,". . must often do so in defiance of the evidence provided by problem-solving. (They) must, that is, have faith that the new paradigm will succeed with the many large problems that confront it, knowing only that the older paradigm has failed with a few. A decision of that kind *can only be made on faith* " (p.158).

He says that rivals may each ,". . hope to convert the other to his way of seeing science and its problems (however) neither may hope to prove his case"(p.148), since they cannot, ". . make complete contact with each other's viewpoints"(p.148). And they cannot make contact because there is, ". . *incommensurability* of the pre- and post- revolutionary normal-science traditions" (p.148). Again, ". . in some areas [scientists] see different things. . . before they can hope to communicate fully, one group or the other must experience the conversion. . the transition. . cannot be made a step at a time, forced by logic and neutral experience. Like the Gestalt switch, it must occur all at once (though not necessarily in an instant) or not at all" (p.150). And also, in connection with comparison of paradigms: "Like the issue of competing standards, the question of values [for comparison] can be answered only in terms of criteria that lie outside normal science altogether" (p.110).

---

41. Several of the arguments which follow are derived from Scheffler (1967).

How good is Kuhn's argument? How good is any argument which rejects the traditionally accepted reasons for scientific change (increased simplicity, logicality, observational evidence, subsumption of what has gone before as a limiting case, descriptive range, predictive capacity, unanticipated but consistent implications, unanticipated practical applications) and replaces them with faith, conversion, Gestalt switch, personal decision, shifts of merely professional allegiance? And which, by blurring the distinction between discovery and justification, applies these not merely to the generation of new ideas, but also to their verification?

Kuhn seems to be maintaining, always by implication, and often explicitly, not merely that propagandizing is important, but that it is only propaganda which is. It all sounds much more like religion than like science. Or as Scheffler pointedly says, "The picture is rather like that of an epidemic" (p.78).

What should immediately be noticed is how much of what Kuhn writes and how much of any similar argument seem to be irrelevant. For even if we granted the truth of his claims about the way scientists come to have belief in a new paradigm, surely it is *not* the case that they appeal to conversion, Gestalt, faith, etc. in *defending and justifying their new beliefs* [42]. Even if scientists do experience Gestalt conversions and insights, in no way do scientists themselves think that these experiences are worth any weight in trying to convince their presently unmoved colleagues. And scientists certainly act as if they believe that theories can be rationally compared. Tycho certainly believed that the sun revolved around the earth, and Kepler believed that the earth revolved around the sun. Stahl believed in phlogiston, and Lavoisier believed it a fiction.

In the quotations above, Kuhn assumes that because issues of paradigm cannot be solved by resort to what he calls "normal science", the two scientific schools cannot be resolved by rational means. But this claim begs the question. Because Kuhn has actually *defined* "normal science" as science which assumes some paradigm, it is hardly to be wondered at, that the meta-level of *choice* between paradigms cannot be conducted within the domain of that which he has called "normal science".

Again, to what extent do we have to accept Kuhn's claims about incommensurability? Perhaps he is just confused. One question to ask is, In what way can incommensurable paradigms be *competing* ? To argue that they are competing is by the logic of the issue to place them within some *common* ground or framework. Indeed, the logic of the concept of competition indicates that at least two people or at least two points of view must desire the *same* thing, or be seeking the same allegiance, and that possession by one rules out possession by the other (Dearden, 1976). Competition is always competition over something. If there is not even agreement over what the supposed facts of

---

42. Although Kuhn is not doing scientific research, but producing Cluster 3 theory, there is also a nice irony here. For he supposes himself to be describing what actually happens in scientific research, and is trying to persuade the reader in a *rational* manner of this, and to believe his account because his account is a *rational explanation* of the supposedly a-rational happenings in scientific research.

the case are, then what are the two paradigms supposed to be competing over? Thus the logic of the situation is that there is after all some common ground (which is to be expected from commonsense, and from the accounts of scientists themselves). And if there is common ground then there is some possibility, despite Kuhn's denial, of a transition being made,"one step at a time". Consider what happened re the conflict between stabilism and drift, at the Lamont Laboratory of Columbia University. Although there was sharp disagreement and the Lamont geologists were scathing in their criticism of drift, the term, "incommensurability" hardly applies either in relation to the meanings of the concepts and the arguments they used, or in relation to the methods of appraising arguments and evidence. Ellen Herron, a graduate student at Lamont in the 1960s reports that, "Doc Ewing's philosophy that the oceans were permanent features was the party line at Lamont" (quoted in Glen, 1982, p. 313). Yet the very same Ewing, together with others such as Heirtzler and Le Pichon moved from being sarcastically contra critics to powerful proponents within a year– *and they explained the reasons for their change of view.*

Second, we should recall the point from White: if there were no way of describing, supporting, or refuting observation, by both using the particular theory to describe the observations and using some other type of description (e.g the categories and words of commonsense), scientists would rarely be in a position to explain that which they originally intended to explain. Thus, if chemists could speak only of combustion in oxygen, they would be unable to explain what they first tried to explain, namely, why things *burn*.

Third, as Newton-Smith says (1981, p.108), it is ironic that Kuhn, who claims to be describing what *actually* has happened in science, says that, for instance, Einstein and Newton are incommensurable, because scientists consider that they are commensurable, and also, in some respects incompatible. But then, as Toulmin would argue (1972; 1983), Kuhn would seem to be providing the wrong account of what has indeed happened in science.

Fourth, as Scheffler argues in relation to works of art (1967, pp.82-83), even if scientific theories *were* incommensurable, that would not stop certain sorts of comparison being drawn. Just as the art critic does not have to translate one work of art, say, a Picasso, into another, say, a Rembrandt, in order to make comparisons and comments of a rational sort, neither do persons who compare two scientific theories *have* to make translations: they can take a step forward, move to the meta level and compare the two ways of explaining in relation to any aspects of the physical world they consider relevant and important. Perhaps this fourth point is begging the question? If scientists on Kuhn's account of the situation are assuming incommensurable paradigms of "normal science", perhaps it may be thought that they will assume differently on the meta level also, stepping up to different meta or evaluative positions. But why should this be assumed? For there is an important distinction to be drawn between criteria for working within a paradigm and criteria by which different paradigms may be contrasted and compared. As Scheffler says,". . it is simply gratutitous to suppose that each paradigm 'dictates' such second-order criteria" (1967, p.85). Indeed, when Kuhn writes, in connection with disagreement over which problems scientific research should be concerned to solve, that, ". .the question of values can be answered only in terms of criteria that lie outside normal science altogether" he appears to contradict his own claim, and to agree with the important difference between criteria within a paradigm and criteria for judging between paradigms. And it is surely to beg the whole question to assume, as he does, that these external criteria, or these questions "of values" must be non-rational ones.

Moreover, when Kuhn says that,". .a scientific theory is declared invalid only if an alternative candidate (theory) is available to take its place", he is being much too strict in the application of the words, "declared invalid" and "take its place". As an historian of science he should be well aware that change from one view to another, such as the change from Ptolemaic-Aristotelian to Copernican-Galilean cosmologies is often an exceedingly slow evolution, even in the same human being. Contra Kuhn there *is* often change *one step at a time*. There is often a sort of "grafting" of one aspect of the new view onto some aspect of the old: Tycho's partly-sun centered theory is an example of an aspect of Copernicus being grafted onto Ptolemy. H.H.Hess grafted his sea-floor spreading hypothesis onto the rejected idea of continental drift, to produce a more powerful explanation. Often the older view lives on as a kind of limiting case within the new, or as an approximation of the new.

Equally, the phrase, "declared invalid" is ambiguous as between entire rejection, limitation as to application, unsuitable as a general guide to research, unsuitable as a practical tool, and so on. "Rejecting" a view can have numerous aspects. And when Kuhn disclaims, ". . . the methodological stereotype of falsification by direct comparison with nature" he is caricaturing the procedures of falsification. Falsification certainly happens, but not in this way. For instance, the successive falsifications by Crick and Watson of their tentative structures for DNA were never achieved in such a manner. Moreover, as was also pointed out in the case-study of DNA, and in the discussion above on observation, although a single test may not be sufficient, nevertheless, failures, challenges, falsified predictions can gradually amount to overwhelming proportions. Again there is a fine example in the overthrow of Aristotle-Ptolemy. Furthermore, it is naive to think that falsification by observation is the only important way in which scientific views are rejected. Considerations such as simplicity of the alternative, its economy, its conceptually greater range, even its metaphysical implications may at times be seen as relevant.

Also pertinent is the fact that Kuhn reinstates the significant distinction between discovery and justification. He does this in the very way in which he describes some of the reasons for change. It is noticable that despite Kuhn's preferred emphasis on Gestalt change, faith, conversion, rhetoric, persuasion, and incommensurability, he *also* talks about the significance of prediction as a criterion, and of anomalies as problematic for the older paradigm. But these two criteria are traditional rational explanations of the way in which matters evolve, Scientific Theories are compared, scientific improvements come about, and the later theory (or paradigm) is justified. Kuhn also writes, without being aware of the destructive effect of these words upon his overall thesis, that scientists will be reluctant to adopt a paradigm unless convinced that (1) it will resolve some important and generally recognised problem which can be resolved in no other way, and that (2) it promises to keep intact, ". . a relatively large part of the problem-solving ability that has accrued to science through its predecessors" (p.169). But of course these conditions are precisely the sorts of conditions which show that it is deliberation, objective control, critical assessment, and public and replicable tests which count.

Newton is better than Aristotle, plate tectonics are more powerful science than stabilism: through scientific research, paradigms, and research programs, and traditions do not just change, they also progress. And Scientific Theory continually improves.

Conclusion

So what has been said in this chapter, about scientific research?

As shown in the various case studies, scientific research uses several different Clusters of theory to provide its inspiration, its working presuppositions and its precision of conception and application. Its concepts are abstract— abstract concepts which relate in complex ways to form the propositions of the Scientific Theory in particular, but also relating in one way or another, or at one time or another, as a crucial part of theory in the other Clusters necessarily used by researchers (Clusters 5 (Hypothesis) and 6 (Presupposition)). Scientific research does not involve generalizations from experience, but rather universal claims from conceptualization.

The Scientific Theory used by and developed by scientific research involves three different kinds of connection between concepts. Rational connection is purely conceptual and relates abstract concept to abstract concept. Empiricist connection is empiricist or "sensible" and relates empiricist concept of observable to empiricist concept of observable. Abstractive connection straddles the abstract and the empiricist and relates abstract concept to empiricist concept of observable.

In being abstract, scientific concepts and propositions can enter into precise relationships with one another by way of conceptual definition and analysis, and mathematical calculation. The abstract nature of Scientific Theory also makes possible that paradigmatic scientific device: the use of ratio scales for measurement.

Contra the subjectivist claims of writers such as Kuhn, scientific research makes possible the continual progress of Scientific Theory.

It is now time to consider the history and nature of empiricist research on teaching, to see how such research fares in comparison with scientific research.

CHAPTER III

THE PRECURSORS OF MODERN EMPIRICIST RESEARCH ON TEACHING

Modern empiricist research on teaching, like the scientific research described in the previous chapter, did not spring out of nowhere, fully-formed in the 1960s and 1970s, but evolved both from earlier versions of statistical research in psychology and pedagogy, and earlier versions of and the philosophy of empiricist research. Some writers (e.g.Travers,1983) might place the origins as far back as Horace Mann's empiricist surveys of the schools of Massachusetts, the first of which occurred in 1838. And if we see such research as part of a natural development of more general aspirations for a science of man, then its origins are earlier still, being found in the European Enlightenment.

Whatever we may decide about the above question, it is clear that the modern empiricist approach is an outgrowth of several interrelated historical progenitors, amongst which are at least the empiricist philosophy of Locke and Hume, J.S.Mill's empiricist views of natural and social science, Francis Galton's psychometrics, Karl Pearson's correlation statistics, Edward Thorndike's scaling, Ronald Fisher's agricultural research and statistics, and of early empiricist studies in the schools such as that of Rice and others. And there are probably other sources. What follows is a quick attempt to provide some information about a representative selection of precursors whose work can be seen as leading to present-day ideas about that which is claimed to be a scientific approach to the study of teaching. What is said here links with what has already been suggested about the differences between scientific and empiricist endeavors.

J.S.Mill

Hamilton suggests that Mill's writings, "..give access to some of the epistemological leaps (or breaks) that went into the early formulation of Western social science methodology"(1980a, p.4). True. It is ironic, but it can be argued that Mill's systematic attempts to improve the "moral sciences" as he called the social and political studies, along the model of the natural sciences, led to an entirely new endeavor. It is as though Mill deliberately ignored the nature of the methods and discoveries of Galileo and Newton and all the other founding fathers of science. Although in my view he failed to produce an account of scientific research, *he did produce an account of systematic empiricist procedures.* What is more, the sophisticated development of his ideas by his successors has not led to scientific research, but to an ever more subtle, statistical, empiricist approach.

Mill inherited the ideas of the "philosophical radicals"— his father James Mill, Jeremy Bentham, and their supporters. The psychological basis of philosophical radicalism was "associationism", a more psychologically detailed version of the empiricist connection of Locke and Hume, mentioned in Chapter I. As Hamilton puts it,

> Associationism assumed, first that mental images were a reflection of external stimuli and, second, that association (or contiguity) of different external objects helped to create the internal (i.e. mental) association of the images of such objects. By such chaining procedures associationism could account for the gradual creation of an educated mind out of a series of initially independent (or atomistic) external sensations (1980a, p.5).

Philosophical radicalism transferred smoothly to ethics— human beings seek pleasurable sensations and association of these; these being natural to human beings, there is thus a large identity of human interests. The latter moves also took Philosophical Radicals into the economics of the free market where buyers and sellers shared an identity of interest (Hamilton,1980a, p.5).

Mill, a second generation philosophical radical extended such ideas, explicitly setting out to contribute to matters of government the same precision and concensus that could be obtained in natural science. Working mightily for thirteen years on the task, he produced an interrelated system of ideas both with a new name "Utilitarianism" and a new manifesto. These are detailed in his *A System of Logic* (1843/1967) (his first book), still (unfortunately) widely regarded as, ". .the most enduring essay on the method of the social sciences which has ever been written," (Fouer,1976, p.86; quoted in Hamilton,1980a, p.60). Here Mill merged mediaeval logic, associationist ideas of sense experience, assumptions about the regularity of the world, and his own ideas of inductive methods.

Mill noted the existence both of rational connection between abstract concepts and and empiricist connection between empiricist concepts (i.e. the assumed regularity of the world) as discussed in Chapter I. But he did not grasp the matter of abstractive connection of abstract concept and empiricist concept of observation which is at the heart of the success of Scientific Theory and research, and which makes possible the interpretation to the empirical world. Thus in his system of logic he kept rational connections and empiricist connections separate. In order to include mathematics in the domain of science he was therefore constrained to argue[1] that we arrive at mathematical knowledge through general-

---

1. As Urmson says, "It should be admitted at once that Mill's account of mathematical knowledge has satisfied nobody. Mill never properly distinguished between pure and applied mathematics, and confused the errors of counting and measuring with those of calculation" (1960, p.270).

ization (empiricist generalization, in the terms of this book). Even so, his innovative system of empiricist sorting of causes does not use mathematics. This is to be expected, given the absence of abstractive connections in his system.

Mill's concepts refer to classes of observable objects, rather than to the abstract concepts embedded in Scientific Theory. Logically they are empiricist categories. In so far as he also calls "temperature", "mass" and "energy" concepts, without distinguishing between the kinds, the issue becomes greatly confused (see Chapter VI).

Mill's concern was with propositions and he was interested in linking his empiricist concepts in propositions about observed regularities. "Induction, then, is that operation of the mind by which we infer that what we know to be true in a particular case or cases, will be true in all cases which resemble the former in certain assignable respects" (Mill,1967, p.188). Induction is, ". .a process of inference; it proceeds from the known to the unknown; and any operation involving no inference, any process in which what seems the conclusion is no wider than the premises from which it is drawn, does not fall within the meaning of the term" (Mill, p. 205). As Medawar suggests (1980, p.119), that view might have been very well if Mill had not also claimed that induction was an exact and logically rigorous process, capable of doing for empiricist reasoning what logical syntax does for the process of deduction. For Mill says that, "The business of inductive logic is to provide rules and models (such as the syllogism and its rules are for ratiocination) to which, if inductive arguments conform, those arguments are conclusive, and not otherwise" (p.120).

In his book Mill, confounds Hume's sceptical view of the logic of induction with the axiom that nature is uniform. "The uniformity of nature principle. . rescued the problem of induction and, at the same time [supposedly], raised it to the same level of certainty as deductive logic" (Hamilton,1980a, p.6). Nowadays it is normally agreed by scholars that Mill rescued induction by presuming the existence of induction. But as was shown in earlier chapters, the context of discovery is concerned with connections between observed regularities only in the sense that these sometimes prompt a move to abstract conceptions.

Mill assumes that the universe is a compound of confounding causal regularities, which overlap one another in ordinary observation. Thus if we could sort out all the confounding regularities for a particular phenomenon, we would on his view be able to isolate the real cause. His system of logic therefore attempts to provide a set of "canons" for systematic sorting. In these canons of experimental procedure he attempts to bring inductive logic closer to deductive logic: through his methods of agreement, difference, residues, and concomitant variation. These provide a sequence of empiricist manipulations by which the basic "atomic" causes are to be separated from the plural partial causes. It is not surprising that Mill has nothing to say about theory, because he believed science to be the results of his empiricist sorting methods. Mill's canons in their modern statistical dress such as analysis of variance (ANOVA) and factor analysis have special importance in empiricist research where they still form the basis for the establishment of "findings". And what Scientific Theory calls a "law", Mill incorrectly believes are examples of the natural relationships to be discovered by his methods. Nevertheless something like Mill's canons do appear at times to be made use of by scientists in their critical work: but only when there is a Scientific Theory which is guiding an hypothesis. The canons themselves are conceptually empty and inadequate as a guide. Mill's ideas, though sophisticated in their own tantalizing way, were actually a scientific regression: to empiricist methods.

It is revealing for my argument that the contemporary empiricist researcher does not expect that he will uncover invariant relationships despite his use of modern descendants of Mill's methods. He only expects "findings" which are more or less probable. This is in stark contrast to the laws produced by Snell and Galileo and Newton, who, living hundreds of years before Mill, did not arrive at their laws and theories by empiricist sorting methods, but by brilliant conceptual leaps to the abstract level. As this and the following chapters attempt to make clear, empiricist researchers on teaching have been gathering immense masses of data and using billions of dollars without isolating a single scientific law. This itself should be good empirical evidence that there is something not quite right with this claimedly scientific method.

Though Mill is not associated with a specific grand empirical investigation as is Durkheim, or with devising statistical methods, as are Fisher and Pearson, his work has been of immense significance in modern empiricist research. With its social and philosophical origins forgotten, Mill's general logic of enquiry has been,". . transformed into a technology as unquestionably accepted [by empiricist pedagogical researchers] as it [is] vigorously applied" (Hamilton,1980a, p.7).

Alexander Bain, James Sully and Others

It has just been argued that present-day empricist pedagogical research has adopted Mill's ideas of methods as its norm, without realizing their empiricist limitations and misconstrual of the nature of scientific research. But quite explicit claims for a science of teaching are at least as old as the writings of the Scots, Alexander Bain and James Sully. In 1879 Bain's book *Education as a Science* was published. Anticipating to some extent Process-Product researchers such as Nathaniel Gage, Bain saw education as an art based upon science (1879, p.1) though he construed pedagogy as a *deductive* science drawing upon the findings of other sciences, rather than an observational one. He was however shrewd enough to grasp that a pedagogical science would be an extremely complex matter. James Sully too, in his *Teacher's Handbook of Psychology* (1886) argued there should be a search for, ". . a body of well-ascertained truths . . from which the right and sound methods of training the young may be seen to follow" (quoted in Travers,1983,p.78).

Monroe and Engelhart in their misnamed,*The Scientific Study of Educational Problems* (1936,p.453) list the following pre-1900 American papers and books which argued in favor of the possibility of a science of pedagogy: Payne (1876), Jerome (1882), Bain (1884) (U.S. printing of text mentioned above), Payne (1886), Scripture(1892), and Findlay(1897). Some of these views are not of pedagogical science as itself an inductive or empirical science in an original sense, but similar to Bain's view, as deductive, or as something derived from the various requirements of pedagogy added to the contributions of various other sciences. The existence of such work nevertheless attests to the long history of the hope that pedagogical study can be scientific.

By 1910, the editors of the inaugural issue of the *Journal of Educational Psychology* were unambiguously writing,

> ..we believe that the time is ripe for the study of schoolroom problems in the schoolroom itself and by the use of the *experimental method*. Educational practice is still very largely based upon opinion and hypothesis, and thus will it continue until competent workers in large number are enlisted in the application of the *experimental method* to educational problems... We propose to maintain a high standard of *scientific* worth (italics added) (Bagley, Seashore and Whipple, 1910, pp.2-3).

The Statistics of Francis Galton

In 1883, Galton published his *Inquiries into Human Faculty and its Development,* which drew upon his anthropological, anthropometric and psychological studies of the previous decade. So fundamental are Galton's century-old ideas to present-day pedagogical research that several of his arguments require detailed stating. He writes:

> The object of statistical science is to discover methods of condensing large groups of allied facts into brief and compendious expressions suitable for discussion. The possibility of doing this is based on the constancy and continuity with which objects of the same species are found to vary. That is to say, we always find, after sorting any large number of such objects in the order (let us suppose) of their lengths, beginning with the shortest and ending with the tallest, and setting them side by side . . . their upper outline will be identical. Moreover, it will run smoothly and not in irregular steps. The theoretical interpretation of the smoothness of outline is that the individual differences in the objects are caused by different combinations of a large number of minute influences... Whenever we find on trial that an outline of the row is not a flowing curve, the presumption is that the objects are not all of the same species, but that part are affected by some large influence from which the others are free; consequently there is a confusion of curves. This presumption is never found to be belied (Galton,1883, p.49).

Furthermore, Galton writes,

> There is no bodily or mental attribute in any race of individuals that can be so dealt with, whether our judgment in comparing them be guided by commonsense observation of by actual measurement which cannot be gripp-

ed and consolidated into an ogive with a smooth outline, and thenceforward be treated in discussion as a single object (1883, p.52).

As Hamilton (1980b) points out, to be noted is Galton's portentious assumption based on the earlier claims of the Frenchman, Quetelet, (1) that physical attributes are distributed according to a Gaussian curve, (2) that so are psychological features, and (3) that in this statistical analysis he has discovered an invariate *empirical* relationship: "This presumption is never found to be belied". But it is arguable that it was never found to be belied because measures which did not fit, e.g. measurements that showed a bimodal distribution, were rejected on the a priori grounds that such discontinuous variation indicated the presence of more than one type of entity being measured. Relabelled by Galton as the "normal curve", this view has become an unchallenged axiom of procedure in empiricist research on teaching.

Psychometrics, norm-referenced testing, procedures for ranking, percentiles, correlation and factor analysis all derive from Galton's imaginative statistical innovations in the 1870s and 1880s (Hamilton,1980b, p.157). Thus, many of the statistical underpinnings for the hopeful science of pedagogy, underpinnings which, without critical challenge have been continually refined and developed since that date both in the social sciences generally and in pedagogical research in particular, were established in Britain rather more than one hundred years ago. Transplanted and developed, they took even firmer root in the empiricist soil of the United States. But their Millian and Galtonian origins had been largely forgotten.

Karl Pearson

At century's end, Galton's work was enthusiastically taken up by Karl Pearson, who further developed the statistics of the bell-shaped normal curve (Galton preferred to show it as an ogive), the chi-square test and the multiple correlation coefficient. All such developments were adapted to pedagogical research, increasing the confidence that a true science of pedagogy was emerging.

Like Mill, Pearson believed that empiricist endeavors were scientific endeavors. Indeed his methods are so different from the great scientists of the previous centuries that Willer and Willer suggest that, "Pearson's work represents the first clear separation of systematic empiricism [empiricist endeavors] from the older scientific tradition" (1972, p.44). On Pearson's view, scientific laws are a sort of empirical summary— ". . brief expressions of the relationships and sequences" (Pearson,1957,p.82), or alternatively, as he calls them, "generalizations" (1957,p.86), and thus scientific progress is supposed to be coterminus with ever-improving generalizations.

Mill's views of the relation between mathematics and science were alluded to above. But Pearson argued that because experience and observation never disclose an exact equivalence, mathematical laws misrepresent reality by suggesting a higher level of relationship than can by justified by experience (1957, p.99). However, this idea is itself *the result of taking an empiricist view of scientific research*. Such a view becomes otiose if we construe science as having theory consisting of (one or more) levels of propositions which involve rational connections between abstract concepts which are related abstractively to the observational world. *Pearson confused general concepts with abstract concepts and empiricist generalizations with scientific laws, believing that it was his empiricist generalizations which were the essence of science.* But with empiricist concepts embedded in propositions which are empiricist generalizations, it is impossible to move from empiricist observational concepts to precise, mathematically stated laws, or to use ratio measurement, so of course *statistics have to be used* : which was what Pearson did.

In a program, in effect diametrically opposed to the core of all recent philosophy of science, to the arguments about the effects of the several Clusters of implicit and explicit theories presented in Chapter I, and to the approach to scientific research detailed in Chapter II, Pearson wished to end the dominance of concepts over experience and facts, and replace it with the dominance of experience and facts over concepts. Thus he writes, "The classification of facts and the formation of absolute judgments upon the basis of this classification . . essentially sum up the *aim and method of modern science* [sic]" (Pearson,1957, p.6) (his italics). Again, says he,

> The classification of facts, the recognition of their sequence and relative significance, is the function of science. . .*Let us be quite sure that whenever we come across a conclusion in a scientific work which does not flow from the classification of facts* , or which is not directly stated by the author to be an assumption, then *we are dealing with bad science* [sic] (italics added) (1957, p.9).

In response, Medawar succinctly says, "Poor Pearson! His punishment was to have practised what he preached, and his general theory of heredity, of genuinely inductive origin, was in principle quite erroneous" (1984, p. 28).

Moreover, because there cannot be exact equivalence or identity in nature, "You cannot get exactly the like causes" says Pearson (1957, p.154), empiricist generalizations (which Pearson calls "laws") have to be stated in probabilities and relations represented by a correlation coefficient. Pearson claims that no phenomena are causal, but all are contingent, and thus, ". .the problem before us is to measure the degree of this contingency, which we have seen lies between the zero of independence and the unity of causation" (p.174). Therefore, ". .*the fundamental problem of science* is to discover how the variation in one class is correlated with or contingent on the variation in a second class (italics added)" (p.165). This would have been a real surprise to any of the great scientists discussed in Chapter II.

But as every student of methods of empiricist pedagogical research learns and in most cases thereafter completely ignores, correlation is not causation. Thus no matter how subtle, Pearson's correlation statistics could not provide a means of producing true empirical association, i.e. laws. Indeed for there to be science, the association of observables must be decided *before* the determination of the extent of such a relationship, otherwise we are back with mere correlation. "Pearson's method is incomplete in that he ignored the problem of determining association entirely and as a consequence did not provide a method that could result in generalization," (Willer and Willer, 1972, p.49).

This critique of Pearson is significant. For as will become clear from the examples of the next chapter, important parts of empiricist research on teaching consist of updated versions both of his views and his techniques: on his view (but using my terminology) scientific research consists in the endeavor to relate empiricist categories to other empiricist categories at levels of correlation.

Rice, Thorndike, Taylor, Ayres, Bobbitt and the Efficiency Movement

The first quasi-experimental study of teaching was that of Rice who in 1897 published *The Futility of the Spelling Grind*, which compared teaching in a variety of schools, a study which began in 1895 and involved over 30,000 children. It is an early example of the attempts to discover "the best way" to teach and thus a clear precursor of current empiricist research, especially of the Process-Product kind. It did not however make use of elaborate statistics.

Rice saw his task as an empirical investigation of the results of teaching. His results were ambiguous. In some schools mechanical teaching fared better than progressive teaching, in other schools the opposite. Rice suggested that no direct relation exists between methods and results, but that rather, results depend upon the ability of the teacher using a particular method (Rice,1914, p.88). He concluded that,". .nothing can take the place of that personal power which distinguishes the successful from the unsuccessful teacher" (1914, p.99). His one seemingly rigorous finding was that the results of spelling tests were substantially the same whether ten or forty minutes were spent each day on drill.

Conceptions of a science of pedagogy have often been closely tied to attempts at measurement, and since the turn of this century many educationists have made the unjustified assumption that bringing scaling[2] into pedagogical research was synonymous with making it scientific. By 1912, the idea of a pedagogical science based squarely on scaling the (assumed) results of teaching was rapidly gaining ground (Tom, 1984, p.19). That year, at the National Education Association annual meeting, Ayers claimed that, "The proposition underlying this entire mass of discussion was that the effectiveness of the schools, the methods, and the teachers must be measured in terms of the results secured"(1912, p.301). By World War I, the development of standardized achievement tests was a vast industry.

In 1912, E.L.Thorndike foresaw a wonderful future for experimental research on teaching, writing,

> Education. . is just beginning to give promise of quantitative knowledge, of descriptions of facts as numerically defined amounts, and of *relations or laws*

---

[2]. Called "measurement" by its advocates, both then and now.

in terms of rigid, unambiguous equations. The changes that take place in intellect and character are coming to be measured with *the same general technique,* and we may hope with the same passion for clearness and precision, *which has served the physical sciences* for the last two hundred years (Italics added) (1912, p.289).

Ignoring the lessons which might have been learned from a close study of what had happened in the development of the natural sciences, Thorndike's exemplars of scientific desiderata were scaling and statistical analysis. He brought to the task the new techniques of such statisticians as Galton and Pearson (Travers,1984, p.279). And it was at the 1914 first conference on educational measurements (i.e. scales) that Thorndike promulgated his sweeping claim that has been so influential in empiricist pedagogical research that if a thing exists in some amount, it can be measured.

Though Thorndike was not himself a supporter of F.W.Taylor's "scientific management" (see below), his proclivities for measuring and testing fitted neatly into that way of thinking. It was a way of thinking which reached flood proportions even in the schools, by the mid-1920s (Callahan,1962). It consisted of unfavorably comparing schooling and business enterprise, ". . of applying business-industrial criteria (e.g. economy and efficiency) to education, and of suggesting that business and industrial practices be adopted by educators" (Callahan, p.6). Moreover, however misleading or restricted the use of the word, "scientific" may have been in this context, the movement helped to convince people that a scientific approach was not just acceptable but was essential to pedagogy. As early as 1908 the U.S. Commissioner of Education had highlighted the idea of efficiency and its related idea, standardization, in his annual report (Travers, 1984, p.126). Taylor and those who followed him made the extraordinarily myopic claim that in every field of endeavor there was,". . always one best method for doing a particular job and this best method could be determined only through scientific study" (Callahan, p.25).

A new surge of criticism was directed at schools, ". .especially large ones, which might be suspected of gross managerial inefficiency" (Callahan, 1962, p.47), and throughout 1912 there was often vicious criticism, in articles such as, "Our Medieval High Schools— Shall We Educate Children for the Twelfth or the Twentieth Century?" which asked such questions as, "Why should [taxpayers] support inefficient schoolteachers instead of efficient milk inspectors?" (quoted in Callahan,1962, pp.47-48).

As an example of the cult of efficiency in schools, and an intriguing precursor of recent research on Academic Learning Time (ALT)[3], in 1912, Mitchell suggested an elaborate system of book-keeping by teachers, pointing out that as much as four minutes were wasted by pupils waiting their turn to be assigned academic tasks, and saying that this, ". .must give way to an ideal of timesaving . . to the end that maximum results may be attained under pressure of

---

3. See Chapter IV.

time and with economy of material" (quoted in Callahan, 1962, p.102). In 1913, Hanus, addressing the annual meeting of American school superintendents on, "Understanding Principles of Scientific Management" said there should be more research so that schooling could be based on, ". . verifiable data which any technically informed person can appeal to." He was especially encouraged by the fact that educators were he said, ". . no longer disputing whether education has a scientific basis; we are trying to find that basis" (quoted in Callahan, pp.66-67).

The following year, Bobbitt, whose ideas were to become very important for pedgogical study, argued in the twelfth yearbook of the influential National Society for the Study of Education that to be scientific there should be precise specification of educational products. He wrote for instance," . . the ability to add at a speed of 65 combinations per minute, with an accuracy of 94 % is as definite a specification as can be set up for any aspect of the work of a steel plant" (quoted in Callahan, p.81). Moreover argued Bobbitt, measurable standards could be established even in the more intangible subjects such as history and literature. Though difficult, such standardization was still possible, ". . for every desirable educational product whether tangible or intangible" (quoted in Callahan, p.84). To decide the best methods, it was merely necessary to examine current alternative practices in different schools. Thus with respect to handwriting says Bobbitt,

> . . suppose each of these groups of schools to be measured in the first week of the school year by the Thorndike or the Ayers writing scale as to quality, and tested by the stop watch as to speed. If they are then measured again at the end of the year in the same way, it is possible to determine which of the modes of distributing the sixty minutes of time for teaching writing is the superior (quoted in Callahan, p.87).

Though eighty years have passed, this is similar to claims made by present day Process-Product empiricist researchers.

A decade later, Bobbitt had changed his interests to teacher preparation, but had not changed his outlook. Teacher preparation was to be vocational. In a manner curiously like that of the movement for Performance Based Teacher Education (PBTE)[4] fifty years later that, he writes, "The method generally agreed upon for discovering the objectives of any vocational school is well known. The first step is activity analysis." The investigator should,

> . . go where the teachers are performing all their tasks as they ought [sic] to be performed. He will then list the two hundred or five hundred or five thousand tasks which the competent teacher accomplishes in his work. The abilities to perform these tasks then, are the fundamental teacher-training objectives— the abilities to do the jobs are the objectives. There are no others [sic] (Bobbitt,1924, p.188).

---

4. see Chapter IV.

And in his 1922 booklet, *Curriculum imaking in Los Angeles* Bobbitt lists some 550 such abilities. With respect to this list, Bode, a contemporary critic pointed out that there is a pedagogically-destructive, systematic ambiguity between Bobbitt's specific abilities and his general ones. He writes that Bobbitt lists,

> . . such skills as putting up shelving, putting on doorknobs. . . renewing washers in faucets, gluing, soldering. . But at other times the abilites rival the 'faculties' in general. . for example, the ability to protect oneself from social, economic, and political fallacies. . [and the pedagogical ones of Bobbitt?] and the disposition to "do one's best whatever the circumstances" (Bode,1924, p.181).

But despite the pertinent criticism of people such as Bode, attempts to discover the basic activities of teaching by similar empiricist methods continued throughout the 1920s and 1930s, as typically evidenced in the researches of W.W.Charters (see below). In the years between 1911 and 1925, as a concomitant of such supposedly scientific management, thousands of surveys of schools and school districts were made, appropriating business and industrial terms and drawing parallels between such concerns and pedagogy. Such surveys encouraged the adoption of standardized achievement tests, and rating scales to assess teachers.

In the first issue of *The Journal of Educational Research* , January 1920, the editor, Buckingham writes, "Since scales and standardized tests began ten years ago, to afford new instruments of precision, a new language has developed. . statistics have come upon the scene, providing a powerful method of analysis" (1920, p.2). However, he criticizes the abstractness of the results, and continues with comments which resemble those of some recent critics, "Some of us wonder whether all this is necessary— whether a writer may not be scientific without being impractical, or profound without being obscure" (p.2).

Thus by such early dates, many of the key aspirations, procedures and problems, which have continued to sustain the quantitative, standardized, empiricist attempt to develop scientific pedagogical research throughout the rest of the century, had been clearly established, expressed, or encountered.

Ronald Fisher

Fisher approached the problem of determining association through a method of statistical testing developed from the theory of errors, based on studies in mathematical probability of the seventeenth and eighteenth centuries. With a splendid dismissal of the kind of scientific work discussed in Chapter II, and drawing from his own statistical work in agriculture, Fisher argued that statistical procedures and experimental design are two aspects of the same issue (Fisher, 1935, p.3).

He saw his method as a way of drawing inferences, ". . from observations to hypotheses; as a statistician would say from a sample to the population" (p.4).

He argued that we can in a manner of speaking, infer from the particular to the general: "We may at once admit that any inference from the particular to the general must be attended with some degree of uncertainty, but this is not the same as to admit that such an inference cannot be absolutely rigorous" (p.4). The rigor Fisher is referring to is that of *careful calculation of the degree of uncertainty*. Thus, as Willer and Willer so astutely put it, "*Rigorous uncertainty* is the desired result of Fisher's process of associational inference (italics added)" (1972, p.50). In order to achieve this he produced some massive statistical moves. Thus he,

> . . strengthened Pearson's chi-square test by introducing the concept of "degrees of freedom"; he provided a rigorous mathematical proof of Student's t-test; he identified and formalised the z-distribution (which allowed significance tests to be performed with correlation coefficients derived from small samples); and finally, he developed the technique known as analysis of variance as a means of separating the effects of different "causes" (Hamilton, 1980b, p.161).

Fisher developed his techniques in agricultural research. In agricultural field experiments, a common systematic design had consisted of three treatments, A, B and C, distributed across an experimental area according to the preconceived plan: ABC: ABC: ABC. "Though such a design sought to eliminate environmental inhomogeneity, a gradient of environmental condition that existed, let us say, from left to right, would affect treatment A to a greater extent than the others," (Fisher-Box,1972, p.171). But it had been well established by empiricist methods that such gradients in the conditions of soil existed everywhere and often over short distances. Fisher's insight, was to realise that if the several positions for treatment were *assigned in a random manner*, then the laws of chance, would make it possible for statistics to provide relatively objective *statistically quantitative estimates* of the error due to any variability in the soil or in the biological materials being tested.

Fisher did not claim to be looking for scientific laws, but, basing his account on his own agricultural research, looked instead for, ". . the principles which are common to all experimentation" (Fisher,1935, p.11). He thereby ignored the four centuries of scientific research since Copernicus in order to draw from his own approaches. Taking Gosset's work as a starting point, Fisher approached the problem of drawing inferences from small samples and as a result produced his comprehensive theory of statistical hypothesis testing for agricultural experiments, by way of his tool of the "null hypothesis". In Millian terminology, his design involves the canon of the method of difference deriving from assignment of objects to two or more groups, manipulating one or more, assessing the results, and deciding by way of Galton's normal curve and the null hypothesis whether there is a statistically significant difference. This is an approach which has become central to the observational and experimental work of empiricist research on teaching.

It is interesting to note what Cronbach has said on these matters:

"Design of experiments" has been a standard element in training for [empiricist pedagogical researchers]. This training has concentrated on formal tests of hypotheses— confirmatory studies— despite the fact that R.A.Fisher, the prime theorist of experimental design, demonstrated over and over again in his agricultural investigations that effective inquiry works back and forth between the heuristic and the confirmatory. But since he could offer a formal theory only for the confirmatory studies, that part came to be taken as the whole (1982, pp.ix-x)

But what Cronbach says can be construed in a different way. It is a nice coincidence (and irony) that he uses the word, "heuristic" in the above quotation, because Fisher's starting points were indeed Heuristics in the sense discussed in Chapter I (i.e. Theory Cluster 5, Type 7). What was missing in Fisher's procedure was precisely the thing still missing in empiricist pedagogical research: explicit acknowledgment of the importance of Theory (Clusters 3 (Explanation), or Cluster 5 Type 8 (Scientific Hypothesis), or Cluster 9 (Science), in other words the kind of theory which might have turned his heuristics into true scientific hypotheses. But Fisher was not interested in scientific Hypotheses (Type 8), or in Scientific Theory. He was not testing Hypotheses drawn from or developing into Scientific Theory. His Heuristics were possibilities with respect to the effectiveness of alternative treatments.

That does not mean that Fisher himself did not get some pertinent results *in agriculture*. But Willer and Willer seem to me correct when they argue that what Fisher failed to realize was that in agriculture there was already organization— inbuilt Scientific Theory already at work in helping him to reach usable results. This aspect is missing from empiricist research on teaching, except for the commonsense of Empiricist Theory from which it draws its Heuristics. As the Willers say, for Fisher's agricultural research the cogency of the "rigorous uncertainty" was inbuilt, because,". . his experimental objects, such as strains of seeds, had been previously manipulated into separate homogeneous strains." Thus, "One strain of seed could be considered to be essentially identical to another because of their careful development following genetic laws." So the associations which Fisher found and which proved agriculturally useful, ". . were at least partly dependent on scientific laws and thus were actually not inductive at all, *but simply demonstrated an expected [relationship]* . Fisher could argue for the sameness of his experimental objects because they were developed according to the dictates of genetic theory (italics added)" (Willer and Willer, 1972, p.55). In contrast, in empiricist research on teaching, of either the observational or experimental sorts, the objects of manipulation are not products or propositions of already established Scientific Theory, but empiricist categories and generalizations.

Lest there be misunderstanding, let me repeat that Fisher himself achieved useful results in agriculture. But he also wrote books, such as *The Design of Experiments* (1935), on how to achieve results. *And what was missing from the books was the part played by the suppressed Scientific Theory* in his agricultural experiments. Thus it is his advice on how to proceed which has been misleading. And it is precisely this which empiricist researchers on teaching have followed. Contrariwise, I would argue in agreement with Willer and Willer that, "While Fisher had the perfect control provided by the use of laws, [empiricist pedagogical researchers] can only randomize the selection of subjects and hope that it is relevant." In short, "Whereas Fisher's experiments merely confirmed Darwinian theory and Mendellian laws, [empiricist research] is left in the circumstance in which there are no such laws to confirm" (Willer and Willer, 1972, p.50).

So I am arguing that Fisher failed to grasp the key role that Scientific Theory was already playing in his statistical research. Instead, as Willer and Willer point out (1972, p.56), Fisher believed that the successes he achieved resulted entirely from the statistical design— a mistake which has been routinely made ever since by pedagogical researchers who adopt his methods.

W.W. Charters and A.S.Barr

W.W.Charters is another researcher representative of the movement aimed at bringing science and efficiency into teacher training. In 1918 he argued that there was a great need for researchers to identify the difficulties which teachers encounter and in order to remedy these, there was a need to collect suitable curative methods, evaluate the efficiency of these, and publish the resulting findings about, ". .excellent methods of teaching", (quoted in Tom,1984, p.21). Like Bobbitt before him, Charters argued that the curriculum for teacher training should be based on particulars, and in 1919, he suggested that such particulars should be stated in behavioral terms. He believed that, ". . while science through controlled experimentation would one day produce 'the right answer to everything' (sic), [at present] we must rely on the judgment of experts to determine the most effective methods " (quoted in Tom, p.22).

What Bobbitt called "abilities", Charters called "functions". By 1924, Charters was applying functional analysis to teacher training. As Capen writes in his Introduction to the eventual study,

> . .the critical inquiry into first, the training of pharmacists, and now, the training of teachers, are functional studies . . . A functional study tries to determine what the professional practitioner does under modern conditions of practice. From the objective record of what he does it attempts to derive the determination of what he must know and what he must be to perform these duties effectively (Charters and Waples,1929, pp.xv-xvi).

In 1925, Charters received a three-year grant of $42,000[5] from the Commonwealth Fund, to develop an improved curriculum for teacher training. In the Preface to the resulting volume, he and his co-researcher, Waples write as follows. Their ideas, with slight change of terminology might easily have come

---

5. Close to $1,000,000, in today's values.

straight out of a late 1970s PBTE manifesto:

> A radical reorganization of the curricula of teacher-training institutions is demanded by a variety of conditions. . .curricula have been developed without clear definition of objectives and with no logical plan of procedure. Sponsors of the project have based their support on the hope that a comprehensive description of the duties and traits of teachers might provide the necessary basis for determining systematically what teachers should be taught (1929, p.v).

S.P. Capen, in his Introduction to the book said, "If the standards are to be precise however, if they are to be galvanic rather than repressive, they must be based on renewed and searching study of the educational process they are designed to regulate" (Charters and Waples,1929, p.xiv). Charters and Waples found three years insufficient time both to do the analysis and to produce the precise and galvanic standards. But their, *The Commonwealth Teacher Training Study* managed to provide the analysis. Their survey was massive. Tyler (of later curriculum fame), the statistician for the study, reports that as the lists of teachers' activities arrived from those collaborating, ". . each activity was typed on a card. The number of specific activites reported was staggering, and eventually more than a million cards were assembled"(1953, p.44). Charters and Waples reduced this mass to a grand total of 236,655 submitted activities (1929, p.21). Out of these, they drew the materials upon which the standards would have been based. There are some eighty-three traits (1929, pp.223-244) and 1,001 activities (1929, pp.257-472), together with rankings of importance and difficulty of learning. "It is as complete as present techniques can make it" (1929, p.23), said Charters and Waples. Typical listed activities are: 100: Teaching pupils to gather reading materials, 201: Filing records, reports, and correspondence about marks, 302: Developing pupils' interest and attention in the performance of acting courteously toward others, 504: Grouping pupils, 605: Selecting points to which to make excursions, 807: Securing action on decisions arrived at with the Superintendent, 908: Establishing cordial relations with community social organizations.

The first ten traits are: accuracy, adaptability, alertness, ambition, animation, appreciativeness, approachability, attractive personal appearance, breadth of interest, and calmness. Each trait was sub-divided into trait-actions. I choose two at random. Number 23, dignity, had the following trait-actions: a. does not pretend to be going to cry whenever boys misbehave, b. does not apply rouge and powder before pupils, c. does not sit on the top of desk and swing feet, d. does not answer pupils back in the same saucy way that they talk to him, e. does not whine about things that do not suit him, f. keeps purely teacher-pupil attitude toward pupils, g. plays the role of host in a charming manner, h. does not make silly remarks that do not suit the situation, i. does not gush over children.

Number 30, forcefulness, lists: a. makes people care for what he says, b. puts energy into his teaching, c. holds the attention in conversation, d. dominnates any circle by his presence, e. makes pupils feel that he knows his subject, f. makes parents see his point of view, g. exerts authority when he should [sic], h. requires attention from pupils when making assignments or when giving instructions, i. gets people to think his way.

Tom reports that reception of Charters's work was mixed. M.E.Haggerty pointed to the failure to synthesize all these findings into actual courses for teacher trainees, wondering whether, ". . this last and necessary step is not more difficult and baffling than the report reveals" (quoted in Tom, 1984,p.25). Indeed, Tom says that he could find only one example of such an attempted course, at Ohio State University, to which Charters moved in 1928. Tyler reports that Judd and Morrison ridiculed such specificity, saying that Charters,". .was expressing a mechanical conception of curriculum development that missed the main point of education, namely, the process of generalizing learning" (quoted in Tom, p.24). But a decade later Charters's faith in the ultimate victory of scientific methods in pedagogy was unshaken, for he writes, "When *scientists have constructed measures for educationists as accurate as those of engineers*, teaching wastes will be cut and curriculum engineering will become a routine of curriculum construction (italics added)" (quoted in Tom,1984, p.26).

A.S.Barr, a director of supervision for the Detroit Public Schools in the 1920s also wanted to place teaching on a solid scientific base, but rather than rely upon Charters's technique of surveying the judgment of expert opinion, Barr, like present-day empiricist researchers attempted to study behavior directly in classrooms, for he had been concerned about the,". . subjective character of current supervisory procedures"(quoted in Tom,1984, p.26). Not unlike current researchers who strive for "low-inference" scales, Barr wished to reduce or replace *all* inference, thus making supervision of teaching more specific. And precisely like present-day empiricist Process-Product researchers, he wished to make the study of teaching scientific in order to demonstrate which behavior or characteristics of teachers were most significant for good teaching. His study compared two groups of forty-seven teachers. (The logically necessary, non-scientific, inferential nature of the initial moves to identify the good and bad categories seem to have been overlooked by Barr.) The groups were chosen by way of,

> . . superintendent nomination, state inspectors' evaluations, and Barr's own classroom observations . . . stenographic reports, observation schedules, interviews, and related procedures were used to gather data on about forty teacher behaviors or characteristics, including posture, student assignments, teacher questions, supervised study, length of pupil responses, the average number of hands raised per question, and so forth (Tom,1984, p.27).

The results, under the title, *Characteristic Differences in the Teaching Performance of Good and Poor Teachers of the Social Studies* were published in 1929, the same year in which the Charters Commonwealth Study had appeared. In it, Barr drew hundreds of comparisons, both particular and general, between good and poor, such as the twenty-five most frequently used responses by teachers to questions, and the way in which teachers allocated classroom time. With respect to the latter, he concluded that, ". . teaching performance is highly variable. . . Good teachers function successfully within a wide range of time expenditures" (quoted in Tom,1984, p.27). So Barr found differences, but not anything which fell into neatly separate categories. He concluded that, ". . he had not identified any factor that might be viewed as critical, that is, a factor so important no teacher could succeed without it" (quoted in Tom, p.27). Barr concludes that good teaching, ". . is probably the result of many small matters well done" and that, ". . the performance of teachers is so variable as to make it next to impossible, in the absence of further evidence, to say that an observed practice is wholly good or wholly bad" (quoted in Tom, pp.27-28).

Such ambiguity did not stop Barr from writing *An Introduction to the Scientific Study of Classroom Supervision* (1931) and another book with Burton and Breuckner, *Supervision* in 1938 in which he summarized parts of the 1929 study. By the second, 1947 edition of the latter, Barr had come to acknowledge that there were some puzzling questions to be asked about this search for optimum features of good teaching, features, ". . true for all purposes, persons, and conditions" (quoted in Tom,1984, p.28). By 1961, in his contributions to the Wisconsin Studies he clearly states that his 1929 study showed that, ". . good teachers cannot be separated from poor teachers in terms of specific teacher behaviors (*there is an appropriateness aspect* to teacher behaviors that must be taken into consideration)(italics added),"(quoted in Tom,1984, p.28). Like some present-day Process-Product researchers, Barr frequently summarized the current status of research on effective teaching— seven times for the *Review of Educational Research*. And the Wisconsin Studies into teaching ability (nowadays ignored) which were conducted between 1940 and 1960 were largely the work of Barr's graduate students.

As Tom notes, "A lifetime of inquiry, including [the supervision of] over one hundred doctoral dissertations, seems like a large price to pay to find out that the original question needs redefinition" (1984, p.29), or, it might be argued, cannot be answered. Yet Barr's continuing optimism is shown as late as a 1958 paper, which shows that while he grants that past efforts to measure and predict teacher success have "met only moderate success", he believes that the problems facing future researchers are soluble (Barr,1958, p.695). His view, like that of leading, present-day researchers such as Gage, Walberg and Berliner is that what is really needed is careful attention to a number of procedural issues. By 1960 however, he is less accepting, writing in a memorandum, "Can behavior be considered in isolation or out of context? I think not. Can behaviors be divorced from purposes, persons and situations. I think not." He continues, "The tabulation of behaviors out of context may be misleading. I believe this is important. Study this carefully" (quoted in Tom, 1984, p.30). Indeed.

Conclusion

The empiricist ideas of Mill, Galton, Pearson, Thorndike and Fisher have had a profound impact upon modern ideas of what is assumed to be a scientific approach to research on teaching— an approach which until recently was taken as normative. In the following chapter I therefore describe several recent examples of such supposedly scientific research. When the early studies are compared with these more recent ones, it will be seen how in the early work, men such as Rice and Barr were making many of the same problematic assumptions which continue to be made by present-day empiricist researchers. Moreover, it will be seen how so much current research may be construed, in spite of the improved statistics, as certainly more sophisticated, but fundamentally the same approach as was adopted in the earlier work. A description of recent research will also allow me to show in Chapters V, VI and VII, that in its assumptions and procedures all such empiricist research remains very different from scientific research.

CHAPTER IV

THE GENERAL FORM OF EMPIRICIST RESEARCH ON TEACHING

The previous chapter has provided an account of some highlights from the long history of empiricist research on teaching. Seen in its light, recent and current empiricist research may be construed as merely the latest in an extended series of attempts to produce improved understanding of teaching by way of claimedly scientific methods.

Not all empiricist research approaches its subjects in the same way. A representative selection of research which falls under such classifications as observational, experimental, statistical, correlational, questionnaire, *ex post facto*, cluster analysis, factor-analysis, meta-analysis, or various combinations of these will be considered in this chapter. Much in its favor, is the fact that this research has been carried out in real school classrooms (rather than, say, in laboratories or with groups of captive undergraduates). It has been argued that as a result the implications are straight-forward.

An important variety of such research has been termed, "Process-Product". The "process" in Process-Product research is those things which teachers do in classrooms– their organization, the procedures they follow, the techniques they use, what they say to pupils, and so on. The "product" is changes in the learning of pupils. What the processes are is established by observation of teachers, either in normal or in experimental situations. What the products are is established by the performance of pupils on standardized achievement tests. The significance or otherwise of the changes is established by the use of the well-established statistical tests produced by Pearson, Fisher and others, such as tests of correlation, t-tests, F-tests, chi-square tests, and a panoply of increasingly-sophisticated statistical maneuvers. Such techniques are believed to make the research task quite specific. A key aspect (decomposing an activity or entity into its supposed component parts) seemed to have worked successfully in other areas such as in industrial training and in scientific research, so, say many present-day empiricist practitioners, as did Bobbitt, Charters and others before them, why not in teaching? As Anderson, Evertson and Brophy say (1979, p.193), an explicit aim of such research has been eventually to promote the adoption of those features of teaching which Process-Product and other forms of empiricist study claim to have discovered to be the most effective.

To be representative, and also to demonstrate the world-wide practice of such research on teaching, the studies which follow are drawn from The United States, England, New Zealand and Germany, from a range of different kinds of empiricist work, from both single studies and clusters of interrelated studies, and from reports in several important books and journals.

(1) Experimental Process-Product Studies at the University of Canterbury (New Zealand) (1970-1973)

What the empiricist research community has considered to be a significant series of Process-Product studies was carried out in the 1970s at the University of Canterbury in New Zealand, by Hughes(1973) and Nuthall and Church (1973). These studies continue to represent well the experimental approach to the study of teaching, and are often quoted in the literature.

In the the three 1973 studies by Hughes, lessons on animals were taught to Grade Seven pupils. The participation by pupils and the reactions of the teachers to the pupils' responses were experimentally manipulated. Hughes's first study involved three "treatments" of participation by pupils: random response (questions were addressed at random to the pupils), systematic response (questions were asked in accordance with the seating arrangement of pupils), and self-selected response (questions were given only to those pupils who volunteered answers). The results showed no statistically significant difference between the groups, and no significant relationship between the pupils' rate of response (voluntary or demanded) and their adjusted achievement scores.

A second study was more manipulative. Randomly selected halves of the classes were asked all the questions. Again, overt participation by pupils was not found to make any statistically significant difference to their achievement.

A third study considered the teachers' reactions to the responses by pupils. In the "reacting" group, the pupils were provided with frequent praise and support for correct answers, together with occasional mild reproaches or encouragement when they failed to provide the correct answers. Pupils in the other group received merely a statement of the correct answer. The "reacting" group improved significantly more than the other.

The studies performed by Nuthall and Church were,

> .. intended to explore the possibility of conducting controlled experimental studies of teaching in a normal classroom context. They were designed to provide information about how such studies could be carried out and to provide a sample of the kind of data which such studies might produce (1973, p.11).

In their first study, one group of teachers emphasized conceptual understanding, using open-ended questions, less direct instruction, and maximizing their use of logical connectives. The other group tried to maximize scores on achievement tests through emphasizing the factual content. Basic conceptualizations in their approach were designated as "verbal moves" and "episodes". Typical examples of verbal moves were: open questions, closed questions, repeating the question, modifying the question, extending the question, address moves (calling on the pupil by name), structuring or orienting, prompting, repeating the pupil's answer, saying "right" or "wrong", and so on. Nuthall and Church write:

> The basic unit of analysis is the "verbal move". A verbal move is a unit of discourse which has a single identifiable verbal function (in either the control or content dimensions). For instance: a teacher question is a verbal move, a pupil response to a question is another kind . . a teacher comment on a pupil response is a further kind of verbal move . . During class discussions, verbal moves occur in independent clusters, which we call "episodes". An episode consists of a single content-oriented teacher question, and all of the verbal moves made by teacher and pupils which are associated with [it] (1973, p.13).

The different approaches were judged to be unrelated to the scores for factual knowledge or for higher-level conceptual understanding.

A second study taught science concepts to ten year olds. The variables were coverage of content, and the behavior of teachers. The amount of content taught, the degree of redundancy and the amount of time spent were manipulated, as was the amount of questioning contrasted with direct instruction. Achievement was deemed to be more closely related to the coverage of content than to the method of teaching. But contrary to the above study by Hughes, Nuthall and Church found that there was a statistically significant difference in the amount of learning of those pupils who were involved more overtly in the lesson.

Leading researchers such as Brophy and Good have seen these Canterbury studies as significant. In their paper, "Teacher Behavior and Student Achievement" (1986, p.333), they conveniently summarize the results thus:

> (a) content coverage determines achievement more directly than the particular teacher behaviors used to teach the content; (b) younger students need to participate overtly in recitations and discussions, but older ones may not require such active participation; (c) questions should be asked one at a time, be clear, and be appropriate in level of difficulty so that students can understand them (most of these will be lower order); (d) teacher reactions to student response that communicate enthusiasm for the content and support (or if necessary, occasional demandingness) to the students are more motivating than matter-of-fact reactions; and (e) teacher structuring of the content, particularly in the form of reviews summarizing lesson segments, is helpful (1986, p.333).

(2) Questionnaires, Observation, and Factor and Cluster Analysis: Bennett's (1976) Teaching Styles and Pupil Progress (England)

Neville Bennett was dissatisfied with the conventional distinction between progressive and traditional teachers. He wanted to develop a more subtle set of categories of "teaching styles", to determine whether the different styles affected pupils differently, and to find out if different types of pupils performed better under particular styles.

Because it was impossible to take a random sample of every school in England and Wales, the study began with the development of a questionnaire that was administered to all teachers in Grades Three and Four in the more than 800 primary schools in Lancashire. Factor analysis was used to try to ensure a minimum of conceptual overlap between the variables and that groups of variables were not given undue significance. Typical of the variables were: pupils have choice in where to sit; pupils are not allowed freedom of movement in the classroom; teacher expects pupils to be quiet; pupils given homework regularly; arithmetic tests given at least once a week; above average integrated subject teaching. For various statistical and practical reasons, the original intention to have a more subtle range of styles was changed, and the study finally considered merely three styles: formal, mixed, and informal. The responses to the questionnaires were cluster-analyzed, and a total of thirty-seven teachers were chosen to represent the three styles. The initial classification of teachers resulted from the responses to the questionnaire, but the researchers later confirmed their decisions when they observed these teachers.

The study lasted one school year. Using standardized tests at the beginning and conclusion of the period of observation, the progress of the children was measured in the areas of English, reading and mathematics. They also used a series of personality tests (called "inventories"). The results of these were also cluster-analyzed to develop a set of personality types, in order to determine if certain types of pupil were affected in different ways by the different styles.

With respect to general effects, the study concluded that the children who were taught in the formal style made the most progress in mathematics and English, and the second greatest progress in reading. Those taught in the informal style made the second most progress in mathematics and the least progress in reading and English. Those taught in the mixed style made the most progress in reading, the second most progress in English, and the least progress in mathematics. The single class that was judged to have made the most progress generally, was taught in an informal style.

The "Aptitude-Treatment-Interaction" effects (ATI's) are summarized by Bruner in the following way:

> Most pupil types progress better under more formal teaching. . . particularly the insecure and neurotic pupil: he seems able to attend to work better, and harder, informal setting . . . for the unstable child, the informal setting seems to invite time-wasting activities . . . the "unmotivated", rather neurotic child was found to work four times as much at his studies in a formal setting than in an informal one . . . the informal class seems to increase favourable attitudes towards school– but it also increases anxiety. Interestingly enough, the ones who suffer most in scholastic performance in informal class teaching are the ablest students. Inadvertently, informal classrooms are academic levellers. They also hurt most the less well-adjusted student, who, it was thought would be most helped (1976, pp. ix-x).

(3) Teachers' Use of Grading Instruments (1985): Helen Suarez DeCasper's Questionnaire Research of the Use of the Georgia Eighth Grade Criterion-Referenced Tests of Basic Skills by Ninth Grade Teachers

The Georgia Basic Skills Tests measure students in reading, mathematics, and problem solving and are among requirements for graduation from the public high schools of Georgia. Because the Georgia Department of Education wished to evaluate the effectiveness of teachers' use of various forms for the assessment and placement of students, this survey examined how Grade Nine teachers used the forms which reported the results of these tests of basic skills for Grade Eight. (The majority of Georgia students change to a different school after Grade Eight.) The several forms are: (1) for Individuals: Individual Student Report, Student Label; (2) for Schools: Student Achievement Roster, Summary of Student Reports, Item Analysis Forms; (3) Support Materials: Teacher's Interpretive Guide, Objectives and Assessment Characteristics.

A sample of 800 teachers (from 141 schools in ninety school districts) were sent the instrument. Of these 572, or 72% responded. In order to be representative of the state as a whole, these had been selected by way of two-stage cluster sampling with the first stage using sampling with probabilities proportional to size.

The instrument was a questionnaire developed from a consideration of the information required, a study of similar questionnaires, veting by the Georgia Department's experts and a pilot study of thirteen teachers in two Atlanta schools.

Information was sought about the following questions: Which of the forms are considered the most useful? How useful is the information from the Individuals Report Form and the Student Label? Which forms are sent to teachers, are available for their use, or would be helpful to teachers? What are the opinions of teachers re the assistance they need or the assistance given to them in using the results of the tests? How useful is the information in the Alternative Individual Student Report Forms A, B, and C for the various instructional decisions teachers make? How do teachers rate the Report Form, Alternative Summary of Student? Which teachers make most use of the Forms?

Typical of the instructions, and specific questions asked were the following: the age of the teacher; the number of hours earned by the teacher in training courses in "undergraduate or graduate courses in testing, educational research or statistics"[1]; whether the teacher, committees of teachers, counselors, principals, curriculum specialists, or others, decide "about the curriculum to teach", "about assigning students to classes", "about accelerating students", and "about remediating students"; whether, in making decisions about grouping students within the class, "the reports, comments and grades of other teachers", "criterion-

---

1. The unjustified assumption that the terms, "testing", "educational research" and "statistics" are reasonably similar in meaning, should be noticed.

referenced test-scores of students", "results of placement tests developed for curriculum use", "results of the teacher's own tests", and "the teacher's observations of the children's work", were "not important", "slightly important", "important", "very important", or "critical"; whether teachers would like assistance or instruction on administering tests, interpreting the results of tests to parents, determining the relationship between the objectives of the tests and the topics in the textbooks, and so on.

To investigate the differences between the means of responses to the three different report Forms, one-way ANOVA statistical tests of significance were performed for each application across the three Forms. The SPSSX program MANOVA was used to peform the analysis.

Representative results were: one hundred and thirty teachers (24%) reported no course work in statistics, educational research, or measurement; their degrees were earned between 1944 and 1984; the teachers expressed strong support for the report forms; teachers rated the forms as important in assisting them in making particular decisions about instruction; 84% rated the Alternative Student Report Form (Form C) as most useful in comparison with the two forms presently used; less than 20%reported that the forms were sent to them.

The responses to the usefulness of the information in the Alternative Individual Student Report Forms A, B, and C for the various instructional decisions teachers have to make, showed lower ratings for A, and highest ratings for C in all applications. The means of scores for all instructional situations for Form A were near to 2 (slightly useful), for B near to 3 (useful) and those for C were near to 4 (very useful). This preference was also strongly reflected in the teachers' responses to the open-ended questions.

Teachers rated the CRT scores as important when making decisions such as planning instruction, grouping students, remediating or accelerating students.

More than half said they would like assistance in using the results of the tests. Strongest need for assistance was expressed by language arts teachers.

(4) Pre-test, Post-test, Control Group Experiment: Study of Reading, by Anderson, Brophy and Evertson

Brophy and Evertson performed an experimental study of instruction in Grade One reading, (Anderson, Evertson and Brophy,1979) which involved teachers' using an approach which consisted of twenty-two principles for managing, and instructing the class as a whole. The principles were derived from the earlier observational Process-Product work of Brophy and Evertson (1974a, 1974b, 1976, 1978). The principles were organized into a manual, and teachers were trained to use them.

Classes from nine schools, predominantly middle-class, were randomly assigned to three groups: ten classes received the special instruction in reading and were periodically observed for a year. Seven classes received the special instruction but were not observed. Ten classes formed the control group, did not receive the special instruction, but were observed. The treatment-unobserved group allowed for assessment of possible effects of the presence of an observer on the effects of treatment, while inclusion of classroom observation in both treatment and control classes allowed for the assessment of treatment implementation and Process-Product relationships, in addition to effects on achievement (adjusted for entry-level reading readiness). Brophy and Good report that the following conditions were found to be the most effective:

> More time was spent in reading groups and in active instruction, and less time was spent dealing with misbehavior; transitions were shorter; the teacher sat so as to be able to monitor the class while teaching the small group; lessons were introduced with overviews; new words were presented with attention to relevant phonics cues; lessons included frequent opportunities for individuals to read and to answer questions about the reading ; most questions called for response by an individual rather than the group; most responses resulted from ordered turns rather than [from] volunteering . . . most incorrect answers were followed by attempts to improve the response through rephrasing the question or giving clues; occasional incorrect answers were followed by detailed process explanations (in effect, reteaching the point at issue) ; correct answers were followed by new questions about 20% of the time rather than less frequently ; and praise of correct responses was infrequent but relatively more specific (although the absolute levels of specificity of praise were remarkably low, even for the treatment teachers). Group call-outs were associated positively with achievement for the lower-ability groups, and negatively for higher-ability groups (1986, p.345).

Following the 1979 study, the principles for teaching reading were revised, and a sample of the key ones as stated by Anderson, Evertson and Brophy (1982) is now listed:

General Principles
1.Reading groups should be organized for efficient, sustained focus on the content.
2.All students should be not merely attentive but actively involved in the lesson.
3.The difficulty level of questions should be easy enough to allow the lesson to move along at a brisk pace and the students to experience consistent success.
6.Although instruction takes place in the group setting, each individual should be monitored and provided with whatever instruction, feedback, or opportunities to practice that he or she requires.
Specific Principles
2.Academic focus. Successful reading instruction includes not only organization and management of the reading group itself . . but effective management of the students who are working independently. Provide these students with appropriate assignments; rules and routines to follow when they need help or information (to minimize their needs to interrupt you as

you work with your reading group); and activity options available when they finish their work (so they have something else to do).

6. Transitions. Teach the students to respond immediately to a signal to move into the reading group (bringing their books or other materials), and to make quick, orderly transitions between activities.

11. Ask questions. In addition to having the students read, ask them questions about the words and materials. This helps keep students attentive during classmates' reading turns, and allows you to call their attention to key concepts or meanings.

13. Minimize call-outs. In general minimize student call-outs and emphasize that students must wait their turns and respect the turns of others.

17. Wait for answers. In general, wait for an answer if the student is still thinking about the question and may be able to respond. However, do not continue waiting if the student seems to be lost or is becoming embarrassed, or if you are losing the other students' attention.

18. Give needed help. If you think the students cannot respond without help but may be able to reason out the correct answer if you do help, provide help by simplifying the question, rephrasing the question, or giving clues.

20. Explain the answer when necessary.

(5) Research with Meta-statistics:
Gage (1978) and Walberg (1986) (USA)

Gage's meta-statistical study, reported in his, *The Scientific Basis of the Art of Teaching*, (1978) was a statistical examination by Gage and his assistants, of four important, relatively large-scale, observational, correlational studies of teaching of the Process-Product kind (Gage,1978, p.38), namely those of Soar (1973), Brophy and Evertson (1974), Kaskowitz (1974), and McDonald and Elias (1976). They analyzed the results of the four studies looking at the definitions of variables and the correlations of the variables with the achievement of pupils in reading.

They took into account the averages and variabilities of the measures of teachers' behavior. The importance of the average level and the variability, usually disregarded in research reviewing, can be seen in the study by Brophy and Evertson, which dealt with a teacher behavior labeled, "Teacher reactions to wrong answers: criticism." This variable correlated .61, that is, positively and strongly, with adjusted pupil achievement in reading classes of higher socioeconomic status. Presumably, this correlation might imply that teachers would do well to criticize pupils in such classes frequently. But further examination revealed that incorrect responses amounted to only about 20% to 25% of all pupil responses. Further, the mean for the measure of all verbal criticism was only about 5% to 10%, and its variability (as measured by its standard deviation) was only about 7%. Gage points out that about 8% of 25% gives 2% of all teacher reactions to pupil responses as the mean frequency of criticism after incorrect pupil responses among these teachers. Thus the high correlation really indicates that the effective teachers provided criticism for only a small percentage of the time, and only with pupils of higher socioeconomic status, while the less effective teachers never gave such criticism.

In this way, the researchers carefully studied the detailed information for several hundred variables in the behavior of teachers in the classroom, developing a set of inferences as to how third-grade teachers should work if they intended to maximize achievement in reading and (they believe) also in mathematics (1978, p.39).

In an attempt to find relationships which might prove superior to anything produced by individual studies of teaching, Gage advocated this further statistical work, or, the "meta-analysis" of findings. Meta-analysis applies inductive statistics to collate the collective probability of results from all the studies which meta-analysts judge to have considered the same question. Biddle and Anderson (1986, p.235) say that rather than having to rely only upon, ". . the weak single studies based on small samples," this technique can be used, ". . to gain the greater statistical power of a cluster of studies." Meta-analysis, itself evolving from Karl Pearson's early work, converts the precise probability value of the result of any particular study into a value of the statistic called chi square. Next, the values of chi square are summed across all the studies, and the significance, or probability, of the sum is determined. Thus, Gage claims, ". . in essence, the technique provides an estimate of the statistical significance, or 'non-chanceness', of the whole cluster of independent findings" (1978, p.29)

Meta-analysis may be seen as a sophisticated recent development of the empiricist tradition, teasing out the general implications of individual studies. Especially in the United States, its results have had a considerable impact on the schools and on those who train teachers. As examples of meta-analysis, as well as the work of Gage, that of Rosenshine (1976), Medley (1977), Brophy (1979), Good (1979), Glass, Cahen, Smith and Filby (1982), Walberg (1986), Brophy and Good (1986), Doyle (1986), and Yates, Chandler and Westwood (1987) may be cited. For instance, Glass, Cahen, Smith and Filby examined several hundred studies which related class size and the achievement of pupils to see how these covaried "on average". They concluded that the achievement of pupils declines as the size of class increases. But as Biddle and Anderson questioningly point out (1986, p.235), such meta-analyses, ". . assume the confirmatory assumption of simple relationships among terms, concepts, and methods used." Gage in defence is careful to point out that the findings of his own meta-analysis". . have emerged from research." They are he says, ". . the joint yield, the convergence, the common general finding of years of careful classroom observation, reliable and valid measurement, and sophisticated statistical work" (1978, p.40).

There are two main approaches, the vote count and the effect size (Glass,1977). The vote count calculates the percent of all the studies which are positive, e.g. the percent of studies in which the experimental group exceeds the control group. The effect size is the, ". . difference between the means of the experimental group and control groups divided by the control group standard deviation. It measures the average superiority (or inferiority, if negative) of the experimental relative to the control groups" (Walberg,1986, p.216).

In his own case, Gage lists the findings as a series of "teacher should" statements, saying, "They must be qualified implicitly with all the assumptions as to objectives and context that I have already mentioned" (1978,p.39). The first five statements are:

> Teachers should have a system of rules that allow pupils to attend to their personal and procedural needs without having to check with the teacher.
> Teachers should move around the room a lot, monitoring pupils' seatwork and communicating to their pupils an awareness of their behavior, while also attending to their academic needs.
> When pupils work independently, teachers should insure that the assignments are interesting and worthwhile yet still easy enough to be completed by each third grader working without teacher direction.
> Teachers should keep to a minimim such activities as giving direction and organizing the class for instruction. Teachers can do this by writing the daily schedule on the board, insuring that pupils know where to go, what to do, etc.
> In selecting pupils to respond to questions, teachers should call on a child by name before asking the question as a means of insuring that all pupils are given an equal number of opportunities to answer questions.

Gage claims that the above statements may be summarized by way of the concept of "academic learning time" (ALT)[2], i.e. time during which pupils are actively engaged on set academic tasks. He argues that the statements are significant, because it is not the case that all or almost all teachers are in fact already behaving in this way. Moreover, he claims that the ways would not have survived the research, had they not correlated positively with the achievement of pupils (1978, p.40).

This underlying rationale for meta-analysis may perhaps be summed in Walberg's optimistic and status-seeking claim that,

> Research synthesis explicitly applies *scientific techniques* and standards to the evaluation and summary of research; it not only statistically summarizes effects across studies but also provides detailed descriptions of replicable searches of literature, selection of studies, metrics of study effects, statistical procedures and both overall and exceptional results. . . synthesis is the *fifth of six phases of scientific research* ; it helps consolidate the first four phases of research before wide-scale adoption and evaluation are justified (italics added) (1986, p.214).

Walberg also claims that,

> Research workers and educators must retain both openmindedness and skepticism about research synthesis. Yet the 1980s do seem to be a period

---

2. See also below, pp 126-127.

of quiet accomplishment. In a short time, research synthesis helped sort what is known from what needs to be known about the means *and ends* [sic] of education. . . Although more and better research is required, synthesis points the way toward improvements that seem likely to increase teaching effectiveness and educational productivity (italics added) (1986, p.228).

Before leaving Gage and Walberg, a fundamental, problematic aspect of their work should be mentioned. Garrison and Macmillan correctly point out that, "Attempts to combine results by meta-analysis and other techniques have leavened the results somewhat, but those techniques themselves *bring in other conceptual and methodological problems without adding new empirical content* to the findings" (italics added) (1984, p.17). For instance, ". .at each step of the meta-analysis more empirical content is lost. That is, we are no longer talking about the subject of investigation when we compare different studies of the subject. Rather, we now have a new study, with data that are different in kind from the original. At each step, *we get further and further from the world"* (italics added) (Macmillan and Garrison, 1988, p.53). But what we really need is to get closer and closer to the world (of teaching).

(6) Pretest, Post-test, Observational-statistical Research: Helmke and Schrader's Practice of Mathematics During "Seatwork" (1988) (Germany)

This research was part of a larger "longitudinal" project conducted within the program of the International Association for the Evaluation of Academic Achievement.

The authors say that there had been little empirical research which concentrates upon practice in normal classrooms. Thirty-nine Grade Five classrooms in twenty-four Bavarian primary schools, and thirty-nine teachers (nineteen men, twenty women), the normal teachers of the children were studied. Study of a random sample of 651 other German teachers, showed no significant differences between the thirty-nine and that sample. The curriculum was with minor variations of sequence, the same in all classes.

Their purpose was to expand the existing body of research on the effectiveness of student practice in the classroom, as well as, ". . to support some findings of [earlier] studies . ." Their study provided an empirical analysis of independent practice, rather than of practice guided by the teacher. Independent practice normally occurs during seatwork, when students work independently (usually on written exercises) at their desks with their own materials.

Helmke and Schrader say that previous research had suggested that the following conditions were characteristic of effective practice. First there are established routines of management. "The teachers monitor the class thoroughly and intervene immediately in the case of disruptions," (1988, p.70). Second, the teachers have made certain the children know what is expected of them during such practice. Third the children are intensively supervised. "These teachers move around, interact with students, and inspect students' work frequently" (p.71). Fourth, contacts with pupils are normally brief, ". .usually not longer than 30 seconds" (1988, p.71).

The researchers were faced with the problem of statistically equalizing the entry abilities of the pupils. They say that they were faced with the problem of unequal classroom entry characteristics because their approach involved nonexperimental study. Given this fact, they believed that the residualization of post-test scores was the least problematic method of equalizing. For reasons of clarity, a constant was added to the residualized post-test score of each class, namely, the average gain of the total sample (1988, pp.71-72).

Because previous studies had shown that classroom entry characteristics influence not only later achievement, but also the teacher's subsequent behavior, all of the measures of teacher behavior were statistically adjusted for classroom differences in math pretest achievement and intelligence. So the correlations between teacher behaviors during seatwork and the students' residualized post-test achievement were partial correlations rather than zero-order correlations (1988, p.72).

The teachers were observed by way of a low-inference system (Five Minute Interaction system) which was a modification of Stallings's Multiple Coding system (Stallings, 1977). Observations were taken during nine lessons by each teacher. Coding was based on a sampling of combined time and event, which allowed every new event to be coded. When the event lasted longer than five seconds, coding was repeated every five seconds. Coding was concerned with context, "persons participating in the interaction", and "content of the interaction". The definitions of interactions were as follow :

> *Seatwork* refers to an instructional context in which students work independently at written tasks and exercises . . .
> *Monitoring* refers to supervision of students' work by the teacher without interacting with students.
> *Discreet support* comprises all subject-related teacher contacts with individual students during seatwork, e.g. diagnosing errors . . answering students' questions. . .
> *Public interactions* comprise all teacher utterances directed to single students that are given indiscreetly and are consequently heard by the whole class.
> *Lecturing* refers to teacher presentation or explanation of academic material.
> *Recall questions* are questions that require recall or recognition of factual information rather than higher level processing.
> *Disciplinary events* are all interactions dealing with student misbehavior.
> *Procedural activities* are all comments concerning non-academic procedures and organizational activities (1988, p.72).

The diagnostic competence of teachers was assessed by a "judgment- accuracy approach". The accuracy of teachers' judgments of their students' characteristics was taken as a measure of their diagnostic competence. In this case, students' achievements in a mathematics test which contained content covered during the last three months served as the criterion against which accuracy of teachers' ratings was judged. Before making their judgments, teachers were given sufficient time to familiarize themselves with the test items by looking through the test booklet. After that, the teachers predicted for each student the number of test items they expected the students to solve.

Statistically, diagnostic competence was measured as the product-moment correlation between scores which were predicted for each student and the students' actual scores. This was assumed to indicate the extent to which the teacher was able to predict the relative achievement position for each student of his class.

With respect to the findings, Helmke and Schrader say that about twenty percent of all lesson time was spent on seatwork, with some teachers spending over half of the lesson time, but the frequency of seatwork was not correlated significantly with achievement.

The indicators of inefficient management and of insufficient preparation showed a significant negative correlation with achievement. The frequency of independent seatwork was not correlated significantly with post-test mathematics achievement. In contrast, indicators of inefficient management and insufficient preparation by the teacher were significantly negatively correlated with achievement.

The overall score of teachers' activity during seatwork was not related significantly to achievement. Of interest is that when this overall score was broken into public activities and discreet activities, the first had a nonsignificant negative effect, the latter had a significant positive correlation with achievement.

There were large differences with respect to diagnostic competence. Helmke and Schrader write that while some teachers were informed almost perfectly with respect to their students' level of achievement, others were largely uninformed. To analyze the joint impact of diagnostic competence and frequency of teacher supportive contact on achievement, they carried out a two-factorial ANOVA (with both independent variables dichotomized). The results pointed to a weak main effect of supportive contact. Although the main effect of diagnostic competence on achievement was insignificant, there was found to be a significant disordinal interaction between supportive contact and diagnostic competence. The achievement of students was highest when both the frequency of teachers' supportive contacts with students and their diagnostic competence were above average. They also mention what they call, "a surprising result", viz that those classrooms with teachers who were good diagnosticians, yet who gave their students comparatively little support, showed pronounced decline in achievement. They speculate that such students may become sensitized to their own weaknesses, but do not receive sufficient support in overcoming them (p.73).

In their "Discussion" Helmke and Schrader write,

> . .the interpretation of the data was based solely on (partial) correlations. Thus it is problematic to interpret teacher behavior (e.g. efficiency of management or sufficiency of preparation) as a cause and classroom achieve-

ment as an effect. However...there are two arguments that support a causal interpretation as suggested above. First, the temporal order of the variables was included in the analyses, because the observation of teachers (yielding scores for teacher management, instructional quality, etc.) started after the administration of the pretest (t1) and was finished half a year before the posttest (t2). Second, by using partial correlations rather than zero-order correlations between teacher behaviors and classroom posttest achievement, we took into account classroom differences in the level of intelligence and prior knowledge.

They found that seatwork as such, was unrelated to academic success. In their words, ".. the amount of independent student practice in the context of seatwork has no impact on achievement" (p.74). According to their data, it is not the quantity but the quality of independent student practice that makes the difference. Thus the organization of independent student practice by less successful teachers characteristically involved frequent interruptions through lecturing and questioning. Helmke and Schrader say that this result coincides with the results reported by Fisher and others (1978) and shows that independent practice characterized by either insufficient preparation or by inappropriate task difficulty level undermines achievement (pp.73-74).

The researchers also found evidence which supported the importance of having a smoothly running classroom. They produced significant negative correlations between achievement and the frequency of disciplining and explaining.

They also suggest,

Surprisingly, our results indicate that teachers' diagnostic competence, *per se*, was not related to learning progress. This is in contrast to a finding of Fisher (and others,1978) whereby teachers' accuracy in diagnosing student levels of skill was related to student achievement. Our results, however, do not necessarily mean that teacher diagnostic competence is irrelevant for achevement. Diagnostic competence becomes relevant for achievement only in interplay with specific didactic teaching behaviors (p.74).

They conclude:

.. qualitative, not quantititative, factors are crucial for the effectiveness of [student] practice. Three conditions are considered especially important in designing effective seatwork.. First, adequate external conditions must be provided for independent practice; for example... disruptions should be kept to a minimum. Second, the continuity of seatwork seems to be of special significance. In order to ensure continuity, independent practice must have been adequately prepared [for]...independent practice is of benefit only if students have already attained a minimum level of competency. Explanations and lectures represent unfavorable interruptions, indicating less than optimal preparation.. Third, practice is more successful when a teacher actively circulates around the classroom . . . supervising students' work and giving support to single students discreetly, without distracting other students (p.74).

(7) Performance Based Teacher Education (PBTE) (USA)

In the early 1970s, there began a new movement (closely linked to the findings of Process-Product research) in the preparation of teachers in the United States : PBTE. PBTE has been an attempt to make a body of claims about effectiveness, which are based upon the results of empiricist research, the central element of the professional component in the preparation of teachers. Gage and Winne say that, "PBTE is teacher training in which the prospective or inservice teacher acquires, to a pre-specified degree, performance tendencies and capabilities that promote student achievement of educational objectives" (quoted in Tom, 1984, p.30).

The ratonale behind this movement was simple enough. In a sort of up-dated Barr-Charters approach, it was to identify by rational commonsense analysis of the results of Process-Product research, those activities and behavior commonly used by successful teachers and not used by unsuccessful teachers, and to base teacher education upon them. The results of the program have however been equivocal, and the justification for the claimed activities and behavior to be acquired by student-teachers is even more debatable. For instance, as Tom argues (1984, p.31), the attempt by Travers (1975) to produce a teacher-training program based strongly upon actual empirically-established activities and behavior seems "paltry".

PBTE has nevertheless had enormous influence. Unlike much that has gone before, it actually had an effect on teacher education. Indeed, as Tom points out, PBTE has even been introduced in many places, prior to the establishment of cogent research results. The mere promise of results achieved widespread change. The Florida Department of Education even mandated that such relationships would be discovered. Assuming its members were also of the opinion that PBTE research was scientific research, then *they were mandating that scientific discoveries would be made* ! Andrews (1972, p.156) reports their decision thus:

> By the end of 1974, competencies expected of teaching personnel in elementary and secondary schools will be clearly identified. *Evidence will be available* showing relationships between teacher competencies and pupil learning. Teacher training techniques will be available for use in preservice and inservice teacher education programs which are aimed at the specified competencies. *Evidence will be available* to state policy makers *which shows the extent to which* teacher effects on pupil learning support various credentialling requirements (my italics).

One ironic aspect of PBTE has therefore been the way in which it became established despite the lack of what would normally have been considered to be empirical evidence for it. Intriguingly like some earlier work of Charters and Barr the claims seem more to have been the result of expert opinion about the competencies required for success, than of actual empirical investigation. Thus Gall (1979) found that fewer than ten out of 255 sets of PBTE materials for training teachers, had been subjected to rigorous statistical summative evaluation and that such evaluation rarely included the impact of the training materials on student learning (reported in Tom,1984, p.34).

Nevertheless PBTE has been tenacious. It has metamporphosed into the flourishing teacher accountability movement : "Though on the wane in schools of education [it] is reemerging at the state level in programs for beginning teachers and/or for evaluating teachers for certification, tenure, or merit increases" (Shulman,1986, p.11). This is still strongly the case in 1992. American legislators, reacting to a host of calls for the accountability of teachers, and always ready to seize a seeming, short-term gain, have been impressed by empiricist Process-Product's dual claims of clear criteria of effective teaching and of a scientific base. From Florida, to South Carolina, to Virginia, to Arkansas they have enacted legislation which has forced teachers to apply such findings for tenure and promotion (at least when they are being observed by the supervisor or superintendent).

(8) Observational-statistical Research: Academic Learning Time (ALT) (USA)

A significant variation of the empiricist research programme was the movement into Academic Learning Time (ALT), or Time-on-Task. This was an attempt to shift some of the emphasis from the teachers' characteristics or actions at one end and the pupils' achievements at the other (which tended to ignore what happened in between as if it were a "black box") onto what pupils were doing. Shulman says that in addition, the initiators of the program, ". .felt that variations in some pupil indicator would provide a more sensitive estimate of the effects of teaching than the more distal product of achievement-test performance" (1986, p.14).

Berliner, one of the leaders in the approach explains the idea in relation to the large-scale study known as the Beginning Teacher Evaluation Study (BTES). The researchers modified the basic Process-Product approach, in the belief that,

> . . what a teacher does at any one moment while working in a circumscribed content area affects a student primarily at only that particular moment and in that particular content area. The link between teacher behavior and student achievement is, therefore, the ongoing student behavior in the classroom learning situation. . .What a teacher does to foster learning in a particular content area becomes important only if a student is engaged with appropriate curriculum content. . . content that is logically related to the criterion and is at an easy level of difficulty for a particular student. . . The variable used in BTES is the accrued engaged time in a particular content area using materials that are not difficult for the student. This complex variable is called Academic Learning Time . . . In this conception of teaching, the con-

tent area the student is working on must be specified precisely, the task engagement of the student must be judged, the level of difficulty of the task must be rated, and time must be measured. The constructed variable of ALT then, stands between measures of teaching and measures of student achievement (Berliner, 1979, pp.122-125).

The four components of ALT are allocated time, engaged time, relevance and high success. The meaning of the first two is self evident. A relevant task is defined as a task which will be tested on the achievement tests at the year's end. High success has occurred when a pupil can correctly complete tasks with ease. In short, ALT is, ". . the amount of time a student spends engaged in an academic task that s/he can perform with high success" (Denham and Lieberman, 1980, p.8).

The study provides dramatic evidence of just how much time is spent by pupils in management activities such as passing in papers, waiting between tasks, moving about the room. Thus in Grade Two an average of 19% of the day is spent in this way. In Grade Five it is 17% (Denham and Lieberman, 1980, p.119). Other than the waste of time, the main finding of the work has been that, "Students who accumulate more ALT generally have higher scores on achievement tests," (sic) (1980, p.18), or, "The more ALT a student accumulates the more the student is learning". ALT is thus construed as an immediate and continuing measure of the learning of pupils.

In a sense, as Tom suggests (1984, p.50), these researchers were returning to Rice's original 1897 question about the relationship between instructional time and what was learned. As Tom points out in respect to the conclusion from ALT, "This commonsense conclusion is the inverse of Rice's finding that increasing drill time (a task of inappropriate difficulty ?) does not necessarily lead to added proficiency in spelling" (1984,p.50). There are probably good explanations for the existence of both findings. The more fundamental question to ask in 1992 is : to what degree have conceptualization, adequacy of design, and relevance of such empiricist studies to teaching in the complex actual world of the classroom, really improved over the intervening ninety-five years?

As Shulman notes (1986, p.15), although ALT was initiated as a muted critique of the Process-Product version of empiricist research on teaching, it continues to use its approaches and ideas. ALT is still concerned with observations and statistical analysis of relationships between variables. The research on ALT is representative of recent developments in empiricist research, and may be seen either as a development of the Process-Product paradigm, or else as a new development in empiricist research on teaching, which is coming about because of recognized shortcomings.

(9)Experimental-Statistical Research by Finn and Achilles: a State-Wide Study of the Effect of Size of Class (USA)(1990)

Finn and Achilles begin their report with a statement about the importance for policymakers and researchers of continuing research on the size of classes:

To school personnel, small classes promise to facilitate increased student-teacher interaction, allow for thorough and continuous student evaluation, and provide greater flexibility in teaching strategies. . .the cost of smaller classes is high, requiring investment in both additional teacher salaries and additional classroom space. This investment is unlikely to be forthcoming unless evidence that positive outcomes will be realized is sustantial and consistent (Finn and Achilles,1990, p.558)

Finn and Achilles then say that the findings on such matters are not consistent. They approvingly quote Robinson and Wittebois to the effect that, "Existing research findings do not support the contention that smaller classes will of themselves [sic] result in greater academic achievement of pupils" (1986, p.558). They approvingly quote Slavin regarding the debatable continuance of any effects in later years : "Only randomized experiments can provide a definitive answer to these questions" (1990, p.558).

Their research involved a large-scale experiment in the teaching of reading and mathematics, in the state of Tennessee. From a final sample of seventy-six schools that were large enough to have at least three kindergarten classes, approximately 6,500 pupils and 328 teachers were randomly assigned to 328 large and small classes. Every class was to remain the same type, viz. small, regular, or regular with aide, for four years beginning in the fall of 1985. The project staff randomly assigned a new teacher to the student group in each subsequent grade. Three hundred, forty-seven classes with some 7,100 pupils were used in Grade One of whom some 331 with 6,570 pupils remained for statistical analysis after data were screened for errors in coding. "The reduction in class size is substantial, from an average of about 22 students present . . in the regular classes, to about 15 in small classes, a decrease of 32%"(p.559). The children remained in the classes for two years. At the end of each year they were measured in reading and mathematics by the Stanford Achievement Tests (SATs) and for Grade One by a set of curriculum-based tests developed by the Tennessee State Department of Education.

The means of the measures of outcomes were calculated for each class and also for white and minority students. Two multivariate ANOVAS were performed, a cross-sectional analysis of the whole first year, and a longitudinal analysis of pupils who were in the study in both kindergarten and Grade One.

The following results were typical:

The benefit of reduced class size is seen clearly for every measure. The effect sizes are about one fourth of a standard deviation among students, and range from about one third to two thirds of a S.D. among class means. The difference in classroom pass rates on the BSF curriculum-based tests is 5.2% in mathematics and over 10% in reading. In fact, the univariate contrast values for comparing small classes with the others is significant at $p < .001$ for four of the five measures, and $p <.01$ for the BSF mathematics scale. If the SAT means are referred to the publisher's table of norms, the small-class advantage is at least eighteen months for reading, and thirty months for mathematics on the grade equivalent scale. The effects of augmenting a regular class with a full-time teacher aide are not as large as the small-class advantage, and are not statistically significant in the multivariate model. However, the average performance in teacher-aide classes is higher than [in] regular classes on each measure (1990, p.567)

"The question of why these effects are realized remains largely unanswered," say the authors, "but in light of these findings, is particularly important to pursue" (Finn and Achilles, 1990, p.557). But they hazard the following:

> Three dimensions of school process should be examined. First, teachers' enthusiasm and satisfaction may be enhanced when there are fewer students to teach; this may be perceived by students and influence their own motivation for learning. The Glass, Cahen, Smith and Filby (1982) meta-analysis concluded that, ". . . class-size affects teachers. In smaller classes, their morale is better ; they like their pupils better, have time to plan, and are more satisfied with their performance" (p.65). Teachers in the Shapson et al (1980) experiment indicated that they were especially pleased with their smaller classes, and informal comments from teachers in the present investigation confirm this view. Second, reduced class size may directly impact on teacher-student interactions, allowing for more individual attention in particular ; third, smaller classes may increase the extent to which individual pupils attend to and become involved in learning activities (1990, p.575).

Conclusion

The foregoing studies provide a fair cross-section of empiricist research on teaching. Already, prior to the analysis of the following chapters, it should be clear that such work bears little resemblance to the scientific research described in Chapter II. Where are the abstract concepts? Where is the imaginative extrapolation? Where is the Theory which relates the abstract concepts in powerful, ever-evolving schemes of explanation? So that there can be no misunderstanding, the empiricist view of what it is to perform scientific research on teaching, the view implicit in the foregoing studies, is made quite explicit in the following two chapters. The radical contrasts between such empiricist research and scientific research are then considered in Chapter VII.

CHAPTER V

WHAT IS THE VIEW OF SCIENTIFIC RESEARCH ASSUMED IN THE STUDIES?

It is now appropriate to tease out the view of scientific research assumed in the studies described in the previous chapter.

There are four sources which will now be drawn upon to make clear the views of scentific research assumed in this research tradition[1]. The first two are the ideas of teaching and of scientific research *implicit* in the studies just described. By "implicit" I mean that if these studies are carefully examined, reasonably determinate accounts of the assumptions being made by the researchers about teaching and about research can be produced. The third and fourth sources of our understanding are the *explicit statements* about teaching and scientific research made by leading empiricist researchers in their more philosophical moments. Ideas about teaching will be discussed first, then I shall consider ideas about scientific research.

Views of Teaching Assumed in Empiricist Research

If we consider the examples of Chapter IV, the implicit assumptions about teaching look something like the following. Teaching may be construed as a cluster of observable behaviors by teachers, occurring in normal classrooms of twenty or thirty students.

It is the view of these researchers that just as there are aspects of the universe which we call swimming, talking, or fishing, so there is an aspect of the universe which we call teaching. Just as we can observe people swimming or fishing, so we can observe people teaching.

For purposes of observation, just as one example of swimming by John is sufficiently similar to another example of swimming by Tom, so one example of teaching by Mrs. Johns is sufficiently similar to another example of teaching by Mr. Thomas for teaching to be regarded as a unified and recurring set

---

1. Empiricist researchers have paid little attention to the many perceptive analyses of teaching by philosophers. That this attitude may be changing is evidenced by the two articles by Fenstermacher and Greene included in the most recent *Handbook of Research on Teaching* (Wittrock, 1986). On the other hand, most of the rest of that huge volume just ignores the results of philosophical analysis.

of observable behaviors. For instance, as described above, Helmke and Schrader (1988) studied thirty-nine teachers, Evertson and Brophy (1979) over 550 teachers, and Finn and Achilles (1990) some 659.

Just as swimming coaches believe that for purposes of improvement, swimming can be divided into atomistic observable aspects such as entry into the pool, movements of the arms, movements of the legs, rhythm, velocity, and so on, so such research *assumes that teaching can be divided into discrete, observable, atomistic aspects*, such as exposition, various kinds of questioning, directing, responding, praising, inventiveness, and so on. It is with these "atomistic" features that the claimedly scientific approach to research must deal.

It is believed that there is sufficient observable similarity between the questioning, directing and so on of the different teachers for these aspects to be treated as examples or instances of the *same* thing— as unified and recurring behavior which is observable. The terms, "questioning", "rebuking", "praising", and so on are general conceptions which are assumed to capture the observably similar individual behaviors which fall under them.

There is also an assumption about context. Just as we can observe someone swimming and at the same time ignore whether the swimming is taking place in a pool in London, or New York, or Sydney, so on this view we can observe someone teaching, and *ignore the context* in which it is taking place, i.e. just as swimming or the various atomistic aspects of swimming can be construed as something discrete, observable and separable from the context, so teaching and its various atomistic aspects can be *observed quite separably from the context*. Teaching is construed as involving a cluster or bundle of aspects which can be deployed across but *remain the same thing across variations of context*. As Shulman puts it, ". . the effectiveness of teaching is seen as attributable to combinations of discrete and observable teaching performances *per se* , operating relatively independent of time and place." And again, "Data from fall may be combined with data from spring. Data from a unit on natural selection combines with data from a unit on the circulation of the blood." As Shulman points out, "All these are seen as instances of teaching, an activity that transcends both individual teachers and specific situations" (1986, p.10). Nuthall and Church, in their study considered in Chapter IV, say that teaching, ". . occurs daily, in much the same way, in thousands of different classrooms. However good or bad the teaching may be, it has the kind of regularity which makes it an ideal object for systematic behavioural analysis" (1973, p.10).

But this supposedly scientific research makes another equally significant but very different assumption. Although the teaching by Mrs. Jones, Mr. Thomas and teacher No.30 is recognizably the same phenomenon, it will also *differ in important respects*, in the sense that the questioning by Mrs. Jones may be more effective, the directing by Mr. Thomas may be more organized, Helmke and Schrader's successful teacher may check the pupils' work more often, and Powell's teacher No.30[2] may be more skilful in structuring ideas.

---

2. See chapter IX.

Thus, although the same word, "questioning" is appended to the behavior of the different teachers, the entity itself varies somewhat from teacher to teacher. Such concepts require judgment in order for a researcher to be able to claim that a particular behavior is an instance of that concept. Indeed, researchers are very happy when they can achieve 80% agreement between observers. In short, the research deals with generalized empiricist concepts. But at the same time it is assumed by the researchers, that in being examples or instances of the "same" thing, questioning, directing, being alert, and so on, are suitable for statistical manipulation and can be dealt with as though they were like gravitational forces, or chemical elements, or nucleic acids (where "observer" agreement is 100%).

Moreover, despite their acknowledged differences, these aspects of teaching are considered to be sufficiently similar, that the probability values attached to, say, an "open question" in one study have been combined with the probability values attached to "open questions" in other studies and the statistics themselves have been approached statistically. This technique as already mentioned is termed "meta-analysis". *There is assumed to be an underlying "true score" for the relationship between the behavior of given teachers and a measured outcome of pupils*: "There is a parameter or law which can be estimated. The problem is to get beyond the limitations of particular teachers, particular classrooms, particular studies to a more stable generalization" (Shulman,1986, p.10).

The stress upon observable, atomistic aspects of teaching, the use of pre-conceived general categories, and the statistical measures of relationships between the processes (questioning, directing, being alert), and the products (ability to multiply, recognizing words, writing chemical equations)– themselves statistically assessed on standardized achievement tests, gives a distinctly empiricist and causal flavor to the research[3].

Teaching is seen as something made up of discrete aspects which *cause* particular results in classes of students. *The researchers implicitly assume that an empiricist, causal account, resting on the statistical manipulation of observables falling under general concepts, and of observables reacting upon observables, is sufficient to understand and improve teaching.*

In passing it is worth pointing out the difference between this view and that of a writer such as Thomas F. Green (1971). Green discusses what he calls, "logical acts" and "strategic acts". The first are such things as explaining, concluding, giving reasons, defining, demonstrating, etc. The second are motivating, counseling, encouraging, disciplining, questioning, etc. But Green deals with such features of teaching in a humanistic manner, as categories for intelligent discrimination, discussion and rational balance. Green is not treating them as "objects" precise enough to be scaled, added, subtracted and statistically manipulated.

---

3. Wirth aphoristically suggests that such research rests upon, " . . the assumption that the only learning that counts is learning that can be counted, and the parallel assumption that teachers will be made accountable in terms of test-score results" (1989, p.536).

The emphasis so far has been upon the notion of teaching which appears to me to be implied by the actual procedures used in the research. But such researchers as Gage also make quite *explicit statements about* teaching. Such explicit statements are very helpful, in the sense that they pertinently demonstrate what researchers are *not* addressing. For it will be noticed just how superior is the conception embedded in their talk about teaching from that which their limited methods make it possible to capture. Thus on the first page of his *The Scientific Basis of the Art of Teaching*, Gage says that although he cannot do full justice to an account of teaching because such matters are complex and are really in the domain of philosophers (1978, p.13), he can, nevertheless, make some attempt. He begins with,

> By teaching I mean any activity on the part of one person intended to facilitate learning on the part of another. Although the activity often involves language, it need not do so, nor need teaching rely solely on rational and intellectual processes. We can teach by providing silent demonstrations for our students or models for them to imitate. And we often teach by fostering attitudes and appreciations whose rational components are suffused with affect (1978, p.14).

Gage shows an intelligent understanding of the complex nature of teaching and of its basis in an art. Thus he writes,

> When teaching goes on in face-to-face interaction with students, the opportunity for artistry expands enormously. No one can ever prescribe successfully all the twists and turns to be taken as the lecturer, the discussion leader, or the classroom teacher uses judgment, sudden insight, sensitivity, and agility to promote learning (1978, p.15).

And,

> . .teaching is something that departs from recipes, formulas, or algorithms. It requires improvisation; spontaneity; the handling of hosts of considerations of form, style, pace, rhythm; and appropriateness in ways so complex that even computers must, in principle, fall behind (1985, p.4).

In an article which synthesizes research on teaching since 1976 at the Institute for Research on Teaching at Michigan State University Porter and Brophy report that research on teaching has shown that much of this instruction, ". . results from professional planning, thinking, and decision making by teachers. Good teachers adapt instruction to the needs of the students and the situation rather than rigidly following scripts." They continue,

> . .teaching is highly complex, containing many points for possible breakdown or error. The best teachers negotiate their way through this complexity by attending to each relevant factor. . . The sheer complexity of

the teaching task and of the milieu in which it is conducted— typical classrooms— make it necessary for all teachers to rely on procedures, routines, implicit decision rules, and other simplification strategies that make the task manageable (1988, pp. 74-75).

Thus in their explicit statements, empiricist researchers demonstrate a grasp of the extremely complex and subtle nature of teaching. But given the incompatibility here between the view implied by the actual research and the statements by the practitioners and supporters such as Gage, and Porter and Brophy, it is obvious and perplexing that such understanding is being ignored in the actual research. To take but one example from their writing: where in all the massive amount of research merely sampled in the previous chapter is allowance made for Gage's ideas of improvisation, spontaneity, form, style, rhythm, pace, or appropriateness? The answer is that there has been some consideration of "pace" in some sense of that word in work by Gumo (1967;1969;1982), Kounin (1970) and Merritt (1982), and more recent studies on so-called "wait-time" (Carlsen, 1991), but such studies scarcely begin to grapple with the subtle notion of "pace" which Gage would appear to have in mind. And ALT is perhaps the merest beginning attempt to accommodate for appropriateness.

Notions of What it is to be Scientific in Research

Given that teaching is being construed in the manner just described, what do the researchers believe are *scientific* methods for studying it? As with the views of teaching, the views of a scientific study of teaching can be drawn from two sources, the *ideas implicit* in the studies and the *explicit statements* of leading researchers.

If we examine what the researchers are doing in the examples in Chapter IV, their supposed scientific approach draws from a multifarious variety of somewhat incompatible Heuristics, empirical representations, procedures, and observational instruments, which involve the establishment of statistical relationships between what it is believed teachers are observed to do and what it is claimed happens to pupils as a result of these doings: as shown in performance on standardized achievement tests. *Attempts are made to gain objectivity through making the observational instruments low-inference, since it is naively believed this will make the results reflect as much as possible the way teaching really is in classrooms and as little as possible the personal preferences or intuitions of researchers.* The research proceeds in a cyclical fashion from observation, to more elaborated Heuristics regarding aspects of teaching, to statistical meta-analyses of these aspects, back to observation and experiment, and so on.

The results of both meta-analysis and various experimental findings are fed back into the research system to provide further results, and in various ways the cycle begins again: Helmke and Schrader's research was part of a larger longitudinal project; ALT was itself a considerable program of reflexive research; Finn and Achilles were inspired by earlier research on the size of classes. The researchers construe the scientific approach as miniscule and cumulative. It is believed that the existing stock of scientific knowledge about teaching is slowly added to, and the frontier of our ignorance about teaching is gradually pushed back.

Superficially, this looks like an account of the so-called "normal science" of Kuhn. But even if we accept Kuhn's debatable distinction between revolutionary and normal science, there are two huge differences between his normal science and empiricist research. His normal science takes place against a background of theory (especially Cluster 6, Type 10, and Cluster 9: Observational Presupposition, and Scientific Theory). Such theory provides the origin of hypotheses, gives them meaning, allows them some chance of being acceptable, and ensures that if they are acceptable they will push out the boundaries of the science. But, despite occasional hortatory mention of theory, both the original empiricist pedagogical research, and its meta-analysis, singularly lack theoretical underpinnings. Moreover, the discoveries and the cycle are merely hopeful— they have yet to be produced by empiricist research.

Thus the implicit notion of scientific research which underpins the studies would seem to be the following:

1. Such research is concerned with observable phenomena— observables which can be described and related in various ways. Teaching is made up of such generalizable, claimedly-observable entities. These entities can be approached descriptively. Observers go into classrooms and become familiar with these observable phenomena of teaching by observing what teachers do. The observers develop general categories of observables which are called "variables": variables such as "warmth", "praise", "discreet support", "disciplinary question", "direct question", "open question", and so on.

2. Scientific research is objective. In the belief that objectivity can be achieved, strong attempts are made to produce "low-inference" operational definitions and denotations of the variables. Low-inference definitions aim at a minimum of human judgment on the part of the observer. For example, Good and Brophy (1982, p.73) define two kinds of question in the following way:

> *Process question* : Requires students to explain something in a way that requires them to integrate facts or to show knowledge of their interrelationships. It most frequently is a "why?' or "how?" question. *Open question* : The teacher creates the response opportunity by asking a public question and also indicates who is to respond by calling on an individual child, but the teacher chooses one of the children who has indicated a desire to respond by raising a hand.

These definitions and denotations are then put into scales for observation, with general instructions about observing. To further increase objectivity, observers are trained in the use of the scales, and make an effort to judge the variables in a consistent manner. Such operational definitions of variables must be memorized by the observers until they can record automatically without reference back to the instructions. As Jane Stallings, a leading empiricist researcher, has put it,

All observation systems necessarily include some procedures for training observers. If data are to be useful, they must be gathered systematically and reliably; that is, two or more observers must record the same event in the same way. It is important to start with objective observations (events that can be seen, heard, and agreed upon by two or more people). Subjective events such as feeling or intuition should be avoided in systematic observation unless the observation team shares common assumptions and has spent a great deal of time rating similar events (1977, p.35).

3. The occurrence of the different variables are summed across teachers and periods and the achievement of the classes is assessed. Correlations which link the occurrences of a variable and the achievement of the classes are sought. Where a comparatively high correlation is found, that action is recommended to teachers, and to those who prepare teachers (alternatively this feature may not appear until after stage 4).

4. Where possible, the supposed relationship is to be tested in an experimental situation.

The experiments at the University of Canterbury, those of Anderson, Brophy and Evertson on reading, and those of Finn and Achilles on the effects of class size, seem to be very different from those of Galileo and Lavoisier, or those by Boyle, Davy, Faraday, Pasteur, Rutherford, Michelson and Morley, Jacob and Wollman, and indeed different from all the others in the long history of scientific research. They are of the statistical kind developed by Ronald Fisher in the twentieth century. *Their design is a consequence of the empiricist ontology and the statistical approach. It rests in the statistical testing of Heuristics by way of so-called "hypothesis testing".*

Two representative samples, a control group and an experimental group were drawn from the population to which the Heuristic might apply. With respect to that which he wished to test, the experimenter treated the experimental group or changed the experimental group in the requisite manner demanded by the Heuristic, but not the control group. The achievement of the two groups was assessed and the means calculated before and after the treatment for each group. The change in the means for each group was calculated, and the change in the means compared. At this point the experimenters made use of "the null hypothesis". The null hypothesis argues that the difference in the means of the two groups is not related to the treatment, but to some other feature of the samples. An appropriate statistical test (ANOVA, MANOVA, etc.) was used to demonstrate beyond reasonable doubt that the difference in means was or was not related to the treatment, i.e. the null hypothesis was or was not to be rejected. In the first two Canterbury studies, the null hypothesis was rejected, but not in the third. In the Anderson, Evertson and Brophy research the null hypotheses were generally rejected. In the Finn and Achilles study the null hypotheses were rejected. In effect, the logic of the situation is that it allows the researcher to gamble at stated odds on a substantive hypothesis (Heuristic) versus a null hypothesis (Heuristic), with the precision of the substantive hypothesis (Heuristic) affecting these odds. Thus we have the "rigorous uncertainty", which

THE VIEW OF SCIENTIFIC RESEARCH ASSUMED 137

was mentioned in Chapter III.

In the experiments described in Chapter IV, and the tens of thousands of similar empiricist experiments performed over the last one hundred years in research on teaching, Empiricist Theory of commonsense and Heuristic has been used (and sometimes Theory from Cluster 3), rather than Scientific Theory. For instance, in justifying their research, Finn and Achilles say that the question of class size continues to attract the attention of educational policymakers and researchers, that the findings of observation and experiment are not consistent, and that research findings do not support the contention that smaller classes will of themselves (sic) result in greater academic achievement.

The results obtained in the experiments were of the following kind: Questions should be asked one at a time and be appropriate in level of difficulty (University of Canterbury); reading groups should be organized for efficient, sustained focus on the content (Brophy and Evertson); teachers' diagnostic competence per se, was not related to progress in learning (Helmke and Schrader); the univariate contrast values for comparing small classes with others is significant at $p < 0.001$ for four of the five measures (Finn and Achilles). These are "findings" not scientific results. They are empiricist generalizations, or oughts drawn from empiricist generalizations, not propositions embedded in Scientific Theory. For a concomitant reason they cannot be related one to another in a scheme of rational connection. They can at best be related in a commonsense scheme of empiricist connection of emphases, of statistical correlations, and of general claims and advice.

It will be recalled that Fisher created this approach for agricultural research with seeds, soils, and fertilizers– where it is still used today with important results. In the present context, what should be noted is the profoundly debatable assumption that one variable, or aspect, or style of *teaching* can be tested against alternatives, just as can strains of seed, or types of fertilizer.

5. It is hoped that a theory (the sense of "theory" being left ambiguous) about how teachers should act will emerge.

The second source of notions of scientific research on teaching is the *explicit statements* by researchers[4]. A number of representative statements will be drawn upon. There is a statement from the report of the study by Nuthall and Church (1973) described in Chapter IV, titled, "The nature of scientific research"; there are claims made in Nathaniel Gage's seminal and influential, *The Scientific Basis of the Art of Teaching* (1978) and in several of his articles (1963, 1984, 1985, 1989); there are two articles by Brophy (1976, 1979); there are parts of Nisbett's article "Educational research: the state of the art"(1982); there is Walberg's section "Characteristics of applied scientific research" from his paper "Syntheses of research on teaching"(1986), in the third *Handbook of Research on Teaching* ; there are Kerlinger's claims in his influential *Foundations of Behavioral Research* (1986). (All the following italics are added.)

Nuthall and Church write that in scientific research on teaching:

---

4. Some of the following account repeats summary points already made.

The *first* stage is one in which the phenomenon being investigated is approached descriptively. Observers sit in classrooms, *becoming familiar with the phenomenon, and developing ways of categorizing* behaviour. The *second* stage is one in which an attempt is made *to find inter-relationships between behaviour variables* observed in natural classroom situations *and measures of student learning*. The studies involved are correlational, and the object of them is to indicate the relative significance of the descriptive categories which have been developed in the first stage. The *third* stage is one in which *an attempt is made to provide experimental validation of the relationships suggested* in the correlational studies. An experimental study is one in which behaviour variables are manipulated in such a way that the effects of this manipulaton can be observed on other behaviours, and on changes in pupil learning.

Each of these three stages *feeds information both forwards and backwards*. Descriptive studies provide the background for correlational and experimental studies. But it is equally important that results of correlational and experimental studies should be used to refine and develop descriptive categories and suggest new observational studies...

Constant interaction between the first three stages should lead inevitably to the *fourth* stage. *In this final stage explanatory theory is developed which accounts for the relationships uncovered* in the experimental studies. This theory becomes embodied in the descriptive system, so that the variables which have proved significant in the correlational and experimental studies can be identified by any user of the descriptive system (1973, p.11).

The scheme of Walberg, another leader in the field, is slightly more complex. He says that scientific research on teaching has the following stages:

1. *Identification of valued outcomes.* The pure disciplines such as physics are free to seek the most fundamental phenomena; but practical fields such as medicine, agriculture, engineering and education concern, by definition the efficient attainment of valuable outcomes such as health, yield, and learning.

2. *Identification of plausible factors.* Substantive research *seeks the causes of* valued outcomes. Such causal factors and their likely bearings on effects *may be sought from the following sources: common observation, practical wisdom, policy controversies, and prior research* in the form of case studies, statistical correlations, and experiments. *The use of parsimonious, falsifiable theory as a guide to empirical inquiry* (Popper,1959)— the modern and commonly accepted view in the natural sciences— *is rare in educational and social research.*

3. *Observation of the factors and effects.* The factors may be *experimentally manipulated or only observed in their naturally occurring states*. In either case, the observed covariations of factors and effects are described in case studies or numerically indexed in the form of correlations, regression weights, F-ratios, and other statistics in quantitative research.

4. *Replication of covariations. Relationships replicated with different subjects and in a variety of circumstances can be said to be generalizable*, widely confirmed, or, in Popperian spirit, unfalsified. Replication attempts by different investigators and probing theory in a program of research, however, are rare. *Most research on teaching, for example, draws not on unifying theoretical paradigms and constructs but on an immense variety of somewhat incompatible constructs, empirical representations, procedures, and observation instruments* . . .

5. *Research synthesis* . Comparable *studies are systematically evaluated and summarized* in the form of frequency distributions, means, standard deviatons, and other statistics. Such analysis helps *to make sense of the overall findings and classifications* of them just as statistics are useful in understanding a mass of raw primary data . . .

6. *Adoption and evaluation* . Practitioners can reasonably accept the implications of research findings that syntheses *prove robustly unfalsified*, that are effective considering total costs, and that are without harmful side effects, *although continual local evaluation is needed*. Because research syntheses published in the last several years suggest a number of consistent, worthwhile effects, educational *research may be entering a new stage of scientific maturity and relevance* to practice (1986, p.215).

It may be pointed out that the syntheses mentioned in Walberg's fifth point have been very significant for the impact of research on the schools and on those who train teachers. Much of such synthesis takes the form of the statistical "meta-analysis" already discussed, which attempts to tease out the general implications of individual studies. Such meta-analysis may be seen as a statistically-sophisticated recent development of empiricist research. Walberg claims that, "Research synthesis explicitly applies *scientific techniques* and standards to the evaluation and summary of research" (1986, p.216).

By putting these accounts by Nuthall and Church, and Walberg together it can be seen that their idea of a scientific study of teaching involves the accumulation of what is claimed to be objective empirical-statistical evidence, built up through observational, experimental and meta-analytical research, which in providing a structure of knowledge will somehow generate new hypotheses and new experiments, until the whole field is covered, and that moreover, a theory of teaching, in some sense of the word, "theory", will evolve.

There is a conception of development by a kind of Baconian induction and accretion in much of this. As Nisbett writes,

Most empirical research studies are, to use Ashby's (1958) phrase, 'miniscule analyses'. This is hardly surprising: *all science proceeds by miniscule analysis, and the building of a coherent theory is a very slow process on which we have only just begun*. For the present, the problems which can be resolved empirically are relatively minor, compared with the major issues which require a judgment of values. Nevertheless, in these major issues, *research* has an important contribution to make, in defining objectives, in evaluation, in assembling relevant and adequate evidence on which to base judgments (1982, pp.4-5).

In his 1978 account, Gage is explicit in stating that although he believes there cannot be a science of teaching, there can indeed be a *scientific basis* for teaching. He explicitly says, "The *kind of science* that has occupied mainstream educational research over the last fifty years involves relationships between variables." And again,". . . *a scientific basis consists of knowledge of regular non-chance relationships*. . . .the relationships need merely to be better than those that would occur by chance. . . . this means that correlations should be different from zero (italics added)," (p.20). He writes further that,

> What they want to know about— especially those who are *interested in a scientific basis for the art of teaching*— are Process-product relationships. They want to know whether the teacher's thinking, behaving, acting— in short, teaching — in one way is demonstrably better. . .than teaching in another way. *If the answer is yes, we have a basis for improving teaching* and the training of teachers. *If the answer is no, we are left without a scientific basis* for the art of teaching (italics added) (1978, pp.23-24).

In recent times, Gage's claims for scientific status have become a little more modest. In a recent paper he admits that his sort of research is scientific, ". . .only in the methodological sense of using tools to minimize bias effects, error, and unreliability, and to ward off the many kinds of threats to internal and external validity identified by Campbell and Stanley (1963)," (1989, p.259). Thus he claims that his Process-Product version of research,

> . . has been scientific in the sense that it has *used the methods developed in all sciences for avoiding self-deception and increasing the validity* of its results. Researchers have sought objectivity, reliable descriptions and measures, public and communicable procedures, and minimization of their biasses. *They have used devices that would make the results reflect as much as possible the way things are in classrooms* and as little as possible their own preferences (italics added) (1989, p. 253).

On this view of scientific research on teaching, some sort of theory of teaching is supposed gradually to *emerge* from the accumulated statistical findings. Thus this approach is largely inductive and *non*-theoretical (in several senses of the word). Such a non-theoretical approach is implied or explicitly stated by all of Gage's writings until 1989, and stated explicitly by Nuthall and Church. Until recently then, the empiricist view has been that research can be proudly non-theoretical. Thus in 1984 Gage wrote, "At the least, a scientific basis consists of scientifically developed knowledge about relationships between variables. . "(1984, p.89). And again,

> . . this conception of a scientific basis says nothing about theory, nomo-

logical networks, systems of postulates and axioms, or hypothetico-deductive relationships. This emphasis on relationships between variables does not deny the desirability of systematic theory; it merely means that, however desirable, *systematic theory is not indispensable to any valid conception of science* (italics added) (1985, p.7)

Except for the belief in the potency of meta-analysis, Walberg's account also is admittedly non-theoretical. As the various findings are produced, says Walberg, general features begin to emerge— claims of a supposedly quasi-lawlike character, which fit the seeming facts of teaching and relate them statistically. In discussing "effect sizes", one of the key techniques of statistical meta-analysis, Walberg has the following to say:

Effect sizes permit a rough calibration of comparisons across contexts, subjects, and other characteristics of studies. The estimates, however, are affected by the variances in the groups, the reliabilities of the outcomes, the match of curriculum with outcome measures, and a host of other factors, whose influences, in some cases, can be estimated specifically or generally. Although effect sizes are subject to distortions, they are the only explicit means of comparing the sizes of effects in primary research that employs various outcome measures on nonuniform groups. *They are likely to be necessary until an advanced theory and science of educational measurement develops ratio measures* that are directly comparable across studies and populations (1986, p.216) (italics added).

The supposedly non-theoretical nature of the research has been somewhat modified in the 1989 paper by Gage and Needles. There, in response to recent philosophical criticism, they make claims that theory is to some extent involved in research. They say that it has made use of (a) what he calls, "formative hypotheses". His example from research is, ". . differences in teacher questions influence students' intellectual abilities in the subject matter being taught," (1989, p.263), and they claim that "Process-product research has produced many such formative hypotheses." It has also he says made use of "elementisms" (e.g. structuring, soliciting, responding), and "conceptual constructs" such as Academic Learning Time (ALT). These claims will be considered and rejected in the critique provided later.

Though in his book, *Foundations of Behavioral Research*, Kerlinger does not discuss research on teaching as such, for years the book's recommended approaches have been construed as normative in hundreds of universities and research institutions which are concerned with such research. The following are typical of Kerlinger's statements about what is going on when scientific research is being carried on:

Scientific research is systematic, controlled, empirical, and critical investigation of natural phenomena guided by theory and hypotheses about presumed relations among such phenomena (1986, p.10).

And,

First there is a doubt, a barrier, an indeterminate situation crying out to be made determinate. The scientist experiences vague doubts, emotional disturbance, incohate ideas. He struggles to formulate the problem, even if inadequately. He studies the literature, scans his own experiences and the experience of others. Often he simply has to wait for an inventive leap of the mind. . With the problem formulated . . the rest is much easier. Then the hypothesis is constructed, after which its empirical implications are deduced. In this process the original problem, and of course the original hypothesis may be changed. It may be broadened or narrowed. It may even be abandoned. Last, but not finally, the relation expressed by the hypothesis is tested by observation and experimentation. On the basis of the research evidence the hypothesis is accepted or rejected. This information is then fed back to the original problem, and the problem is kept or altered as dictated by the evidence (1986, pp13-14).

And,

Another point about testing hypotheses is that we usually do not test hypotheses directly. . we test deduced implications of hypotheses. Our test hypotheses may be: "Subjects induced to lie will comply more with later requests than will subjects not induced to lie," which was deduced from a broader and more general hypothesis: "Increased guilt leads to increased compliance." We do not test "inducement to lie" nor "comply with requests." We test the relation between them in this case the relation between lying (deduced guilt) and compliance with later requests (1986, p.13).

It will be noticed that Kerlinger has (merely) mentioned experiment. Most empiricist researchers say little about the *nature of* experiment. Given their standardized, statistical notions, this is hardly surprising: on their approach, learning how to make use of the statistics *is* learning how to experiment. And on these matters Fisher has already said it (1935). As he so clearly puts his key axiom: "Every experiment may be said to exist only in order to give the facts a chance of disproving the null hypothesis" (1935, p.16). And taking Fisher's view as Gospel, Kerlinger adds, "It should be remembered that a hypothesis is never really proved or disproved" (1986, p.20). In empiricist experiment he is correct. Indeed, in relation to the research, say, on the effects of class size the empiricist debate never ends. Nevertheless, Kerlinger's claim would have surprised Galileo and Lavoisier ! But, of course, Kerlinger is talking statistics in relation to testing of the empiricist research hypothesis (Heuristic) by the statistical null hypothesis (Heuristic). In contrast Galileo and Lavoisier were doing scientific research, and though they might have agreed that the proof of an hypothesis raises complex epistemological difficulties, they would certainly have argued for the cogency of disproof of hypotheses. Indeed, there just are myriads

THE VIEW OF SCIENTIFIC RESEARCH ASSUMED 143

upon myriads of instances of disproof in the history of scientific research, and Chapter II provided some typical examples.

Though she is discussing empiricist sociology, the essence of what we find when we examine the nature of empiricist experiments on teaching is clearly stated by Willer:

> The statement of results, when general, is in terms of the connection of the class of objects manipulated to the class of objects supposedly affected, and is therefore an [empiricist] generalization. The judgment of connection is concerned with the deviation of effects from chance expectation and usually uses a test of statistical significance. The judgment of connection has nothing to do with [abstractive connection to the observational world] but is instead completely [empiricist]. The . . generalizations which result from these experiments are tendency statements, not theoretic laws, and thus are not useful for scientific explanation. These [empiricist] generalizations are isolated from one another because their [empiricist] character does not allow them to be connected in a rational scheme. Although this information may have immediate practical or technical value, it does not have theoretic value or explanatory use. It certainly has nothing to do with science (1971, p.123).

In contrast with what was claimed in the first part of this chapter re the notion of teaching, where it was argued that the rationalized accounts of teaching failed to fit the notion of teaching implied by the research, there does seem to be a closer equivalence between the notion of scientific research implied by their actual procedures, and the notion of scientific research which the researchers claim to be employing.

Conclusion

The claim to scientific status is of considerable significance for empiricist research. Perhaps the most important result of the claim is that, because of the kudos attached to scientific research in our kind of society, if the results are accepted as scientific they take on an acceptable face. Politicians, Local Educational Authorities, school districts and state and national departments of education, who and which are looking for advice and quick results can be impressed by claims to science. The Florida legislation (p.125, above) which nonsensically states that "evidence *will be* available" from research is an example of such kudos.

In providing this explicit account of the notion of scientific research on teaching, the stage has been set for the specific critique of the chapters which follow. There the great differences between empiricist research and scientific research will be detailed, and the two recommendations as to how empiricist research might be rescued will be discussed.

CHAPTER VI

PRELIMINARY CRITIQUE OF THE PRESENT EMPIRICIST APPROACH

Prior to demonstrating in the following chapter the immense gulf between scientific research and empiricist research on teaching, and prior to the positive suggestions of the last two chapters, it is necessary to provide some initial critique of the notions with which researchers are working, so that it becomes quite clear why changes are required. This critique will occur in three sections. Initially there will be a detailed examination and rejection of the claims Nathaniel Gage has made in his various apologia for his Process-Product version of empiricist research— claims typical of those made by empiricist researchers generally. In particular there will be a consideration of his published views on the place of theory in research. Secondly there will be a consideration of the difference between the general concepts of empiricist research and the abstract concepts of scientific research. Finally there will be a discussion of whether some empiricist research has been carried on when philosophical and/or commonsense analysis were required. These three critiques will point the way to the later more positive suggestions re the conciliation between Empiricist and Scientific research.

Critique of the Apologia by Gage

Because Nathanial Gage has written so much, and is one of the leading empiricist researchers, perhaps the doyen of those who follow the Process-Product version of this approach to research on teaching[1], both setting the style of the research, training disciples, and providing much of the philosophical rationale to support it, it will be his work and writing which are mosty adverted to in this section. The critique will be mainly of his book of 1978, *The Scientific Basis of the Art of Teaching*, his article of 1984, "What do we know about teaching effectiveness?", his monograph of 1985, *Hard Gains in the Soft Sciences: the Case of Pedagogy*, and his article (with Margaret C. Needels) of 1989, "Process-product research on teaching: a review of criticisms". Because Gage's arguments are influential in the research community, it is important for the sake of any improved research that they be rejected, especially

---

1. Gage received the 1988 Award for Distinguished Contributions to Educational Research, of the American Educational Research Association.

the misconceptions about theory, if there is ever to be either a possibility of a scientific basis for pedagogy or of an improved empiricist approach. It is worth repeating a point made in the Introduction, viz. that what follows is not an attack on Gage. The point is that while I respect his aims and his complex attempts and those of other researchers in this tradition, I believe that the tradition itself is fundamentally wrong-headed and inadequate.

Gage's own attitude to teaching would seem to be inconsistent. On the one hand he has always shown an intelligent understanding of the nature of teaching and of its basis in an "art". As mentioned in the previous chapter, in 1978 he wrote that, "No one can ever prescribe successfully all the twists and turns to be taken as the lecturer, the discussion leader, or the classroom teacher uses judgment, sudden insight, sensitivity, and agility to promote learning" (p.2); and in 1985 he wrote that, ". .teaching is something that departs from recipes, formulas, or algorithms. It requires improvisation, spontaneity. ." (p.4).

On the other hand, despite the immense human complexities which he has himself agreed feature in teaching, Gage emphasises that there can indeed be a *scientific basis* for teaching. His view, like that of empiricist researchers generally is as we have seen, that such a scientific basis rests in statistical relations between variables. He has in the past, claimed that his conception of a scientific basis says nothing about theory, and indeed that systematic theory is not indispensable to any valid conception of scientific research.

So Gage's view like that of other empiricist researchers is that a scientific basis consists in the establishment of statistical relationships between what teachers are observed to do and what (it is claimed) happens to pupils as a result of these doings, and that concomitantly systematic theory (it is unclear what Cluster is being alluded to) is not indispensable. As might be assumed from what I have already written about the significance for scientific research of theory of various kinds, these are unacceptable claims.

Contra such views, and as will be clear from what was written in Chapter I, I argue that the presence of theory from several of the Clusters is far from being dispensable if an enterprise is to lay claim to scientific status and to production of scientific knowledge. It should be clear from the accounts of research provided, that consistent with Gage's views, empiricist research largely ignores theory, and in the present chapter I shall try to show that empiricist researchers generally and Process-Product researchers in particular, misunderstand this theoretical requirement (especially Clusters 5, 6 and 9) in several ways, ways which in turn lead to false claims to scientific status.

In its features over the last twenty-five years and in its nineteenth and twentieth century ancestral versions from Rice onwards, empiricist research appears to be a very different engagement from the work of Galileo, or Newton, or Lavoisier, or Crick and Watson, or Wilson, or from the work in any of the established sciences. As a key inadequacy, until very recently empiricist researchers do appear to have been unaware of the manifold significance of theory (in many of the senses) in science, and even now continue to misunderstand their nature. They are for example still largely unaware of the effects of theory (Clusters 6 and 9) on what is observed, and thus on what can count as a fact.

Gage, in drawing an analogy between teaching and medicine, says that the physician makes use of, ". . a thousand variables that have been identified and related to other variables by scientific methods" (1978, p.17). But this is a misdescription. First, it is misleading to call that with which the physician works and that with which the empiricist researcher works by the same term, "variable". As will be made clear in the second section of this chapter, "matter-of-fact reaction", "pupils have choice in where to sit", "misbehavior", "clues", "system of rules", "diagnostic competence"— typical concepts and conceptions from Chapter IV, are epistemologically radically different from the concepts of medicine-- "respiration", "red blood cell", "plasma membrane", "protein", "virus". Secondly, the nature of the "relating" of these different kinds of concepts is itself different. Gage's use of the kudos-seeking phrase, "by scientific methods" is ambiguous. Though some relationships in medicine have been established initially as statistical correlations, *which await explanatory Theory* (Cluster 9), most connections are first embedded in such Theory, which *suggests the hypothesis to be tested* and also (if they happen to be interested in correlations) why and where a correlation may be found. And even those first established by correlation do not normally have to wait long before such theoretical explanation becomes possible. Moreover, the relationships which have been established initially as correlations were not just chosen at random, but on the basis of theoretical hints. There seems to be a kind of excessive inductionism hiding here, one which ignores most of the valuable instruction we have received over the years from Sir Karl Popper.

Macmillan and Garrison also properly point out that Gage does not compare educational and medical practice or research, but rather compares the *use of research findings* in the two fields, particularly, ". . the use of comparable statistical results as the basis for practical recommendations" (1987,p.39). They say that there is a failure of analogy insofar as medical researchers and medical practitioners share the same (generally causal) assumptions about the world, whereas, as alluded to above, teachers view their work intentionally but empiricist researchers ignore intention (in any pedagogically-significant sense). Again, they indicate that medical practice became potent only after it had been able to make use of biomedical research. (Though they do not say so, there is also a disanalogy here. For such research is not research into medical practice, but into that which sustains medical practice; in contrast, Process-Product work is research into teaching as a practice.)

Moreover, and very importantly, both medical and natural scientists are *very unsatisfied* when they can state that the relationship has been established only, as they say, "empirically" (i.e. by empiricist methods). Until there is some theoretical explanation (drawn from Cluster 9), they do not regard with equanimity what has been "established". They say such things as, "It's *only* an empirical relationship we have found— we do not yet know why it occurs." As is the case with the medical efficacy of zinc, they recognize that the finding exists in some kind of limbo.

Gage's discussion of the significance of the (quite low) correlations between smoking and lung cancer is an example of the misunderstanding of the relationship between correlations and theory (Clusters 6 and 9), indeed of theory generally. He claims that even weak correlations may be important, mentioning that in the 1967 study by Lilienthal, Pedersen and Dowd, the correlation between cigarette smoking and lung cancer was only .14 . And yet, says he, although the coefficient of .14 is somewhat lower in magnitude than that which might be found for any single variable relating the behavior of teachers and the achievement of pupils, ". . on the basis of such correlations, important public health policy has been made and millions of people have changed strong habits" (1978, p.21).

On the contrary, the medical doctors and the health departments of the scientifically-advanced countries around the world, which advise people not to smoke, did not and do not base their advice upon the statistical *correlations* which have been established by statistical analysis of the hospital data that compare cancer deaths and illhealth in those patients who smoke with deaths and illhealth in those who do not— even if they consider such correlations to be additional evidence. The advice is based upon a whole range of interrelating medical theory (Clusters 6 and 9) of nicotine poisoning, of the presence of tars, and of the myriad of other empirically obvious (such as the actual pathological lungs taken from the cadavers of heavy smokers) and theoretically obvious (Cluster 9) relationships between smoking and diseases such as heart disease, emphysema, influenza, increased kidney damage amongst diabetics who smoke, and so on. Such Theory shows why cancer and the other problems *are to be expected* . Such Theory also suggests *ways in which to research into cures* for cancer. The point is that we should be extraordinarily surprised if there were no correlation. In fact, we should realise that we must have made a mistake in the calculations. The correlation does not itself persuade or convince, or provide us with any sort of medical power. Indeed, we may well search for reasons to explain why the medical correlations listed by Gage are so *low* .

Gage's 1989 rejoinder to such claims for Theory is to say, ". . many medical experiments and treatments have proceeded without theoretical bases, knowledge, or even hypotheses concerning the treatment's modus operandi. A classic example is the experimentation and treatment with aspirin, discovered in 1890 but given a theoretical basis only in 1970" (Gage and Needles, 1984, p.284).

In part this rejoinder is appropriate, but it is also misleading. It is appropriate because the practice of medicine like the dyeing of cloth, the brewing of beer, and the construction of cathedrals, has indeed been, as discussed in Chapter One, until the present century, in large part a craft. But the rejoinder is misleading because as also argued in Chapter One, *such crafts have not made use of the results of scientific research* . They have been trial and error, rule of thumb, based upon generations of experience— empiricist, not scientific activity. They have been empiricist "technologies" not scientific technologies. Not being Theory-driven (Cluster 9), not embodying abstract concepts, they could not be scientific. The Scientific Theories which have explained their successes have come centuries later.

But the rejoinder is also partly false, at least to the extent that there have always been medical theories, even though such theories (straddling Clusters 1, 5,

6 and 9) might be later shown to be unacceptable, and even though it is exaggerating matters in many cases to construe them as Scientific Theory. Mediaeval physicians used astrology and cast horoscopes of particular patients. Physicians had always worked with general theories such as Galen's teleological theory of bodily functions, and the ancient Theory of Humors, while Versalius's teaching on anatomy was available from 1543, and Harvey's Theory of the Circulation of the Blood from 1628; there was the Iatrophysics of the seventeenth century; and so on. There was also always what was known as the "principle of sympathy", or the "theory of like cures like", e.g. that Jasper was of value in preventing hemorrhage because of its red color (Lyons and Petrucelli, 1987).

But the overriding significance of the rejoinder is that it *weakens* the case for his research. Gage, in alluding to the practice of medicine as not being theory-driven surely does not wish to suggest that it is empiricist craft practice which is the true analogue for his Process-Product research? For researchers like himself place great significance on the claim rejected in this book, namely the claimedly *scientific* nature of their results.

There are further problems. Gage writes,

> Body temperature, red cell count, cholesterol content, blood pressure, respiration rate– these few variables that even laymen recognize– have acquired their medical significance from scientific methods. But the physician uses the variables in what must be considered an artistic way as he approaches each new patient (1985, p.17).

Here there is confusion between the *establishment* of scientific claims and the *application* of established claims. Scientific knowledge of the way the human body works is applied by the physician in seeking cures. The physician applies his knowledge in an artful way. But the physiology of temperature and its effects, the knowledge of the function of red blood cells, the awareness of the effects of cholesterol, the explanation of the dangers of high blood pressure are not themselves based upon correlations. They are based upon well-argued physiological, biological, biochemical, and chemical theory (Cluster 9) which explains how and why such matters are related.

Now it is true that when a new drug is being tested, it is common for it to be *compared with* some existing drug being used for similar purposes, and the comparison is a statistical one. But the initial Scientific Theory which suggested that this new substance might be helpful was not based on correlations. It was upon established biological and chemical theory. (Otherwise why test this drug rather than a thousand others?) We do *not* make use of such Theory because there are *high statistical correlations* between the way strands of DNA separate, or the way mercury and oxygen combine ! The chemistry of mercury and oxygen, and the biology of cells do not rest upon correlations, but upon the universal conceptual propositions of Scientific Theory. Priestley and Lavoisier made no use of statistics; and Crick and Watson used none in their research.

Macmillan and Garrison have shown that the attempt to draw an analogy between research on medicine and Process-Product research on teaching confronts a further difficulty when the researchers try to convert their supposedly non-theoretical findings into practical recommendations for teachers, i.e. into theory, Clusters 2 or 4 (Compared with Practice, or Practical Theory). As they say,

> The causes of biological and medical conditions and changes are seen as the same by researchers and practitioners and the conversion from research to practice is almost direct . . . In fact, researchers and practitioners share, generally speaking, the same theory [several Clusters] . . .This is not clearly the case with current research on teaching and pedagogical practice (1987, p.39).

Surely they are correct. The supposedly non-theoretical and explicitly non-intentional point of view of empiricist research on teaching, scarcely matches the specifically intentional, theoretically-conglomerate, personal pedagogical theory (Cluster 4) of practitioners in the classroom. As they say, such incommensurability cannot be bridged by merely, ". . turning the results of teaching research over to practical artists (teachers) for they, wise only in the ways of intentionalty, may not understand how to use the results, or worse still, may misuse them" (1987, p.40). Moreover, given the gap, what reasons does a teacher have to make use of the empiricist findings– other than being impressed by the scientific pretensions, or being forced by university researchers, or supervisors?

Gage also provides an inadequate analogous argument from engineering. He suggests that engineers, like medical doctors, have a strong scientific background, normally in physics and chemistry, but that in their actual work on bridges, engines, or communication systems, ". . they also rely on artistry as they balance the claims of competing considerations" (1978, pp.17-18).

Once again there are problems. First, the physics and chemistry which the engineers apply *do not rest upon statistical correlations*. They consist of rock-hard Scientific Theory (Cluster 9): coefficients of linear expansion of metals, precise calculation of the forces set up by such expansion, application of Hooke's Law of elastic limits, the laws of friction, and so on. Second, even if engineers are using artistry when they decide which variables to count as most significant in their calculations, they are *not using artistry* when they *calculate* such limits of tolerance and stress, when they allow for safety margins of error in calculating how much a bridge can hold, and when they apply the above physical theory– they are using cold mathematical calculation. Third, Gage has *confused creation* or discovery or invention in engineering, *with application*. This is clear when he uses Wanlass's description of how he (Wanlass) *invented* an energy-saving electric motor, viz., "It's almost a creative process, like art or music" (1978, p.18) to support Gage's claim that engineering is artistic. But Wanlass's invention of the electric motor is a different matter from the carefully calculated use of that motor in engineering projects.

In order to shore up the case for his research, Gage, as many other empiricist writers have done before and after him, points to the *complexity* of the interactions between variables in teaching. He rejects the possibility that these are so complex that practical applications are not possible, saying that research on teaching is no more complicated in this respect than in a science such as physics. In physics, he claims,

> . . many relationships are about as firm as anyone could wish [sic]. *Yet, considered one at a time, these relationships hold only under laboratory conditions and not in real life.* When it comes to real-life phenomena, physical scientists must apply many of these laws in combination. . . But physical scientists do not denigrate their main effects simply because interaction effects also occur. They simply take more laws into account (1978, p.19) (italics added).

Similarly, Gage (1963, p.120) has written (presumably alluding to Galileo),

> Physicists did not attempt to predict all aspects of the undulating motion of a feather as it fell in an everyday breeze. They refined the problem into the prediction of the fall of a frictionless body in a vacuum and derived laws from this phenomenon. Such laws, combined with other laws applying to similarly pure situations, can account for the vagaries of the motion of actual objects under field conditions.

Though Gage might express these matters differently were he writing today, these quotations still describe some typical present-day empiricist confusions. First, as demonstrated in Chapter II, in becoming established, the laws of physics have related abstract concepts by way of various kinds of theory. So in the strict and limited sense in which they are *created* as conceptual abstractions, they do not apply in real life, for by definition they are idealised. But they are quickly shown to apply in real life, by way of the deductive observational inferences involving ratio measurement and mathematical calculation, in the abstractive connection. Second, it is incorrect to suggest that, ". .considered one at a time, these relationships hold only under laboratory conditions and not in real life." What has been confused is the *discovery* of the relationships under laboratory conditions and the *existence* of the relationships in the empirical world. It was precisely because such relationships were indeed active in the empirical world that they were able to be discovered by laboratory study. It was precisely because such relationships were indeed active in the empirical world that they could be adopted as a basis for scientific technology.

And it is true that physical (and other) scientists,". .do not denigrate their main effects simply because interaction effects also occur. They simply take more laws into account." But the reason they do not denigrate their main effects (to use Gage's term) is because they *do indeed have main effects* ! They have, for example, the laws which describe the way oxygen combines with other elements, Hooke's Law, the Laws of Friction, the Laws of Thermodynamics, Doppler's Principle, Ohm's Law, Huygen's Principle, Pascal's Principle, Bragg's

Law, Chargaff's Rules, the Genetic Code, and thousands of others. Moreover they can certainly, where judged necessary, "simply take more laws into account", (e.g. combine the Laws of Reflection with Snell's Second Law of Refraction, or Newton's Law of Gravity with his Laws of Motion), because they *have so many established laws*. But in any presumed scientific basis of teaching established by Process-Product research there are *no well established main effects* equivalent to such main effects in physics as the above-mentioned laws, and the thousands of others of Scientific Theory. The nearest equivalents which have come out of empiricist research remain either moderate statistical correlations, or claims to a significant difference at some stated level of statistical significance.

Moreover, (a) in many cases in physics and other natural sciences, as, say, in the case of Snell's Laws, the complicating variables can be ignored, with miniscule effect on accuracy, and (b) even where there are substantial interactions, natural scientists still know how to make allowances within particular limits and with various degrees of acceptable error stated prior to the fact. To be carefully noted is that neither the establishment of such well-established physical laws, nor the acceptance of the results of the application of such laws within degrees of acceptable error is the same thing as the tentative establishment or the tentative application of regularities through correlations and statistical tests. *In the first case we have well-established hard laws applied within limits of error. In the second case we have tentatively-established correlational and statistical relationships, the application of which does not even have arguable limits of error*.

All this poses problems for the claimed scientific status of empiricist research.

So the claims of such research are contestable and are based upon tests of a statistical kind, or upon low correlations. There would thus seem to be a sense in which the results are not even a collection of facts. At *best* they are a collection of facts: for in what sense are statistical claims and correlations facts? Perhaps in the sense that, as a reader of one of my conference papers put it, "It is a fact about our society at this particular stage in our evolution that these variables, thus defined, are thus correlated." Indeed this very claim points to the confusion in empiricist work and its doubtful status as scientific research. *For just what counts as a fact depends upon the theory (Clusters 5, 6 and 9) we bring to its conceptualization. So even if the variables are defined in some sense of the word, "definition", their lack of an explicit theoretical basis would seem to make their meaning indeterminate.*

As we have seen, Gage has in the past explicitly claimed that systematic theory is not necessary for activity to be scientific. (Which of the above senses does he have in mind?) He writes, "At the least, a scientific basis consists of scientifically developed knowledge about relationships between variables," and also that, ". .systematic theory is not indispensable to any valid conception of science" (1984, pp.87 and 93).

Contra Gage's views of how to be scientific, I would argue that the presence of various Types from several of the Clusters of theory is far from being dispensable. But these matters are obviously exceedingly complex. Paradigmatic scientific endeavor in physics, chemistry and biology draws its hypotheses (Cluster 5, Type 8) from Cluster 9. Such endeavor also presupposes theory Types 9 and 10. But where such established sciences proceed without needing to make such presuppositions explicit, fields such as empiricist research on teaching, which are still striving to be scientific need to make their theoretical presuppositions absolutely explicit, so that they can be fully examined, critiqued, and perhaps changed or rejected. For example, empiricist researchers are still unaware of the effects of theory Cluster 6 on what is observed, and so on what can *count as* an observation. Thus Gage's point claiming that what we require is the establishment of statistical relationships between what teachers are observed to do and what happens to students is far from straight-forward; indeed it begs the question. For what teachers are observed to do and what (it is claimed) happens to students as a result, depends upon the Presuppositional Theory which the researchers bring to their observations. Just what is the teacher doing? Just what is happening to Johnny? Because teachers perform actions and not just movements[2], normally it is the case that what is going on is *not* just observable, but depends upon the teacher's own construals and intentions and upon Johnny's state of mind and body. What is scored as praise on the observer's observation scale may be, as will be considered in more detail later in this chapter, exaggerated, superficial, calculated, and so on. And what is happening to Johnny depends both on how he construes the "praise", and on whether he is tired? lazy? drugged? anxious? short-sighted? dyslexic? autistic?

So, in general what is to be made of the view espoused by Gage that scientific research can do without theory? Garrison suggests three interpretations: (1) such researchers hold tacit theories, in which case they ought to think about them and then make them explicit, or (2) They hold them explicitly, but deliberately refrain from stating them, or (3) "They pack structural concepts that properly belong to theory into their methodology where they are hidden from their view as well as from ours"(1988, p.24). And as he cogently continues,

> The problem with all three of the above alternatives is that the hidden theory that must be present in order to conduct the experimental inquiry need never face the tribunal of sensory experience. These hidden theories most commonly expose themselves in ex post facto theorizing wherein the tacit, withheld, or methodologically compacted theory is revealed. When research findings are only interpreted after the fact, the interpretations are typically never tested. That's not science either (p.25).

My own view is that Garrison's first possibility best describes the Gage situation. For carefully to be noticed is that there *is* an underpinning theory to such research and that it is of the Clusters 6 and 8 varieties. It is Presuppositional theory of a behaviorist kind, together with ideas and points of view from the commonsense and ordinary language of Empiricist Theory about

---

2. See discussion in Chapter VIII.

teaching. Concepts such as praise, or high success, which are considered to be pedagogically significant, have not themselves resulted from the research, or from the statistical methodology. They have been drawn from suppressed Empiricist Theory of teaching, which has evolved over hundreds of years of practical experience. The remainder of the inexplicit, underpinning theory is Cluster 6, Type 9 (Ontological Presupposition) which involves the belief that statistical tests and correlations are superior to everyday experience and understanding, or to commonsense observational connections and relationships, or to the subtleties of ordinary language description and explanation of Empiricist Theory.

This lack of awareness of the significance of theory from various Clusters would also seem to be the reason for such beliefs as,

> Rather than seek criteria for the overall effectiveness of teachers in the many, varied facets of their roles, we may have better success with criteria of effectiveness in small, specially defined aspects of the role. Many scientific problems have eventually been solved by being analyzed into smaller problems, whose variables were less complex (Gage,1963, p.120).

This latter claim is true, but superficial. For as argued above, the scientific problems which have been solved in this way, were embedded in Scientific Theory. Such Theory provided meaning, and (as Kuhn quite properly emphasized in his talk of paradigms) pointed the way to a solution; moreover, the Theory linked one problem and the solution of that problem to another problem and its solution— continually.

And in Gage's 1984 quotation above, it should be noted that the words, "scientifically developed knowledge" are merely hortatory, or honorific. They add nothing of substance to the phrase, ". . about relationships between variables." Gage is here repeating the interminable claim of himself and other empiricist researchers on teaching that their research is scientific.

The claimedly non-theoretical nature of such research has been somewhat modified in the 1989 paper. There Gage and Needles agree that in several ways theory is involved, saying that Process-Product has made use of (a) what they call "formative hypotheses". The example from the literature is, ". .differences in teacher questions influence students' intellectual abilities in the subject matter being taught"(1989, p.262), and he claims that "Process-product research has produced many such formative hypotheses." These would seem to be theory in some sense of Cluster 5, Type 7, but they are scarcely theory-driven in the manner in which scientific hypotheses are theory-driven. It has also, he says, made use of "elementisms" (e.g. structuring, soliciting, responding), and "conceptual constructs" such as Academic Learning Time (ALT). To concede that some variety of theory is being used is a very positive move, but as will be suggested below, the examples provided add to rather than reduce the confusions.

As an example of this confusion, ALT may be considered. ALT at first sight could be seen to be the empiricist concept most likely to meet the requirements of abstract concepts in the scientific sense. But does it? Gage and Needles claim,

The theoretical explanation of the effectiveness of many of classroom management practices is that they enhance ALT. These practices keep teachers and students from wasting time, and hence reducing ALT. . . many findings of Process-product research can be explained by their connection with ALT. That theoretical construct in turn, derives from the importance of time in the cognitive processes (1989, p. 265).

But to use the appelations, "hypothetical construct", or "conceptual construct" (Gage and Needles, 1989, p.263) to dignify the somewhat tautologous notion of ALT is most misleading. ALT is not an abstract concept in the way that, say, "rectilinear propagation" of light, "frictionless plane", "hydrogen bond", or "transform fault" are. It does not by positing an abstraction which clearly relates to other abstract concepts, thereby help to explain the phenomena of teaching in the way these make possible the explanation, understanding and prediction of the phenomena of light, the behavior of bodies in motion, the structure of DNA, and the occurrence of earthquakes. For example, as already mentioned, the bold conceptual leap which posited the rectilinear propagation of light, suggesting that light might be conceived as something which both travels and travels in straight lines, eventually helped provide an explanation of the multitudinous phenomena of reflection, refraction, shadows, twilight, and so on. Snell, making use of these abstract concepts was able to perform a set of experiments which produced the precise mathematical relationship now known as Snell's Second Law of refraction.

From Chapter IV it will be recalled that ALT is a composite measure of the four components of allocated time, engaged time, relevance, and high success, and that it is an attempt to gauge the success of teaching without having to wait for the later administration of achievement tests. But it is immediately seen that of these four components only the first is precise and measurable scientifically, with the following three components depending upon the personal judgments of observers. Even if there is a high inter-observer agreement, say 80%, that would still leave gross error in application. In contrast, with, "rectilinear propagation", "frictionless plane", "hydrogen bond", or "transform fault", there is no such ambiguity, or error.

Moreover, ALT attempts to be a measure of learning without paying attention to the effects induced by changes in content. As Confrey writes, "I am suggesting that learning is influenced profoundly by the subject matter which is to be learned and that, unless the nature of the subject matter is considered, our progress in understanding learning will be minimal" (1981, pp.88-89).

Confrey suggests that a proper attention to content involves, ". . the consideration of the structure and purposes of an academic discipline and how topics are selected and presented." It allows teachers,". .to share our intellectual heritage." It, ". . promotes students' introspection. They become aware of themselves and their roles in the world" (1981, p.89). It may be added that a particular subject matter such as physics, or economics, brings with it a specific pattern of ideas, and has concepts which are related in a logical order. Even a subject such as American history has a loose structure for one cannot understand much about how the United States of America came to be as it is today, without knowing something of American ideas of "starting from scratch", of the war between Britain and the Colonies, of the significance of the Constitution, or of the "Age of Jackson". To ignore the special nature of the content and its peculiar structure is to ignore the most significant thing about the learning. Confrey points to other problems of ALT:

> . . it draws one's attention away from the quality of learning to the quantity of time spent learning. There is semantic ambiguity in discussions about learning which makes it deceptively easy to consider learning as a quantity and hence as a simple function of time. For example, one talks about "learning more", or "increasing learning". Learning however, is also referred to as a quality when one talks about something being "easy", or "difficult" to learn. The Beginning Teacher Evaluation Study encourages one to think of learning primarily as a quantity. . . A simple cumulative measure of engaged time deters consideration of such issues. ALT ignores the subtleties of learning, reducing it to a single uniform porridge (1981, p.90).

Confrey also points out that the ALT studies were performed on reading and mathematics in a debatable manner, which construed them primarily as skills, rather than as conceptualizations. What is more, ALT is accumulated in relation to items to be tested on standardized achievement tests. Yet achievement tests are notorious for testing information, and (some kinds of) skills, rather than conceptualization, understanding, or critical awareness. Furthermore ALT ignores that immensely significant qualitative learning which good teaching of a subject makes possible: the special excellences of honesty, perseverance, finesse, creativeness in relation to that subject. Again in such subjects as social studies, ". . . the content to be taught (e.g. an understanding of social interactions) and the way in which the class is conducted will influence each other profoundly" (Confrey, 1981, p.91), and it might be added that such can equally be the case with the teaching of science.

In short, instead introducing scientific precision, the concept of ALT remains an empiricist term which is sytematically ambiguous. So ALT is certainly a new *term*, but no abstract concept falls under it. It is merely a banal, jargonistic, *re-description* of things human beings have known for a million years, namely that to maximize learning, a person must *concentrate* on *the task* for a sufficient *amount of time*. The single addition to this million year-old understanding is that of "high success". And even that is a matter for debate. Whose criteria are to be used? Is it not conceivable that, depending on what is being learned, a continual sequence of "medium success" might be equivalent, or even more useful in the long term? Moreover, with the learning of matters such as propensities, attitudes, and values the very notion of ALT may be inappropriate.

Nobody should be surprised to learn that proponents of ALT found that teachers were often reluctant to impose practices which beliefs about ALT seemed to imply. Fenstermacher reports that:

> On occasion the reluctance became outright resistance. The diaries kept by the various members of the research team. . provide ample indications of frustration, conflict, and misunderstanding . . What emerges clearly is the ambiguity of the role, uncertainty over whether the researchers were the bearers of rules or the presenters of evidence. There is an undercurrent of mixed metaphor: the researcher as therapist on the one hand (presumably concerned about the subjective beliefs of the teachers), and the researcher as management consultant on the other (presumably prepared to stipulate the rules) (1979, pp.171-172).

Suffice it to say that all of Gage's examples are very different from the theory-driven hypotheses and the abstract concepts which have made possible the long, successful history of physics, chemistry, biology and geology. So once again, like the claim to "scientifically developed knowledge" above, the claims of Gage and Needles about formative hypotheses, elementisms and conceptual constructs are hortatory and status-seeking.

Gage's apologia are unconvincing. What he has shown is not that his Process-Product work is scientific, but rather that in using careful observation and sophisticated statistics, it, like all present-day empiricist research is sophisticatedly empiricist in a statistical way. But as will be suggested in the last two chapters, there are additional and perhaps better ways of being sophisticatedly empiricist. The nature of such empiricist concepts will be further examined in the following section.

Abstract Concepts cf. General Concepts

The careful observation, the use of experimental and control groups, and the sophisticated statistics of the various studies considered in Chapter IV show that such empiricist research is both subtle and systematic.

But the concepts of such research are no more subtle or sophisticated than those of any other empiricist endeavor. As the earlier discussion of Empiricist Theory and Scientific Theory emphasized, there are profound differences between abstract concepts and general ones.

The concepts, of empiricist research on teaching, like those in all empiricist endeavors are general-- they are drawn from the individual cases of observable empirical phenomena, or as Manicas and Secord say, are concepts, ". .that are more or less readily available to direct empirical confirmation" (1983, p.403). Thus, as pointed out in Chapter V, although the same word(s), say, "discipline question" may be appended to the behavior of the different teachers observed in a particular study, the entity itself varies from teacher to teacher in a way that, say, "volt" or "tectonic plate", which are themselves embedded in Scientific Theory, do not. Moreover such research deals with concepts which require subjective judgment in order for a researcher to be able to claim that a particular behavior is an instance of that concept. Unlike the case of "specific heat of fusion" or "five grams of mercury" there is agreement between observers only up to a point. In being generalized empiricist concepts they are only superficially like the abstract concepts of scientific research, such as "ellipse", "mass", "velocity", "mercury", "transform fault", and so on. This essential difference has escaped most empiricist resarchers. Thus Kerlinger is unaware of the questions he is begging, when he writes, "A construct is a concept with the additional meaning [sic] of having been created or appropriated for special scientific purposes. Mass, energy, hostility, introversion and achievement are constructs. They might more accurately be called constructed types or constructed classes" (1986, p.4). But the precision and usability of "mass" and "energy" are radically different from the others in Kerlinger's list. Examination of the studies considered in Chapter IV reveals such empiricist general concepts as, "rules", "formal style", "mixed style", "reacting group", "frequent praise", "frequent opportunities", "interesting assignments", "actively involved", "high praise rates", "brisk pace", "monitoring", "verbal move", "episode", "focus on content", "random response", "rate of response", "observer presence", "process question", "adjusted achievement score", "academic needs", "clear feedback", "task relevant feedback", "non-evaluative learning environment", and so on.

The meaning of such concepts is *either acquired from ordinary language and everyday experience of the Empiricist Theory of social life, or these common meanings of ordinary language are given a specific jargonistic or statistical twist*, so that they have the appearance of being abstract concepts, as in for instance, "verbal move", "episode" and "adjusted achievement score"; or, *an abstract concept from natural science or technology is being used metaphorically*, as in "clear feedback". As is not the case with the abstract concepts of scientific research, once the jargonistic or statistical twist has been exposed, understanding the term poses no problem. Moreover, the term does little or no work that is not already done or could not be done by ordinary language equivalents. Little or nothing new is being contributed.

Whereas the meaning of "force", or "compound", or "cytosine", or "tectonic plate", is learned only through mastering the Theory (Cluster 9) of the particular science, to grasp the idiosyncratic usage of such empiricist concepts does not require the mastery of a body of Theory in which they are embedded. They are not *ontologically* separated from the world in the way in which those already listed, together with such common scientific concepts as "acceleration", "mass", "oxygen", "electron", "light ray", "specific heat", "relative density", and fifty thousand others are separated from the world, or the way in which the equations and laws which relate these abstract concepts in mathematical terms are separated. *At best* they are empiricist shorthand, rather than abstract concepts; *at worst* they unnecessarily complicate matters, when ordinary language would do the job more simply, more clearly, and with more cogency, power and precision[3].

Moreover, lacking precise meanings the concepts used in the same piece of empiricist research commonly overlap in meaning. In such cases, not only can they not be related by the mathematics used in scientific research, but it follows

---

3. See the masterly discussion of its potency in Austin, 1962, 1970.

that the results of all such statistical manipulation are to some extent equivocal.

To further illustrate these points, I shall consider examples from several of the studies described in Chapter IV. Some of the more usable (though still rather commonsensical) results of such research are listed by Gage in, *The Scientific Basis of the Art of Teaching*. It will be recalled that he lists seven "teacher should" findings. These findings, Gage claims, are, ". . the joint yield, the convergence, the common general finding of years of careful classroom observation, reliable and valid measurement, and sophisticated statistical work" (p.40). If years of very careful work have resulted in such banal advice, we may with justification wonder about the aptness of the methods used. But the present point is that the limitations of such research may well derive from the fact that the research uses general empiricist concepts such as,"pupils", "rules", "academic needs" "monitoring", "seatwork", "awareness", "independently", "easy enough", and so on. Thus the concept, "pupils", refers to John and Jane and Tom and Tammy considered as a whole despite their manifest differences and individuality. The concept "communicating" refers to facial expressions, to bodily movements, to complex sentences, to simple sentences, to partly-finished sentences, to leading questions, to exclamations, and so on. Such generalized concepts are very different from scientific concepts.

The Bennett study, *Teaching Styles and Pupil Progress* fares no better. It will be remembered that Bennett attempts to compare three teaching "styles: "formal", "mixed" and "informal". Once again these are empiricist concepts. For he has divided the myriad of relevant features such as teaching practices and tasks, the values, habits, and so on, into merely three observationally-based "styles".

Given the imprecision of such concepts, any research which attempts to clearly distinguish the comparative efficacy of these three "styles" of teaching is *necessarily* going to have problematic results. As Barrow puts it, "If one operates with very broad categories, such that many facets of performance are contained within each one. . . then it is almost certain that there will in practice be no clear instances of any category" (1984, p.192). The problem is that with such general categories, numerous of the above tasks, values and so forth will fit in each of the three "styles". But there may be thousands of different features and dimensions involved, many of which could be particularly significant, but will find equal representation in each of the three empiricist categories. This results in the antithesis of the precision of theoretically-embedded, abstract concepts.

To make the point, let us consider just ten features. The similar use of sessions of critical questioning by teachers can fall into any of the three styles. So can their use of praise. So can an unusual ability to work an overhead projector. So can a propensity to mark all written work immediately. So can a teacher's moral belief in the abiding worth of each individual pupil. So can a teacher's charismatic personality. So can his patience with children. So can a pleasantly modulated voice. So can depth of knowledge of his subject, history, or mathematics or whatever. So can his ability to make things clear to learners. All of these features and perhaps thousands of others may have significance in any particular teacher's success. Thus we have here the necessary imprecision of immense conceptual-overlap between the three general concepts. Any supposed causal relationships derived from such categories will be about as useful as (to adapt Barrow's example) would be such relationships derived from dividing teachers into three categories of hair color: those with red, blonde and black hair. So why practise *sophisticated statistical analysis* with these three categories as a basis ? Even were the "styles" able to be refined to say ten or a dozen, the situation would scarcely improve because there are just so many dimensions involved.

Thus the very concepts which Bennett seeks to relate are themselves ambiguous and vague empiricist ones. As Bennett himself agreed more recently,

> The major problem with this approach was that the styles themselves were composed of groups or bundles of teacher behaviours and activities, making it impossible to ascertain the impact of any individual teacher behaviour on pupil outcomes. . .as such, the notion of teaching styles, in itself, cannot provide an adequate explanation of differences in pupil outcomes (1988, p.21).

Much the same problems face the experimental studies of the faculty of the University of Canterbury, in which the participation by pupils and the reactions of the teachers to the responses of pupils were experimentally manipulated. Once again we have general empiricist concepts such as "random response", "systematic response", "praise", "verbal moves", "open questions", "prompting", "content coverage", and so on. There is both imprecision and conceptual overlap. For example, a "modified question" overlaps with the notion of a "repeated question"; a "modified question" is not conceptually distinct from an "extending question"; "structuring" and "orienting" are not conceptually distinct from "prompting"; and so on. Moreover, as argued above, there is no difficulty in learning or in applying the idiosyncratic meanings of concepts such as "verbal move" and "episode", once they have been quickly explained. But more important: such concepts do no useful work. Unlike the use of "oxygen" or "light ray" or "tectonic plate" which respectively have helped mankind to understand chemical combination, the phenomena of light, and earth movements, "verbal move" and "episode" may even obfuscate rather than clarify.

There is a further very important difference. One of the fruitful approaches of science has been to break down the totality of the empirical world into small, isolated aspects, or metaphorically one might say "atoms". Such atomistic aspects are then researched into, while the rest of the irrelevant empirical world is ignored or kept as constant as possible.

This is a feature of Kepler, Galileo, Lavoisier, and Crick and Watson, and the plate-tectonic theorists, and a feature of scientific research generally. For instance, Lavoisier took meticulous care to isolate his combustion experiments from the surrounding empirical world, so that he was dealing only with oxygen and mercury and the heating of these. *But what must be emphatically stated is that such "atomism" occurs within the confines of theory* (several Clusters, but in particular, Scientific Theory). It is precisely such theory which makes both the "atomism" and the search, meaningful and achievable.

As has been shown in Chapters IV and V, as part of their attempts to model their approach on that of natural science, empiricist researchers have also broken down teaching into a system of supposedly conceptually separable "atomistic" factors or variables, which they have then dealt with as though these were distinct scientific entities.

Atomistic isolation in scientific research has proved immensely fruitful. But the belief that something essentially similar can be done in research on teaching, or any other variety of empiricist research is naive. The view that we can isolate a teacher's questioning or a teacher's praise from the rest of that which is occurring in the classroom at that time, and try to relate these in various ways to other variables, would seem at best to be questionable. The reason for this doubt has to do with the distortions and changes which occur to such factors or variables when they are thus atomized in the observation of teaching.

When we are dealing with such matters as praise, the situation requires exceedingly subtle conceptual analysis prior to any empirical research. Had that occurred, researchers might have realized that their research was problematic. An action such as the praise given by a teacher is embedded in a human context, and in being so it is inseparably involved in a myriad of meaningful human relationships, from which it cannot be separated without destroying the special nature of the situation. Practical social actions and activities such as those of teaching are contextually *holistic* and cannot be broken into atomistic parts for consideration as atomistic without destroying the special nature, even the very existence of those actions and activities. Harré writes,

> A smile, for example, can mean many different things depending on all the other actions which precede and accompany it. . .So there could never be experiments on the effect of smiling, in which the smile was taken as an independent variable, its effects on others as the dependent variable and the situations in which smiling occurred fixed as parameters (1988,p.16).

In empiricist research this is precisely what has been done with human actions logically equivalent to smiles. For example, in Process-Product research on teaching, praise has been isolated from the infinite other dimensions of teaching and attempts have been made to show empirical regularities in its effects on the achievement of pupils.

But this is simplistic, because praise always occurs in a *context, in a complex network of meanings* , for both teacher and taught. Praise *to be praise* depends upon the context allowing it to be so. To mention but one dimension of this totality: praise is *experienced* by the pupil *as* for instance, something fitting or appropriate, or unfitting or inappropriate. Moreover, unmerited praise can, for example, breed a variety of problems, from casualness in the recipient, to cynicism in the child who sees its being given to his undeserving classmate.

Praise can be exaggerated, superficial, calculated, repetitious, habitual, emphatic, sarcastic, ironic, and so on through perhaps an infinite number of describable situations. It can also be given by teachers because they know that a supervisor or a researcher interested in praise happens to be in the classroom. McNeill tells of the fatuous situation where a gifted high school teacher of Advanced Placement courses praised the students consistently (as required by the regulations) throughout the lesson which was being observed by the supervisor. After the lesson, her students, "..who were accustomed to the teacher's probing questions and impatience with gilbness, asked, 'What were you *doing* in there today? Are you feeling all right?' " (McNeill,1988, p.485). My friend, who teaches chemistry in a New York City high school says that after a while the students with a knowing look, sarcastically or jokingly reverse the situation, and say, "Now wasn't that a *good* answer I gave Mr X?"[4].

The above features make it exceedingly difficult to deal with praise as an isolable, unitary, particular entity. For this reason, it may not be possible sensibly to study praise as a dimension of teaching, in an atomistic, empiricist manner. "To attempt analytically to do away with [the] background and treat human acts as though they are object-like entities is a methodological error, because it [is] to remove the phenomenon being studied" (Packer, 1985, p.1087). Another way of putting this issue is to suggest that in empiricist research, unlike in scientific research, *that which is constitutive of is sometimes confused with that which causes* .

This does not mean that a teacher or a supervisor cannot understand when and how to use praise in the classroom. Certainly teachers can understand the phenomenon of and place of praise in teaching *in a holistic way* as part of our general intelligent understanding of human relationships, derived from our experiences of the subtle human societies in which we have grown up. In a general way we can understand why it is important in a classroom and how teachers should deploy it, because we understand its *meaning* [5]. And if we want to know how best to make it an effective tool in our teaching, then we must get to know as much as possible about the nature of human beings as human beings. This means going to the fertile studies of the novel, drama, history, anthropology and ethnography, and perhaps to humane psychology, not to the too simple studies of too much empiricist research, as it is presently constituted.

One irony of such atomism is that the researcher has to use his general intelligent understanding of human relationships in order to identify instances of praise. There are two levels to this intelligent understanding. Such research is performed with the use of "low-inference" observation categories agreed upon prior to the observation of the teaching. It is claimed that the "low-inference" is required so as to to remove as much as possible inferences which require the exercise of intelligent judgment by the observer, for this is construed as subjective and non-scientific. But first, we should carefully note that this intitial categorizing of what does and what does not count as an instance of say, praise, is necessarily the result of intelligent human judgment. Second, the observers, no matter how well they are trained to construe the empirical world in identical

---

4. Personal communication, April 1990.
5. See the discussion in Chapter VIII.

ways[6], must still use their judgment in deciding what does and what does not count as an instance, in the teaching they are observing. To do this they must make use of the background and context in which the action of praising is embedded. This is the case whether or not the observation of actions of praise are supposed to be of the "low-inference" kind. Outside a particular context it is impossible to know whether the utterance of a word such as, "good" is functioning as praise.

In other words, researchers can develop their presumed objective empirical knowledge about praise in the classroom, or draw up their scales of low-inference "observables" which are to count as praise, only by making use of at least a minimum of contextual understanding. But at the same time, the researchers are trying to isolate praise from its whole context in order to manipulate it. They are construing the very features which give meaning as, ". . secondary and unreliable or irrelevant to the goals of scientific enquiry" (Packer,1985, p.1087). But as is the case with all such general concepts, the more the entity is removed from the context, the less sure can we be that it remains the same entity.

That these claims about context are accurate can be shown by an analysis of some of the actual findings from the Process-Product empiricist research in this area. Following is a list of some of the findings, called "Guidelines for effective praise", as stated in Brophy and Good. Effective praise,

    1. Is delivered *contingently*
    2. Specifies the *particulars of the accomplishment*
    3. Shows spontaneity, variety, and other *signs of credibility* [sic]; suggests clear *attention to the student's accomplishment*
    6. Orients students towards better *appreciation of their own task-related behavior* and thinking about problem solving
    7. Uses students' own *prior accomplishments as the context for describing present* accomplishments
    8. Is given in *recognition of noteworthy effort* or success at difficult (for *this* student) tasks
    11.Focuses students' attention on their own *task-relevant behavior.*
    12.Fosters appreciation of, and desirable attributions about, *task-relevant behavior* after the process is completed (italics added) (1986, p.368).

The features which have been italicized should be noted. They would appear to emphasize what has been described above as *context*. They are the kind of context which makes it possible for the words the teacher utters *to function as praise* . The following are some of the features which they list as characteristic of ineffective praise:

    1. Is delivered *randomly* or unsystematically

---

6.   It may be that such training is itself problematic because of the Ontological Presuppositional Theory (Cluster 6), which it assumes.

3. Shows a bland uniformity that suggests a *conditioned response* made *with minimal attention*
4. Rewards mere participation, *without consideration of performance processes* or outcomes
6. Orients students towards comparing themselves with others and thinking about competing
8. Is given *without regard to the effort expended or the meaning* of the accomplishment
9. Attributes success to ability alone or *to external factors such as luck* or low task difficulty
11. Focuses students' attention on the teacher as an external *authority figure who is manipulating* them
12. Intrudes into the ongoing process, distracting attention from task-relevant behavior (Brophy and Good, 1986, p.369).

What should be noticed about the italicized features is that whenever they are implemented, they help to make any such action *something other than* praise. In other words *these findings are mainly conceptual truths, rather than results of true empirical study*. This point is a good way of demonstrating how a general empiricist concept such as "praise" is embedded in its context, for the researchers *are here teasing out meaning rather than discovering new facts about the world*. Thus it would seem to follow that whereas abstract concepts such as "surface tension" or "momentum" take their meaning from being embedded in Scientific Theory and mean precisely the same thing from one case to another, general concepts such as "praise" or "open question" take their meaning from their social context and their applicability varies from one case to another.

To criticize the general concepts of empiricist research in this way is however not to reject them entirely. It is rather to show how they are unsuitable for use in Scientific Theory, or scientific research.

There may be a place for some of these concepts in pedagogical discussion, and in a Practical Theory of Pedagogy (Cluster 4), for they may describe *very general emphases,* which teachers and educationists may wish to distinguish. Different, loosely-demarcated, general approaches, or "styles" may be more suitable for different personalities and varying contexts. So it may be useful for us to have very general categories with which to mark these differences in emphasis. Equally, it is useful for teachers to think about the different kinds of pedagogical question there may be and their broadly different functions. In other words, *they may be doing a useful empiricist job.* Thus what has been thought by its protagonists to be developing Scientific Theory of pedagogy has been a development within Empiricist Theory (Cluster 8), and/or Practical Theory (Cluster 4). Or as Willer says above, "...this information may have immediate practical or technical value". The gross mistake is made when researchers manipulate such categories as though they are scientific concepts and try to find out in an absolute way which of these styles, or questions, is generally more effective. And this point should be generalized (empiricistically) to all empiricist studies.

Thus the useful concepts of science such as "orbit", "force", "acceleration", "compound", "thymine", and "tectonic plate" are abstract and embedded in Scientific Theory and *derive their meaning from that theory*, whereas say, "episode" and "praise" are general concepts of Empiricist Theory, which derive their meaning from *specific uses of the always somewhat ambiguous and overlapping meanings of commonsense language used in particular contexts* And empiricist researchers delude themselves if they believe that by providing operational definitions for such concepts and training observers in these uses, they can obviate this situation. They delude themselves because such terms still have to be *applied* in a complex and evolving context of ordinary-language meanings.

When do We require Such Research Anyway?

As the discussion of praise has implied, the *empiricist approach has sometimes been used to research into matters which do not require that kind of research*. Practitioners of empiricist research should be very careful to ensure that the matters upon which they are carrying out their exceedingly expensive labors do indeed require such research. Although this is a matter of fundamental significance for research, it has been ignored by empiricist researchers. Yet there exists a sizable literature which shows just what is at issue here, and why it is simple-minded to perform empirical research (empiricist or scientific) on conceptual and commonsense issues. See for example, Wilson, 1972, 1975, 1988; Popp, 1980; Barrow, 1984, 1986, 1988; Scriven, 1988; Egan, 1988; Burgess, 1990.

For instance, few or any findings of the University of Canterbury studies needed to be established *empirically*. Consider for instance the recommendations that questions should be appropriate in level of difficulty so that students can understand them, or that teachers' reactions which communicate enthusiasm are more motivating than matter-of-fact reactions, or that reviews which summarize segments of lessons are helpful. Very similar ideas could be arrived at (and have been arrived at) by conceptual, logical and commonsense analysis. Or consider the finding of Helmke and Schrader that teaching should give, ". . support to single students discretely, without distracting other students." The piece about distracting is pedagogically obvious. Why (unless it were part of some elaborate pedagogical plot) would any teacher deliberately distract students? But the first part of their suggestion is more complex. Such advice may not be sound across contexts. It is at least conceivable that some kinds of students could react more favorably to *public* comments, e.g. comments which are of general help to the class. Again, where there is strong rivalry within a class, other students may be stimulated to work harder when they hear public advice to another student. Such matters are complex conglomerates of normative and empirical issues: matters which depend upon the context and the content being taught. They require the teacher's educated judgment, rather than empiricist research.

As another example, consider the study by Finn and Achilles, on the size of classes. It is germane to ask, What worthwhile knowledge is it possible to establish *empirically* by research into class size, that we do not already *know* by commonsense and philosophical analysis, and through the actual classroom experience of teachers? Such knowledge is similar to that used in scientific thought experiments (see Chapter VII), which itself *provides a foundation* for further extrapolation and empirical experimentation. I once heard a famous psychologist at an international education conference confidently claim that there was no empirical evidence to show that children learn better in small classes than in large. He thought that such evidence concluded the matter. Yet classroom teachers know that if this is indeed the finding of research, then such research *must* be faulty. I *know* from my own teaching in both elementary and high schools, and in universities, that smaller classes do make a positive difference.

But the fundamental question to ask in this context is, Do we need any kind of large scale empiricist research to answer such questions? I believe that we do not need such research, and that we have here merely a matter for commonsense empiricist experience, and philosophical and commonsense analysis. (Finn and Achilles believe we do need empiricist research, writing, "Only randomized experiments can provide a definitive answer to these questions" (1990, p.558).) But fundamentally at fault are the myopic notions of Finn and Achilles and the famous psychologist, of what can count as evidence, viz. that produced by the kind of narrow observational-statistical, or experimental-statistical research continually criticized in this book. And that kind of evidence is not needed in the present case[7], because in small classes the humanness and humaneness of contact and reciprocity between the teacher and the students can take their rightful place— with all that that means for improvement of learning. If empirical research (scientific or empiricist), however sophisticated, contradicts these truths, its findings *must* be in error.

If, however, the teaching is occurring in a situation where humanness and personal relationships can only be minimally brought into play, e.g. in a lecture to large numbers, then large scale research which shows that numbers do not make any difference, may be correct. But once again, we hardly require empirical research (scientific or empiricist) to tell us this. For we already understand why this is so. It is so because the personal contact and reciprocity are removed: there is no room for sharing individual human concerns in a reciprocal way, or for making adjustments for individuals. In such a situation, the size of the class may as well be 1,000 or 1,000,000— as long as everyone can hear and see the teacher. It is interesting that the research by Finn and Achilles contradicts the claim by the famous psychologist, as they write, ". . the benefit of reduced class size is seen clearly for every measure." And at the level of the small numbers of the children in the smaller classes in their study— an average reduction of from twen-

---

7. Moreover, as will be argued in some detail in the last two chapters, such research filters out or ignores the nature of the participants, and their situation. In so doing, it invalidates its techniques. As soon as we include consideration of the features discussed in Chapter VIII— intention, context, the nature of the content being taught, and the nature of the learners— features fundamental to any really adequate empiricist research, we can see why the more narrow approach to research cannot be depended upon.

ty-two to fifteen children, i.e. about 32%, we can see that if their research had found the opposite, or no significant effect, then it necessarily would have been in error. So we return to the question asked above, Just what worthwhile findings can be established *empirically* by research into class sizes? While such research remains in its present narrow empiricist state, the answer would seem to be: for a Practical Theory of Pedagogy, little or nothing of value that we do not already *know*. In other words, we did not require the exceedingly expensive, empiricist research of Finn and Achilles.

Conclusion

This chapter has argued for the following points of view. Empiricist pedagogical research *cannot be theory-free*, because the fundamental epistemological situation is such that theory from various Clusters necessarily influences and orients research. The kinds of apologia offered for empiricist research, as typified by those of Gage, are conceptually confused and need to be rejected. There is *a profound difference between the general concepts of empiricist research on teaching and the abstract concepts of scientific research* and the failure to draw this distinction has contributed to unacceptable atomistic analyses of situations, which actually require understanding in a holistic way. At least some empiricist research has been carried out in an attempt to answer questions which actually required philosophical and/or commonsense analysis; perhaps the point should be put more strongly— *some expensive empiricist research has been carried out in an attempt to answer questions which required no sort of empirical approach, so that the research in such cases has been otiose.*

The scene has now been set for the point by point comparison of the following chapter.

CHAPTER VII

THE MANY CONTRASTS BETWEEN SCIENTIFIC RESEARCH AND EMPIRICIST RESEARCH ON TEACHNG

In the previous chapters a strenuous attempt has been made to show the typical features of present-day empiricist research on teaching and to say something about its ancestry. Occasionally in the account there has been some criticism, e.g. there have been some intimations of problems or allusions to the periodic repetition of ideas, or the absence of theory. Moreover, the reader may have noticed that this quasi-historical approach has at the same time demonstrated that there has been for something like one hundred years, a repetition of assumptions, aspirations, and relatively similar procedures– that modern empiricist research differs from its predecessors mainly in its standardization of empiricist methods and the concomitant increasing sophistication of its statistical procedures.

As examples of repetition may be cited: the continual desire to be and often the explicit claim to be scientific and the relative consistency in the attempts and approaches to the achievement of this aim, from the nineteenth century arguments of Bain, and the studies of Rice, through the work of Thorndike and Ayres in the early decades of the twentieth century, to that of Barr in 1929, to post-World War II efforts in Performance-Based Teacher Education, Academic Learning Time and Syntheses of Research on Teaching ; Rice's intimations of ideas about, and Bobbitt's hopes in 1913 and in the early 1920s of finding the "one best method" of teaching, compared with the broadly similar aims of Gage's research in the 1970s and 1980s, none of which has produced it; the repetition of basically the same conclusions decades apart, e.g. Rice's 1897, "nothing can take the place of that personal power which distinguishes the successful teacher from the unsuccessful teacher," to Barr's 1929 finding that he had not identified any factor in teaching as critical, to which we may add Barrow's 1984 view after having carefully considered the results, that, ". . the research . . is not particularly impressive and yields few clear conclusions"– yet the continual hopeful attempts to discover empirical results which will disprove such conclusions ; the continual allusions which stretch from Thorndike and Bobbitt to the 1980s refinements, that measurement and statistics may not be doing all that is possible, so that what is required is continually improving ways of measuring and of using statistical methods; in short, it may be argued with some justification that the history of empiricist research is one of the continual production of old wine in new bottles.

It may also be worth pointing out that throughout this period of some one hundred years, the empiricist innovations and aspirations have not gone unchallenged[1]. But the point of this chapter is to make explicit much which has so far been largely implicit, namely, the additional wide variety of ways in which empiricist research differs from scientific research.

It is important to understand the logic of this chapter. There may be other significant aspects not here considered, and additional ways of construing what is going on in scientific research. The present claim is merely that the following criteria are in most cases widely acknowledged as significant in the literature of the history and philosophy of science, that they can be seen to be applicable to the procedures and key features of the accounts of Copernicus, Newton, and so on, in Chapter II, and that empiricist research as presently pursued, despite its claim to scientific status, singularly fails to fit most of them.

There are three points which need to be made prior to the detailed examination of differences. The first is that in discussions of research on teaching (and on education generally), there is a disconcerting tendency to slide from the word, "disciplined", to the word, "scientific". But although all scientific research is disciplined, not all disciplined research is scientific: these are far from being synonyms. Precisely this confusion occurs in the following two quotations from Shulman, an otherwise careful scholar, in the authoritative recent publication on educational research of the American Educational Research Association. He writes: "What is important about *disciplined* inquiry is that its data, arguments and reasoning be capable of withstanding careful scrutiny by another member of the *scientific* community (italics added)" (1988, p.5). Two paragraphs later he continues, "First, *scientific inquiries* cannot involve mere recitation of the 'facts of the case'. Indeed, inquiry demands the selection of a particular set of observations or facts from among the nearly infinite universe of conceivable observations... so in *disciplined inquiry* in education there is often lack of concensus about the grounds, the starting points.. (italics added)" (1988, p.5). *Empiricist pedgogical research is disciplined inquiry, but it is not scientific inquiry–* for at least the reasons listed below.

The second point is the common (though usually implicit) confusion in pedagogical discussion and writing between empiricist knowledge and scientific knowledge. The same confusion with which this book is concerned– the confusion between scientific research and empiricist pedagogical research– is a large-scale example of this. *But it is common for there to be equal confusion in the discussion of education and science, and in that of science teaching.* In the following four examples from writing over the last quarter century, the confusions can be seen quite explicitly. They are also typical.

"Science is precisely the process of using one's previous experience as a basis for prediction... Our own experience sensibly used in our everyday life can be

---

1. For examples from early in the century, see Callahan (1982, pp.121-124).

said to be identical with science and scientific method" (George,1967, p.46). Grotesquely untrue. Everyday experience is indeed *everyday* experience, and results in empiricist knowledge. What may result from everyday experience is either (a) commonsense understanding of trial and error generalizations, which work more or less, or (b) questions which puzzle us enough to stimulate some scientific endeavor, i.e. questions which may eventually *lead to* some scientific research. Before science comes into existence, there has to be, as already mentioned, a "rape of the senses", or a "breaking with everyday experience".

The second example is from a 1983 article which claims to tell us how to improve educational research, in which we find: "As developmental and cognitive researchers continue to provide us with insights into how children acquire knowledge, we will simultaneously be provided with insights into how scientists acquire scientific knowledge" (Galfo p. 7). Not so. Knowledge acquired by children is acquired in a childlike way, without the aid of developed theory. In contrast, scientists acquire knowledge as they further develop the Scientific Theory, which is itself already underpinning their work.

The third example is drawn from a 1989 book which explores areas of common interest between the teaching of and the study of science. Its editor writes,

> The key to this [understanding how children learn science] is to perceive science, not as an algorithmic method for "finding out", but as akin to the everyday processes of gaining commonsense knowledge. The way we learn science parallels the way we learn about common things. . . In everyday life we reach agreement with others about events and processes. We develop shared concepts and theories which allow us to talk and discuss situations with one another. We have varying degrees of confidence in how things will behave, ranging from those contexts about which we feel very certain to those where our knowledge is qualified with terms like "maybe" and "probably" (Millar,1989, p.4).

Mostly untrue. The writer's ideas about reaching agreement about events and processes and about varying degrees of confidence apply more to the learning of empiricist knowledge, than to the acquisition of scientific knowledge. They apply to acquiring commonsense knowledge about our relations with others, to everyday assumptions about the world, to our knowledge of and confidence in statistical surveys of opinion, not to our acquisition of knowledge of chemical reactions, of the behavior of DNA, or of the earth's crustal movements, and how such things interrelate in broad realms of ever more subtle meaning. What this writer says applies scarcely at all to students as they learn science. It applies to some extent to the scientific researcher on the fringes of his discipline. (If the acquisition of empiricist knowledge were the acquisition of scientific knowledge, then more of what he writes would be correct.)

The fourth example, from a commercially-successful 1984 manual of educational research says that, "Scientific inquiry is the search for knowledge by using recognized procedures in data collection, analysis and interpretation. The scientific method is usually a sequential research process" (1984, p.7). And again, "Scientific inquiry also implies an attitude towards solving problems. It is a willingness to question, search out the explanations for results, and examine methodology. As an attitude, *scientific* means a willingness to accept new knowledge after critical examination and to modify one's beliefs and ideas in [the] light of this knowledge" (McMillan and Schumacher, 1984, p.7). Hardly. The account is no more representative of science than it is of, say, history or philosophy. While it may describe some necessary features of scientific activity, it certainly fails to describe anything sufficient. It emphasizes merely a rational and disciplined orientation to the solving of problems and thus relates back to my first point above, about the slide between the words, "disciplined" and "scientific". Given the context of their book, it is ironic that the authors say that scientific inquiry involves, ". .a willingness [to]. .examine methodology," for although that is true of scientific inquiry, their book is actually describing empiricist methods of research; and that is precisely what so many empiricist researchers are unwilling to do, when they take their statistical methods as a given, at the same time that they claim to be doing scientific research.

All four claims, though typical of pedagogical discussion about science are either incorrect, or require strong qualification.

Surely the significance of the methods of scientific research is that they are not the same as everyday and empiricist methods. Surely the significance of scientific thinking is that it is not the same as everyday and empiricist thinking. As Phillips so perceptively said over twenty years ago, "There is little point in claiming that scientists use scientific method if everybody uses it" (1971, p.31).

The third point is that although empiricist research on teaching is not scientific research, it still suffers in some of its features from the same sorts of complexities and difficulties as scientific research does. Thus when I am pointing to the scientific complexities in the examples below and then commenting on some similarities in pedagogical research, I am not thereby suggesting that empiricist research is the same as scientific research. It fails to be scientific; but it also suffers from some of the problems and features which make scientific research so complex; and its practitioners are unaware of both of these difficulties.

Now let me make explicit just some of the wide variety of ways in which empiricist research on teaching differs from scientific research.

The Inductivist Orientation of Empiricist Research

It will be recalled that Nisbett believes that, "Most empirical research studies are . . 'miniscule analyses'. This is hardly surprising: all science proceeds by 'miniscule analysis', and the building of a coherent theory is a very slow process." And the view of scientific research espoused by the research studies described in Chapter IV saw scientific research as an accumulation of claimedly objective, empirical-statistical evidence, built up through observational, experimental and meta-analytical research, which would generate new hypotheses and experiments, until the whole field is covered and a theory of teaching somehow slowly evolves. This inductivist-observational view is unfortunately still pervasive: a recent book on science education, whose author wishes to emphasize the importance of scientific theories, at one place still writes, "All scientific investigations begin with the collection of data. It is at this level that science first establishes a concrete connection with the natural world" (Duschl, 1990, p.49). There are many problems with such ideas.

The first is that although, as Kuhn might argue, there may be a sort of natural history and loosely inductive stage prior to the development of a science proper, scientific research does not begin until people move beyond such limitations and begin making the conceptual moves demonstrated in Chapter II. All civilizations, no matter how primitive, do a sort of inductive natural history. In (what was to become) astronomy people noticed certain regular movements in the sky, in (what was to become) physics they noticed that objects fell to the ground in some kind of regular way, in (what was to become) chemistry they saw that particular dyes could be made from particular berries, in (what was to become) biology they noticed that some techniques for breeding were superior to others, in (what was to become) geology they noticed regularities in strata. *But that gave them empiricist, not scientific knowledge.* And the significance of such a stage is overestimated, even though it *may* lead to a scientific research. I say, "may", because although these kinds of regularities were observed everywhere, by African tribesmen and Mongolian herdsmen, it was only in Europe, in particular in the last four hundred years, that such observations led to scientific research.

As Popper points out (1969) if such observations really were important scientifically, then the only thing stopping science from developing would be some variety of misunderstanding in the observations. What scientists would need to do would be to observe ever more carefully, to get the observational facts, and to make sure that they avoided their own distorting prejudices (as Bacon suggested in the sixteenth century).

Only in a trivial way can it be true that scientific research begins with such observation and generalization (with empiricist inquiry), e.g. heavy objects like pieces of wood are seen to fall to the ground and fires are "seen" to ascend to the skies, and the observer seeks answers to why these things occur. This is trivial because that is also the way in which *all* human knowledge-seeking endeavors begin: particular things are observed (or believed) to occur, or particular actions are seen to take place, and questions are posed and regularities are assumed. But this is no more the beginning of science than the beginning of religion, or of art, or of morality.

Second, the significant scientific developments are not miniscule at all, but are *large-scale imaginative, conceptual developments* which involve abstract concepts. The scientific research of such people as Copernicus and Galileo can hardly be called miniscule, while it is fatuous to say that of Newton's mind-reorienting conceptions, or the radical developments of Plate-Tectonics. These developments were vast and theoretical (Cluster 9). And although, say, it was the *observed* perturbations in the orbit of Uranus which suggested there was a problem the only reason these perturbations were observed was because the observers had mastered Keplerian and Newtonian Scientific Theory; moreover, the solution was produced by theoretical work which used Newton's Theory. Neptune was observed where the Scientific Theory said it would be.

Third, it is true that the building of a Scientific Theory takes time: consider for instance the slow evolution of astro-physical theory from the time of Copernicus's first thoughts on the matter, to the publication of Newton's *Principia* (150 years?), or Lavoisier's developing oxygen theory between 1764 and 1783, or the quarter century during which the theory of Plate Tectonics was evolving. But there could be no progress: until Copernicus had placed the sun rather than the earth at the centre of the system; until Lavoisier had rejected phlogiston and developed both a quantitative approach and radical new ideas about relations between chemical elements; and until sufficient members of the geological community became convinced that stabilism was erroneous and some variety of drift was occurring. In short, in the three cases, the significant moves were not "miniscule analysis", but mind re-orienting conceptual leaps whose details were then slowly *created* — leaps which are so challenging that they are often rejected by some contemporary scientists for decades, e.g. Tycho's rejection of a sun-centered cosmology, Priestley's of oxygen theory, and Jeffreys's of continental drift. And although there is an aspect of scientific research which is cumulative, such as the post-war work of Hess, Vine, Wilson and the others with respect to continental drift, or the work on molecular biology post Crick and Watson, carefully to be noticed is that this work is always embedded in and grows out of various kinds of geological and biological theory (Clusters 5, 6 and 9)— in the present instances out of an awareness of the research implications of earth movement, and of nucleic acids.

Fourth, the conservative views and activities of such scientists as Brahe, Priestley and Jeffreys demonstrate an often unacknowledged, but nevertheless misleading aspect of the inductivist orientation common to empiricist approaches, but also appearing periodically in science. The sun does appear to circle the earth; something does seem to go out from substances when they burn; observational commonsense would appear to indicate stablist rather than moving continents. *Only theory of different sorts, involving visionary, creative conceptual leaps can take us past "the facts".*

Fifth, one cannot necessarily build science even on seeming *experimental* data by themselves— for these too are often-enough misleading. Two quotations from Crick are significant. In 1950, Crick's colleagues, including the famous Bragg, had been beaten to the discovery of the alpha helix by Pauling's team. Crick writes,

> This failure on the part of my colleagues .. made a deep impression on Jim Watson and me. Because of it I argued that it was important not to place too much reliance on any single piece of experimental evidence. It might turn out to be misleading, as the 5.1A reflection undoubtedly was (1988, p.59).

In contrast, Crick and Watson beat Pauling's team to the discovery of the structure of the DNA helix, because, as Crick says,

. . every bit of experimental evidence we had got at any time we were prepared to throw away, because we said it might be misleading [Pauling] missed the [DNA] helix because of that . . evidence can be unreliable, and therefore you should use as little of it as you can. . and when we confront problems today we're in exactly the same situation. . people don't realize that not only can data be wrong, [they] can be misleading. There isn't such a thing as a hard fact when you're trying to discover something. It's only afterwards that facts become hard (quoted in Judson,1980, p.169).

In contrast to scientific research, empiricist research on teaching has so far indeed proceeded by miniscule analysis, and its proponents do hope that some kind of theory will slowly emerge. Process-product researchers have hoped for example that their various statistical facts about teaching will somehow be able to be brought together, by logical or meta-analyses. But except for Cluster 5, Type 9 (Heuristic), occasional allusions to conceptions from Cluster 3 (Theory as Explanation) and some kind of implicit and unexamined Cluster 6, Type 9 (Ontological Presupposition), *present-day empiricist research on teaching is largely inductive and drifts theory-less.*

There is No Scientific Method or Art of Discovery As Such

There is no such thing as a "calculus of discovery", a recipe, or a way in which by sticking to say, the supposed rules of the game as enunciated by a J.S.Mill, we can make scientific discoveries. As was implied by the discussion of Lavoisier, the fact that some one hundred years after Newton's marvellous developments in physics, ninety-nine out of a hundred chemists were still thoroughly wrong about the nature of combustion (something which even schoolboys take for granted today) should make researchers pause in their use of the catch-phrase, "the scientific method".

Unfortunately, empiricist pedagogical researchers do still talk about "the scientific method". Consider the following statements from Borg and Gall : "The scientific method widely used in both the natural and social sciences is derived from a system of philosophy [sic] known as positivism" (1989, p.16). "The scientific method that has dominated educational research right up to the present day is based on the principles of positivism" (p.17). "The scientific method is commonly thought to include three major phases. . The first phase is to formulate a hypothesis. . The next . . is to deduce observable consequences . . The third phase of the scientific method is to test the hypothesis by collecting data"(pp.17-18). They have a section titled "The Scientific Method" (pp. 28-30). (However, on p. 31 they write, "Just as there is not a single scientific method.")

The closest thing to recipe research in science occurs at graduate school. But even at graduate school, research does not always "come out". And far from being "recipe", sometimes research by graduate students can be momentous: Crick did not have his Ph.D. at the time of his discovery. Medawar claims correctly that there is, ". . no art or system of rules which stands to its subject-matter as logical syntax stands towards any particular instance of reasoning by deduction. . we can give no rules for the pursuit of truth which shall be universally and peremptorily applied" (1982, p.116). Nagel says, "There is no scientific method as such, but only the free and utmost use of intelligence" (1954, p.5). Woolnough writes, "Science has progressed and continues to progress, by a mixture of insight, personal judgment, inspiration, and personal commitment in a way that contradicts the tidiness of any one method" (1990, p.219). As I said in Chapter II, scientific research can in a sense be construed as a search for hidden liknesses in nature; and it requires not a method, but radical imagination to produce these. As Bronowski says,

> The symbol and the metaphor are as necessary to science as to poetry. We are as helpless today to define mass, fundamentally, as Newton was. But we do not therefore think, and neither did he, that the equations which contain mass as an unknown are mere rules of thumb. If we had been content with that view, we should never have learned to turn mass into energy (1965, pp. 36-37).

But there are no recipes for using our imagination and there are no recipes for doing research; the so-called "hypothetico-deductive" method is not a *method* so much as a mental orientation, an orientation to, "what if?" It is appropriate for Medawar, himself a Nobel Laureate, to emphasize that,

> No layman who reads [Watson's] book [about the discovery of DNA] with any kind of understanding will ever again think of the scientist as a man who cranks a machine of discovery. No beginner in science will henceforward believe that discovery is bound to come his way if only he practices a certain Method, goes through a certain well-defined performance of hand and mind (Medawar, 1968, p.5).

Indeed, scientific research, as a going concern develops in many different ways.

One important tactic which uses the imagination, a tactic much underestimated in discussions in philosophy and methodology has been the thought experiment. Galileo, Newton and Einstein are the three most famous exponents. Several of Galileo's thought experiments were discussed in Chapter II. Newton's initial linking of the falling apple and the moon falling forever around the earth was a type of thought experiment. Huygens, Carnot, Joule, and Bohr were other famous scientists who used them. A few basic points about the logic of thought experiments should be made.

First, all successful thought experiments must make use of some prior information, but as Kuhn says, ". . that information is not itself at issue in the experiment. On the contrary, if we have to do with a real thought experiment, the empirical data upon which it rests must have been both well-known and generally accepted before the experiment was even conceived" (1977, p.241). That is true as far as it goes, but Winchester is even more apposite in pointing out that a thought experiment is,

.. a certain species of metaphysical argument or argument proceeding from our certainties of an ordinary kind about the world in which we live; about, that is, absolute presuppositions . . . Our certainties, I shall argue, are manifold. They are the *things which enable us to get on not only with everyday life but also with science* (italics added) (1991, p.236).

Second, thought experiments help to remove previously unrecognized conceptual confusion. Thus they are involved in the formation of new abstract concepts, and in the separation of concepts presently confused with one another.

Third, in showing how the empirical world fails to conform to the previously held conceptions and expectations, thought experiments propose ways in which theory and explanation must be reformulated. This reformulation is possible because the thought experiment enables the scientist, ". . to use as an integral part of his knowledge what that knowledge had previously made inaccessible to him. That is the sense in which they change his knowledge of the world" (Kuhn, 1977, p.263).

Fourth, in the following sense, as explained by Kuhn, thought experiments also remove theoretical contradiction (Clusters 6 and/or 9), and in so doing, provide us with new knowledge of nature:

. . thought experiments assist scientists in arriving at laws and theories different from the ones they had held before. In that case, prior knowledge can have been . . "contradictory" only in the rather special and quite unhistorical sense which would attribute . . contradiction to all laws and theories that scientific progress has forced the profession to discard. Inevitably, however, that description suggests that the effects of thought experimentation, even though it presents no new data, are much closer to those of actual experimentation than has usually been supposed (1977, p.242).

Resting in a priori axioms of experience and observation, in the certainties of empiricist experience, and in most of the "provisional certainties" of the current Scientific Theory, thought experiments form a continuum with the production of hypotheses. The thought experiments of Galileo and Newton rested in the empiricist certainties that like produces like, that gravity retards objects moving upwards and accelerates objects moving downwards, that forces are required for there to be any change in states of motion, and in the mathematical certainties that when equals are added to or taken from equals, the sums and remainders are equal.

To sum up, thought experiments both clarify and untangle our present concepts and our conceptions of the way the universe works, and also provide us with important new scientific understanding.

Science also develops when one theory (various Clusters) is grafted onto another incompatible theory, as when Copernicus grafted his heliocentric cosmology onto Aristotle's physics. There may be grafting of one theory onto another compatible theory, e.g. Tycho Brahe's planetary astronomy (Cluster 9:

Science) onto the then-accepted earth-centred cosmology (Clusters 6 and 9: Presupposition and Science). There may be continuing use of two different theories (Cluster 9), with the earlier still being used in specific instances, as at the present time in physics in the case of Newtonian physics and Einsteinian physics being pursued for different purposes. One theory (Clusters 8 or 9) may be entirely replaced by another (Cluster 9): Galilean-Newtonian physics eventually completely replaced Aristotle's physics; stabilist theories were replaced by drift theory in its modern garb of Plate Tectonics. There is often a long period of time in which the older theory performs *in continual competition with* the new (e.g. phlogiston versus oxygen theory), a time in which both conceptualizations can account for the observed data equally well (at least in the minds of the proponents).

Medawar (1984, p.18), suggests that,

> A scientist commands a dozen different stratagems of inquiry in his approximation to the truth, and of course he has his way of going about things and more or less of the quality often described as 'professionalism'-- an address that includes an ability to get on with things abetted by a sanguine expectation of success and that ability to *imagine* what the truth might be. A scientist, so far from being a man who never knowingly departs from the truth, is always telling stories in a sense not far removed from that of the nursery euphemism— stories which might be about real life, but which must be *tested very scrupulously* to find out if indeed they are so. A 'story' is more than an hypothesis: it is a theory [Clusters 5 or 6 or 9: Hypothesis, or Presupposition, or Science], an hypothesis together with what follows from it and goes with it, and it has the clear connotation of completeness within its own limits. I notice that laboratory jargon follows this usage, e.g. 'Let's get so-and-so to tell his story about' something or other. There is a slightly depreciatory flavour about this use of 'story' because fancy has to be used to fill in the gaps and some people overdo it[2] (Medawar,1982, p.40).

Several scientists and philosophers of science have indicated the need for competing theories (various senses). Perhaps the most famous instance is advocacy of the so-called Method of Multiple Working Hypotheses. As mentioned in Chapter II, this method was argued for explicitly by the early twentieth century geologist and educational administrator, T.C.Chamberlin. It claims that science progresses most quickly through the simultaneous development of several competing theories (various Clusters) which explain the current facts and phenomena. Chamberlin writes,

---

2. Because, to be successful, they need to provide not just any relationships between abstract concepts, but good relationships— ones which will have some possibility of linking into Scientific Theory and being interpretable to the observational world-- and to provide these is profoundly difficult.

The effort is to bring up into view every rational explanation of the phenomenon in hand and to develop every tenable hypothesis relative to its nature, cause, or origin. The investigator thus becomes the parent of a family of hypotheses and by his parental relations to all is morally forbidden to fasten his affections upon any one. . . The reaction of one hypothesis upon another tends to amplify the recognized scope of each. Every hypothesis is quite sure to call forth into clear recognition new or neglected aspects of the phenomena in its own interests, but ofttimes these are found to be impartial contributions to the full development of other hypotheses. So also the mutual conflicts of hypotheses whet the discriminative edge of each. . . the sharpness of discrimination is promoted by the coordinate working of several competitive hypotheses (1897, pp.839-843).

As mentioned in the continental drift case study, in a parallel fashion Feyerabend has in several places advocated the search for alternative theories (various Clusters). He argues that a scientist must be,

. . prepared to work with many alternative theories rather than with a single point of view and 'experience'. This plurality of theories must not be regarded as a preliminary stage of knowledge which will at some time in the future be replaced by the One True Theory. . . It takes time to build a good theory, and it also takes time to develop an alternative to a good theory (1968, p.14).

Feyerabend is pointing to the deep significance of competing ideas in research. He argues that it is normally the case that until an alternative is available, scientists cling to the current explanation, even if they find inadequacies in it.

Francis Crick provides a useful example, "Olby demonstrates what was not clear to us at the time, that in the absence of our structure Pauling was not inclined to accept the more obvious criticisms of his own" (Olby, 1974, p.ii). The examples of Chapter II seem to support this view. Astronomers went along with Ptolemy until Copernicus and his supporters appeared; physicists believed that Newton had either said all, or pointed the way to it, until relativity and quantum mechanics came along; without the work of Wegener, the earth sciences might still be stabilist. Only through numerous shifts and developments of a logically-disparate sort, does the new conquer in the long term. This was the case with the clash between Ptolemaic ideas and Copernicus. The two opposing camps emphasized different criteria. Priestley argued against Lavoisier's oxygen theory of combustion and remained a supporter of phlogiston until his death; but the younger generation of chemists took up oxygen chemistry. In twentieth century geology, stabilism and continental drift competed for about fifty years (stabilism in the U.S.A.; both stabilism and drift in Europe, South Africa and Australia). Drift (as plate tectonics) triumphed only after the development of new techniques, which provided overwhelming evidence and new predictive capacities. So some of the inadequacies of a particular Scientific Theory can be exposed, only by the development of a rival.

A recent version of the argument for using multiple theories is Siegel's discussion of the importance of pluralism in the teaching of science. This version may be of particular interest to readers of the present book, because Siegel's is a pluralism of epistemology embedded in a *pedagogical* theory (on my view, Cluster 3, as a part of Cluster 4). As he says,

> . . pluralism recognizes the virtue and potential fruitfulness of allowing rival ideas to establish their merits in the free exchange of ideas. . . Philosophically, it recognizes the fallibility of scientific knowledge and the goal of improving our ideas as well as the virtues of freedom of thought and expression. Pedagogically, the conflict of ideas can serve to stimulate students, and to spur them on to deeper understanding of the matter (1991, p. 55).

The importance of alternatives is beginning to apply with manifest force and sometimes embarassing results to empiricist research on teaching, as competing methods are being developed. For empiricist researchers have traditionally believed that there is *a* scientific method and *an* art of discovery (the ones they use), as explained in the recipes of any of the manuals of empiricist research listed in the Introduction. In Chapter I, I wrote that standardized empiricist endeavors have, ". . always been surrounded with very precise, sequential ritual. . there has to be a precise ceremonial which is passed on as a tradition, a ceremonial which fixes the recognition of timing and sequence of procedures." This is a good description of the key experimental method of empiricist research. Following such a recipe is very different from engaging in the expanding, open-ended, enquiries of scientific research.

Empiricist research has also manifestly ignored *thought experiments* . This is perhaps to be expected because the thought experiment is so very different from the hopeful building of a structure from careful observation. Given the revolutionary part thought experiments have played several times in the history of scientific research, this failure to make use of them may also be a considerable mistake in empiricist research. For although empiricist research is not scientific research, it still makes good sense for empiricist researchers to adopt fruitful techniques whatever their source. I suspect that thought experiments may be particularly appropriate for developing the kinds of insights which are so needed if empiricist research is to make itself truly useful to pedagogy. It may also be revealing that empiricist research has largely ignored the insights of analytic philosophy of education, yet analytic philosophy of education is characterized by procedures in some ways similar to the thought experiment.

When those who claim to be doing scientific research in pedagogy grasp the significance of thought experiments in scientific research, both in conceptual development and in the deriving of hypotheses from Scientific Theory, perhaps they will have a new respect for their possible role in the development of any Scientific Theory of teaching. If scientific research on teaching is ever to be developed, it will never be created by looking carefully at the empirical world, but by speculating freely about what regularities are underlying it, or what hidden likenesses are lurking there just below our pedagogical, conceptual horizon. If scientific research and the concomitant Scientific Theory in pedagogy are to become a reality, creative imagination will certainly be required— a willingness to conceive anew what form the pedagogical world might take.

The Context of Discovery is Not Isolated from Other Inquiries

Equally fascinating is the fact that scientific research as an human activity is not self-contained, or completely rational. Many ideas in the context of discovery derive from other fields such as religion, or metaphysics. And there is always Presuppositional Theory of one sort or another. This origin in metaphysics is particularly intriguing in the light of the history of the verifiability criterion of meaning, so popular amongst philosophers earlier this century. For as Popper points out, these logical positivists, ". .were trying to find a criterion which made metaphysics meaningless nonsense. . and any such criterion was bound to lead to trouble since metaphysical ideas are often the forerunners of scientific ones" (1976, p.80). More recently, Kerlinger, in his manual, (1986), which I have already mentioned several times has also produced confusion, in being dismissive of the usefulness of metaphysics for scientific endeavor.

Ideas from other fields help to provide the conceptual leaps and the imaginative extrapolations for the abstract concepts and propositions of scientific research. This really is different from the slow accumulation of observed data. Copernicus was not merely dissatisfied with Ptolemy, but his numerous beliefs about the theology of nature seem to have played a crucial part. For Copernicus there was a wonderful cosmic relationship between the universe and God, with the sun being conceptualized as a visible symbol of this theocentric universe (Wrightsman,1980). Kepler also harbored mathematical, mystical and theological notions about the sun, which moved him to lavish speculation in particular directions. The episode of the five regular solids provides a good example. *The Double Helix* (1980), Watson's report of his work on DNA, is permeated with talk of pairing, sex, and au pair girls, which strongly hints at a pre-conceived preference for a double helix rather than some other combination of helices. But it is not these ideas from other fields which make scientific research powerful and progressive in the context of verification: neither metaphysics nor religion (as Galileo painfully experienced) is scientific.

Such features of scientific research contrast strongly with the plodding, repetitive nature of empiricist research on teaching, which seems to be trapped in an inductive-observational stage. There is little if any originality except for the immensely sophisticated continuing refinement of statistics, there is little variation in methods used, there are no competing construals, there are no striking imaginative insights, there are no thought experiments. Based upon the atomistic breaking down of the activities of teaching into observables and upon the twin hopes of accurate observation of these and statistical manipulation and analysis, such research continues to be substantially the same. There is no stimulation from alternative ways of grasping the world-- use, say, of insights or beliefs from outside teaching.

It is precisely such powerful, expansive, metaphysical and other ideas, which empiricist researchers have quite deliberately ignored in their attemps to model their approach upon what they have believed to be the ways of scientific research. There is even little drawn from other ways of construing what is going on in teaching, as in, say, literature or history; though there is a positive move in the current adoption of ideas from other kinds of research on teaching, as in attempts to use ideas from cognitive psychology.

Because of Their Theoretical Background (Clusters 5,6 and 9), Scientists are aware of a Question, Problem, Puzzle, or Discontent, or see Anomalous or Unexpected Phenomena

Except for those purely documentary or descriptive studies, such as, say, marking the position of a nova, determining an atomic weight, finding a particular refractive index, or examining the anatomy of a new species of beetle (all of which still *assume* specific Scientific Theory), scientific research is both questioning and creative. Embedded in his program or tradition, the scientist wishes to answer his question, solve his puzzle, resolve his anomaly, settle his discontent. Copernicus was affronted by the complexity of epicycles, deferents and equants, and wondered what really was going on in the firmament. Lavoisier was dissatisfied with the contradictions of phlogiston theory, and wanted to know why things burn. Crick and Watson asked what kind of helical biochemical structure could provide DNA with such marvellous genetic properties of replication. Wegener sought evidence and an explanation for drift. Roentgen wondered what could have made the barium-platino-cyanide plate become discolored, finally leading to the unsuspected phenomena of x-rays, embedded in electro-magnetic theory. In all these cases the problem, puzzle, discontent, etc. in the minds of the researchers derives from its embedding in Scientific Theory.

It is because scientific research also requires critical thinking, that the scientist's research must grow out of a thorough theoretical (Cluster 9) background. For, as various writers (e.g. Siegel,1988, 1991) have clearly shown, critical thinking involves both subject-neutral and *subject-specific* reasons and principles.

The attempts of empiricist researchers to show what observable processes can be related to or correlated with what measured products, or, say, to show what typical features correlate with what educational populations are not a problem in any of these senses. Moreover, such attempts *do not grow out of a thorough knowledge of the actual complex subject-matter of teaching* – for as will be argued in the next chapter such research just ignores several of the crucial dimensions of teaching: to become adequate, empiricist research requires embedding in theory in the Cluster 4 sense. Walberg almost admits as much in his claimed account of a scientific approach, quoted above, when he writes, "Most research on teaching, for example, draws not on unifying theoretical paradigms and constructs, but on an immense variety of somewhat incompatible

constructs, empirical representations, procedures, and observation instruments."
In most empiricist research on teaching, theory is supposed to "emerge", rather than be conjecturally created.

But the scientist's question or puzzle is not answered merely by careful observing, or by strict attention to technique. Originality and imagination are required. To use a somewhat dated psychological jargon, scientists are divergent rather than convergent thinkers. In strict contrast, empiricist researchers have mainly been convergent thinkers who use a set of techniques in the hope of finding something important about teaching. I believe it is a forlorn hope. Ironically, the courses in empiricist statistics and research design which I studied at one of the world's most prestigious institutions for educational research in 1987 and 1988 actually discouraged divergent thought. And they were typical of their type.

It may look as if the subject-matter of teaching is different in significant ways from that of the cases whose history was discussed in Chapter II; and of course this will be the case; for instance, teaching deals centrally both with values and intentional actions[3]; but that does not in itself mean that Scientific Theory of teaching might not be produced from scientific research on teaching, which amongst other things would involve true experimentation (see below)– experimentation which both developed and depended upon such Theory. It is the confused conceptions of theory together with an empiricist version of relationships and experiments which have brought about the present impasse in those dimensions of teaching which may be amenable to a scientific treatment.

Misunderstandings which sustain the empiricist view of scientific research are pervasive. A recent book which actually stresses the significance of theories in the teaching of science offers a number of interesting, sound, and usable suggestions. But as already mentioned the author still writes: "All science investigations begin with the collection of data. It is at this level that science first establishes a concrete connection with the natural world" (Duschl, 1990, p.49). On the contrary, all scientific investigations begin with theory of one sort or another, particularly with Scientific Theory from which hypotheses are drawn and tested. And just what the "concrete connection" turns out to be, depends upon this theory. *The "concrete connection" is theory-laden.*

Moreover, techniques however refined cannot be enough. In their research manuals, empiricist researchers *talk* about theory. But that is often all they do. In Chapter I, it was pointed out how confused about theory Kerlinger was. But there are numerous other examples. Consider the account by Wiesma (1985). Wiesma devotes two pages to a discussion of the nature of theory in scientific research– a discussion which makes many good points. But following his discussion *there is not one single example of that sort of theory in education* ! Of course, given the reliance on merely statistical techniques, and the empiricist basis of the approach Wiesma advocates, the situation could be no different– at present there just is no such Scientific Theory in pedagogy (though there are

---

3. For discussion of intention, see the following chapter.

hopeful candidates). My point is that Wiesma fails to notice the extraordinary inconsistency between what he says about theory in scientific research, and his abject failure to be able to produce a single instance of it in what he claims is a scientific field.

The multiple meanings of "theory" also allow much loose thinking in the manuals of research methods. Thus in their section headed, "Scientific Theories", Borg and Gall (1989, p.27), like Wiersma, say numerous sensible things about the nature of Scientific Theory, but they soon slide from Scientific Theory to other kinds. The examples they discuss of supposedly Scientific Theory from education are Piaget's theory of intellectual development and Walberg's theory of educational productivity. I would see Piaget's theory as making epistemological and empiricist claims (Clusters 3 and 8) rather than scientific ones. Walberg's theory has even fewer claims to scientific status. As Borg and Gall point out, Walberg proposes five essential factors of productivity, which,

> . . appear to substitute, compensate, or trade-off for one another at diminishing rates of return. Immense quantities of time, for example, may be required for a moderate amount of learning if motivation, ability, or instructional quality are minimal. Thus no single essential factor overwhelms the others; all appear important (Walberg, 1984, p.22).

Walberg's account is not Scientific Theory. It is but a concoction of commonsense ideas of the empiricist kind— *at best* a rather elementary and banal version of Cluster 3 theory (Explanation). Without knowing anything at all about the methods Walberg actually used, we *know* that if he had discovered the opposite, viz. that immense amounts of time are required for a moderate amount of learning when motivation, ability and quality of instruction are maximal, then his findings just could not be correct. To have this kind of knowledge, "in spite of the research" or "despite the research" is not possible in scientific research.

In developing their argument, Borg and Gall write,

> Suppose further research discovered that one of these five factors, if present only to a moderate degree, produced a high amount of learning, even though the other factors were minimal. This finding would force a reexamination and revision of the theory. Another possibility is that further research would continue to validate the theory for certain educational contexts, but not others. In this scenario, the theory would need to be revised to account for these discrepant findings. In this way, theories change over time and gradually may be replaced by other theories that have greater explanatory power (1989, p.27).

Such suggestions are eminently acceptable. And they certainly apply to scientific research. But they also describe characteristics of *any* genuine rational endeavor. Respect for the evidence, consistency, willingness to re-assess and to change the account are virtues indeed. But they do not turn commonsense into science.

There is Potential Falsifiability

As early as 1629 in his *Novum Organum*, Bacon had pointed to the importance of falsification. Drawing upon the modern arguments of Popper (1968;1969) and Laudan (1977; 1986) and considering the historical development of science, we grasp that the beliefs and claims made within a research program, or tradition, change, because in some complex fashion they have encountered observational falsification, or theoretical difficulties, or inconsistencies. Thus, an important feature of scientific progress is the eventual rejection of proposed suggestions and hypotheses, when they are found inadequate. The 150 or more years of the early revolution in Copernican astronomy is a good example on a large scale; the DNA research of Crick and Watson is a good example of a continual cycle of such changes in a more specific field of research, over a more limited time. The change from stabilism to drift in the earth sciences is another example: it took some sixty or so years at varying rates of progress to show that stabilism was false.

For reasons of the Duhem-Quine sort, apparent falsification may not result immediately in rejection of the claims. The Copernican-Keplerian astronomical research program was continued because of its continuing powerful discoveries and fruitfulness, despite the non-appearance of the implied parallax of stars[4], which on a simple view of falsification should have caused it to be rejected. Naive ideas of falsification of scientific propositions are themselves false. Popperian ideas of falsification in science however are still fundamental, but in a more subtle manner. Falsification and testing are complex interweavings of judgment, observation and theory (several Clusters), as well as comparison with developing theories. The theory-ladenness of observations and the possible temporal inconclusiveness of falsification and corroboration ensure that testing is a subtle business. This,". . is just to stress that decisions cannot be made in any mechanical way" (Phillips, 1987, p.13). And claims fundamental to a program, even if seen as falsified (e.g. those made in Phlogiston Theory or stablist geology) are never abandoned unless there are others to replace them.

A good example of these ideas is provided by the differences between the discovery of Neptune and the "discovery" of "Vulcan" (mentioned in Chapter I). It is true, as Quine has stressed, that the power of seeming falsification has at times been dissipated somewhere else in the theoretical network, by making a greater or lesser *ad hoc* change; but not to just *any* part of the network. Lakatos (1970) and Laudan (1977) have pointed to the importance of the problem-solving capacity of a program. But they have not stressed as much, how this affects where the *ad hoc* move is made. Just which part is adjusted and whether the adjustment continues to be accepted is decided by the greater or lesser "*ad hocness*" of the move, for some moves are more rational than others. Thus, the postulation of a new planet to explain the unacceptable discrepancies which arose between the observed positions of Uranus and the positions it ought to have tak-

---

4. Parallax was first observed with the star 61 Cygni, by Bessel, in 1838.

en, as predicted by Newtonian theory was much less *ad hoc* than in the case of Vulcan. Thus construed, the pertubations of Uranus, the phenomena thus "saved", fitted easily into the astronomical scheme of things, because similar pertubations were known to occur as a result of gravitational forces amongst the other planets. But saving the phenomena by the postulation of Vulcan between Mercury and the Sun was much more *ad hoc*, for the pertubations in Mercury's orbit were quite extraordinary, viz. the precession *of the orbit* at perihelion. Moreover once the calculations had been provided, Neptune had been quickly observed close to the predicted point in the heavens, whereas the scientific community never agreed about the claimed sighting of Vulcan. (And with the new Relativity Theory, Vulcan became redundant.)

Critique of all kinds is central and fundamental to scientific research– though it is often unwelcome to individual scientists (as is critique of empiricist research). While still a graduate student, Crick gave a paper in the regular seminar at the Cavendish Laboratory. The subject of his talk was direct and unmistakable, in general that , ". .they were all wasting their time and that. . almost all the methods they were pursuing had no chance. . . Bragg was furious. . . a little later [at another lecture]. . Bragg turned around to speak to me over his shoulder. 'Crick,' he said, 'you're rocking the boat' " (Crick, 1988, p.50). Further research by the very same scientists at the Cavendish showed that Crick had been correct.

In the philosophical writing there has unfortunately been a persistent ambiguity between falsification of some proposition *within* some particular branch of Scientific Theory, and falsification of the theory (or program, or tradition) as a whole. But the kind of falsification which occurs varies between these different cases. Compare, for instance, the falsifications which proceded within the Crick and Watson DNA research, with the eventual falsification of stabilism in geology. Clearly, the first type is a less complex matter than the second.

Empiricist researchers themselves occasionally make kudos-seeking noises in the direction of falsfication, such as those by Walberg, quoted earlier, but they are unconvincing. Consider his claim: "The use of parsimonious, *falsifiable theory* as a guide to empirical inquiry (Popper,1959)– the modern and commonly accepted view in the natural sciences– is rare in educational and social research." This sounds all right, until we realize that, *such theory is not rare in empiricist research on teaching, it is non-existent.* Walberg also writes, "Relationships replicated with different subjects and in a variety of circumstances can be said to be generalizable, widely confirmed, or, *in Popperian spirit, unfalsified*. Replication attempts by different investigators and probing theory in a program of research, however, are rare," and, "Practitioners can reasonably accept the implications of research findings that syntheses *prove robustly unfalsified . .*" But the italicised phrases are at best hortatory, at worst bombastic and false. No syntheses by meta-analysis of large numbers of studies with statistical findings have ever shown claims to be robustly unfalsified. Meta-analysis has nothing to do with falsification. For what is involved are collations, probabilities, and correlations.

It is Helpful to have Alternative Paradigms, Programs, Traditions, Explanations, Hypotheses, Theories, if a Science is to Progress

In his, "How to be a Good Empiricist", Feyerabend made a powerful plea for the importance of alternatives. As already mentioned, he pointed out that some of the limitations of a paradigm or a theory (several Clusters) can be exposed only by the development of a rival. And, as alluded to in Chapter II, T.C.Chamberlin wrote about "the method of multiple working hypotheses", advocating that, especially in geology, scientists consider alternative explanations for observed phenomena. He writes, ". . the effort is to bring up into view every rational explanation of the phenomenon in hand . . and to give to all of these as impartially as possible a working form and a due place in the investigation" (1897, p.843).

Phlogiston would have gone merrily along, unless Lavoisier had evolved his oxygen theory. In the late nineteenth century many physicists thought their subject was nearing its end– the coefficient of expansion of this, or the refractive index of that, might be refined, but not much else was left to do. Then along came Einstein in 1905, and Rutherford and Planck and Bohr and de Broglie and Heisenberg and Dirac and Pauli and . . .

This idea of alternatives is of deep significance in scientific research, but it is even more so for non-scientific endeavors such as empiricist pedagogical research. For not merely in scientific research, but also in all other endeavors, perceptions of the kind of relationship between entities is controlled by the system of knowledge and belief being used. In an empiricist system empiricist relations are perceived: prayer to the rain-god brings rain; falling rain causes mud; changes in teachers' observable behavior as construed under observational-statistical concepts cause changes in students' learning (as assessed on standardized attainment tests), e.g. a teacher's praise causes improved results. In most circumstances, experiences within the system of knowledge and belief will not be seen as contradicting *the system*. Thus, if a second researcher finds that praise inhibits the learning of students, this will have little effect on the confidence of the first researcher in the empiricist techniques. Doubt about the adequacy of the system occurs when a second system of knowledge and belief is brought to bear on the situation, e.g. when the phenomena are looked at from within a different form of knowledge and belief (say, from morals, or aesthetics, or religion), or when a different variety of Theory (Clusters 3 or 5 or 7) is used to explain. Without this, the empiricist researcher on praise will merely report that, "further research [of his own kind] is required." This is indeed the typical conclusion of empiricist studies of teaching. Experience of empiricist approaches has rarely caused empiricist researchers to modify their systemic views. The automatic move is usually to modify the techniques: with more subtleties in the questionnaire, improved observation scales, more rapid computers, more assistants, more grants from the funding bodies, more millions of dollars.

Doubt as to the suitability of empiricist approaches as presently constituted only became earnest, when alternative approaches began to be adopted: illuminative research, action research, case studies, ethnographic approaches, hermeneutics, *Verstehen* , phenomenology, etc.

There are No Theoretically-Neutral Observations in Scientific Research

Scientific observations are embedded in Scientific Theory and moreover, always presuppose some suppressed Presuppositional Theory (Cluster 6), and commonly they also presuppose similar theories concerning the behavior of the instruments used to make the observations. However, for the reasons provided at the end of Chapter Two, this is not the problem it has often been suggested to be. We should note first, that not all observations and propositions are laden with all theories (several Clusters); secondly, scientific propositions are not necessarily laden with the theories (Clusters 5 or 9) which are under test; and thirdly, even when they are theory-laden with the theories under test, counter-evidence is still in principle perfectly possible.

This theory-laden nature of observation, and this suppressed theory underlying the use of instruments, e.g. of the observation scales, also applies in empiricist research on teaching, and has been particularly ignored. (To adapt the heading of this section: there are no theoretically-neutral observations in empiricist research either.) As already pointed out, observations based upon operational definitions do not escape theory-ladenness either, for they are themselves laden with whatever theory or theories underlie their definition. Moreover, the importance that is still given to operationally-defined observation in empiricist pedagogical research may be seen as a metaphorical coelacanth of the Verifiability Principle– so thoroughly discredited in philosophy of science.

Scientific Research is Open and Continually Evolving

We can see a gradual progression in the ideas of scientists such as Kepler, Galileo, Newton, Lavoisier, Roentgen, Crick and Watson, H.H.Hess, Tuzo Wilson, and a thousand others, who both developed the ideas of predecessors, and produced evolving accounts to match their increasing creative speculations and experience.

Later thinkers subsume much of what has gone before. Kepler used Copernicus's heliocentric theory as the basis of his own conceptions, calculations and observations, eventually modifying it into ellipses. Newton showed that Kepler's Third Law was but a special case of Newton's own laws of motion and gravitation. Einstein showed that Newtonian theory is a limiting case of his own Theory of General Relativity. Science is fruitful and ampliative. Roentgen's work was not merely the isolated discovery of x-rays, but was soon fitted by himself and others into, and radically extended, the theory of electro-magnetic radiation. F.B. Taylor's mid-Atlantic Ridge, from which he assumed that, ". . the two continents on opposite sides of it have crept away in nearly parallel and opposite directions," was shown fifty years later to be a crucial part of the very reason why the continents were indeed moving apart. There are research programs, but even their "hard-cores" can evolve. Nothing is sacrosanct. In a science such as biology or physics, because of the diversity of activities, emphases and interests which co-exist, there is rarely a simple *period* of strict "normal science" as Kuhn claims (though there may be scientists doing mundane research). As Siegel has pointed out (1985, p.102), it is ironic that Kuhn defends relativism in epistemology, but dogmatism in the training of scientists (i.e. the pursuit of Kuhn's supposed normal science). We should recall that when Watson became even more convinced of a particular helical structure and tried to persuade Wilkins of the urgency of the situation, if they were to beat Pauling, Wilkins said that, ". . if we could all agree where science was going, everything would be solved and we would have no recourse but to be engineers or doctors" (Watson, 1968, p.134). This is a most profound observation (and Wilkins too won a Nobel Prize).

In contrast, empiricist research as presently constituted continually refines its observational categories and its statistics continually improve, so in a very restricted sense of the word, it may be evolving— but only towards more and more precision in empiricist techniques of statistical tests involving scaling. Improvement is seen as *more of the same kind.* Such research is introverted and closed. Meta-analysis is the perfect example: statistical results feeding upon statistical results. Meta-analysis fails to add new content. Nor does it expand anyone's theoretical conceptions (Clusters 3,4,5,7 and 9).

To adapt Kuhn, it is somewhat ironic that in empiricist research on teaching there has indeed been a long period of "*normal empiricist*" research, and dogmatism in the training of empiricist researchers. Indeed, the writing of Kuhn may well have helped to encourage this unfortunate situation. Those outside this tradition were for a long time not even seen as doing research. Continuing refinements of the approaches were what was "normal". And those researchers who used these methods rarely challenged the adequacy of their assumptions. Rather they tried to make their observation scales more precise, to improve their observations with these, to achieve greater agreement amongst observers, to improve their means of assessing the effects of statistical experiments, and to make the statistics more sophisticated.

Scientific Research Uses Abstract Concepts Embedded in Theory

This feature of scientific research is of fundamental significance and has already been discussed in detail. Medawar says that science will continue as long as we retain an ability to conceive, ". .what the truth *might* be and retain also the inclination to ascertain whether our *imaginings* correspond to real life" (1980, p.15).

Consider the following passage from Huberman:

> I collaborate with a gifted Swiss sociologist, somewhat in the Bourdieu tra-

dition, who genuinely believes, I think, that data get in the way of good research. He collects data, usually in a casual, participant observation mode, and reads widely, then generates some breathtaking generalizations on institutional factors in schooling that I have trouble refuting. This maddens me of course, so I hear myself saying that he got lucky; he replicated findings of more rigorous empirical work; had that not been the case, he would simply have been another crank. My sociologist friend looks intently at a forest and sees a forest, while I see trees (1988, p.9).

This passage is revealing in several ways. It shows the imperialist assumptions of some empiricist researchers, e.g. "he got lucky" ; "he replicated . . empirical work" ; "he would simply have been another crank". It is also revealing with respect to abstract concepts. For, like Kepler with his five regular solids, the Swiss may have achieved the first stage in scientific research, namely the development of abstract concepts related by rational connection. Data get in the way of good research for him, because data such as Huberman's are empiricist (he sees "trees"). The Swiss uses empiricist data merely as sources for ideas to produce his "forest": not "breathtaking generalizations" as Huberman calls them, but insightful, imaginative leaps. But unlike, say, Kepler with his three planetary laws, the Swiss cannot interpret his concepts back to the observable world by way of abstractive connection involving (ratio) measurement, and so he has not yet developed a scientific form of research.

As already argued, there are no such imaginative conceptual leaps, and no such productions of abstract concepts in present-day empiricist research (despite its continual "sanguine expectation of success" in the future). Both the abstract concepts and the Theory (Cluster 9), which they help to form by being embedded and interrelated, are missing.

The Statistics and Probabilities of Empiricist Research are not the Mathematics and Probabilities of Scientific Research

Empiricist researchers place great store upon the statistical foundations of their work, and many see them as a *sine qua non* of scientific research. But scientific research is not based on such statistical foundations (Nagel, 1962; Meehl, 1967, 1978). To be carefully noted is that scientific research both in establishing its laws, and in applying those laws works within *degrees of acceptable mathematical error stated for the observational interpretation of the abstractive connections prior to the fact*. Natural scientists *make allowances within particular limits* and with various degrees of *predicted* error. The mathematical rigor rests in the rational mathematical connections between the abstract concepts within the propositions of Scientific Theory, and the interpretation to the observational world through scientific (i.e ratio) measurement.

Scientific research does not now consist and never has consisted of the manipulation of propositions of general concepts. The characteristics of scientific research are different from the tentative establishment of general regularities through correlations, or through experiments using statistical tests, or the tentative judgment-based application of the results of those correlations or experiments. Empiricist research on teaching is empiricist research on teaching. Resting upon the "rigorous uncertainty" of the statistical relationships, which link the general empiricist concepts of the propositions of Empiricist Theory, its major concerns remain the establishment of the *existence* of such relationships and the *degree* of such relationships.

To put the general point in a slightly different way: it has sometimes been suggested that empiricist research on teaching is scientific, *because* both the established sciences and it *make use of probability* as a significant aspect of their approaches. However, as Hernandez-Cela (1972) has pointed out, this assumption is misleading. Indeed, there is an hierarchical set of at least four different confusions here.

The first confusion rests upon the view rejected earlier in this book, that scientific research involves inductive generalizations from observed similarities and regularities. The next three are both more specific and more complex. The second claims that even such a system as classical Newtonian mechanics is actually statistical and probabilistic. The third claims that even if classical mechanics is not, classical statistical mechanics is. The fourth claims that even if classical statistical mechanics is not, then sub-atomic physics certainly is. None of these claims is true.

First, because some empiricist researchers believe that scientific research proceeds inductively from observed similarities, they mistakenly assume therefore, that the propositions which link the abstract concepts of science are generalizations with some degree of probability. Thus, *when they learn that physics also uses probability theory, they naturally assume that the claims of the hard sciences are themselves always probable rather than definite.* Early in the century Pearson argued (1911/1957), and later writers such as Carnap (1950), Blalock (1960), Kyburg (1974), and MacMillan and Schumacher (1984) have followed him in claiming so succinctly and erroneously, that all science is probabilistic or that statistical inference underlies all "scientific generalizations". Assuming that the term, "scientific generalizations" makes sense, then this would have been news to Copernicus, Kepler, Galileo and Newton! And how perverse of Lavoisier, Crick and Watson, Wegener, Hess, Vine and Matthews, and Tuzo Wilson to ignore this "truth".

Hernandez-Cela summarizes the key point of the remainder of this section when he writes, "If probability is properly understood as a *theoretical conception* [Cluster 9], it is no more a reflection of observable experience than any other theoretical construct used in science. Since it is part of a [theory] it does not have to be based on facts; it is not intended to be so" (italics added) (1972, p.105).

The main point is that probability in Scientific Theory and in scientific research *applies to relationships between the propositions involving the abstract concepts of the theory* , i.e. to the rational connections. Probability in Scientific Theory and scientific research does not apply to the abstractive connections, which relate the propositions involving abstract concepts to the empiricist propositions about the observable world ; nor does it apply to the empiricist connection which relates the general empiricist concepts within those propositions. The probabilities in scientific research apply to propositions based neither on observation, nor defined in terms of observational categories. Probability in scientific research is a calculating device which relates abstract concepts at some theoretical level. *To imagine that there is an issue to do with the probability of the truth of propositions of chemistry and physics and their interpretations to the observable world , is a category mistake .* In scientific research, unlike in empiricist statistical research, the question as to the probability of the interpreted, observable phenomena is irrelevant. For the relationships between the interrelated propositions of the Scientific Theory and between that Scientific Theory and the observable phenomena, are already established correctly or incorrectly by that very evolving Scientific Theory.

Second, what of the claim that classical Newtonian mechanics uses probability? This is the claim that the phrase, "mechanical state of the system" does not mean the set of *theoretical* variables of the state, but rather the set of *experimentally measurable* values of positions and times. As Nagel indicates, if this were the case, then the theory of mechanics would assert no more than a high statistical correlation (or "relation of probability") between the experimentally defined mechanical states of the system at various times. This would mean that the laws of mechanics should not be construed as strictly universal statements, but as "idealized schematizations" of the true state of affairs, and as, "only relations of probability" (Nagel, 1969, p.284). As Nagel goes on to point out, this view is confused. It confounds two separate issues. *There is the issue of the internal structure of the theory itself— of the relationship between the abstract concepts in the theory. And there is secondly the issue of the manner in which the theory connects with the observable world,* " . . the problem of the precision with which predictions of the theory are actually confirmed by experimental data" (pp.284-285), by way of abstractive connection.

Classical mechanics involve a set of equations which relate specific traits of bodies with particular physical properties. The science of classical mechanics is a systematic study of the relationships which exist amongst a large number of properties belonging to a certain kind or class. But not all such properties require explicitly noting, because there is a small set of variables (the coordinates of position, momentum, and a point-mass) in whose terms variables for other mechanical properties can be defined— so that coordinates of position and momentum constitute the variables of state in mechanics. Thus, if we know the position and momentum of a particle, then its kinetic and potential energies can be calculated. As a result, if the force-function and the mechanical state for a system at some initial instant are given, the equations of motion are able to determine a unique mechanical state for the system for any other time, and hence also deterine the magnitudes of all other "mechanical properties" of the system at this particular time.

The second half of the problem of probability and classical mechanics refers to, ". . the precision with which predictions of the theory are actually confirmed by experimental data". *We can say that of course, actual measurements confirm the predictions of the theory only approximately. But this does not make the theory probabilistic* . After all, any theory which is stated in the form of magnitudes capable of mathematically continuous variation necessarily must be "statistical"— *in that restricted sense*. The numerical values obtained by experimental measurement of such physical magnitudes as velocity do not form a mathematically continuous series. Any set of values obtained in this way, will, in abstractive connection, show some *dispersion* around the values which have been calculated from the theory. But this is a very different situation from that of present-day, empiricist research, where all the data are in one way or another statistically established.

What of the third claim that classical statistical mechanics is probabilistic? Classical statistical mechanics were developed as Scientific Theory to account for the properties of gases. The Scientific Theory (Cluster 9, Type 15) made the assumption that gases are aggregates of large numbers of molecules, the movements of which can in principle be analyzed by Newton's equations. In fact it is not possible to find out the physical state of the system. And even were this possible, it is not possible to predict future states of the system, because of the immense complexities involved in the solution of the vast number of simultaneous differential equations of motion. In order to be able to get around these difficulties, *the theoretical expedient of a statistical approach* was adopted. The imaginative conceptual leap was: even though it is not possible to predict the individual motions of the molecules, it is still possible to predict certain *average* values of magnitudes associated with the individual motions. Thus, the statistical-mechanical state-description is defined by way of the variables of the *statistical* state, not in terms of the state variables of particle mechanics. But this is an imaginative Cluster 9 kind of conception, not an empiricist-statistical one, because with respect to its *own* mode of specifying the state of a system, such statistical mechanics are a strictly deterministic Scientific Theory (Cluster 9, Type 14), not a probabilistic one. Statistical mechanics are not probabilistic, because, if we have the value of the state variables for some initial period, then the Scientific Theory *logically determines a unique set of values* for those variables for any later period.

But what of the fourth claim that at least quantum mechanics are probabilistic? At first glance this might seem to have some purchase. However, the same issues apply here also. *Probability in quantum mechanics is applicable only to the propositions involving the abstract concepts within the Scientific Theory, not (as in empiricist research) to the observable phenomena.* The abstract concepts and their interrelationships are not based upon observation; and they are not defined in terms of observables. In quantum mechanics as in classical and statistical mechanics, probability is used simply *as a calculating device at some theoretical level* , and it is related derivatively to the observable phenomena. Unlike the case in empiricist research, *the question of the probability of the observable phenomena does not arise* . The relationship is established by the abstractive connection between the Scientific Theory and the observable phenomena.

Probability in empiricist research on teaching and probability in scientific research are of entirely different significance.

Scientific Research makes Use of Ratio Scales in its Measurements

As pointed out in Chapter II, because they use mathematics rather than just statistics, and because they use abstract concepts embedded in the propositions of Scientific Theory, scientific researchers are enabled to use ratio scales in measurement; in contrast, because they use statistics and propositions relating general concepts, empiricist researchers on teaching are unable to make use of ratio scales.

Walberg admits this point in his discussion of meta-analysis quoted in Chapter IV. He says that problematic measures, ". . are likely to be necessary until an advanced theory and science of educational measurement develops ratio measures that are directly comparable across studies and populations" (1986, p.216).

Walberg's statement is significant, both for what it admits, and for what it omits. Walberg, himself a leading synthesizer of empiricist research, admits that there is not yet an advanced theory and science of educational measurement, which makes use of ratio measures. That is indeed true. One of the features of Scientific Theory is its use of ratio measures such as centimeters, seconds per second, Newtons, Ohms, degrees Celsius, BTU's, Angstrom Units, milligrams per milliliter, and so on. But carefully to be noted, and what Walberg omits, is the fact that in science this use of ratio measures occurs in the rational connections within and amongst the propositions of the Scientific Theory and in the abstractive relationships to the observable world.

Empiricist research on teaching uses so-called nominal, ordinal (and, it is claimed) interval scales. And the hope is that somehow ratio measurement will evolve. It is a fond hope, because ratio measures (i.e. measurement proper) cannot be used in research until there are propositions which link abstract concepts. So it is merely ingenuous to hope that pedagogical research will develop ratio measures without first developing the requisite type of Scientific Theory. Contra Walberg, it is not an advanced theory of pedagogical measurement which is required. To search for ratio measurement prior to producing the Scientific Theory is to put the cart before the horse. Measurement in science is *a way whereby* propositions which involve abstract concepts are both interrelated and interpreted to the world.

It is thus ironic that an empiricist researcher such as Kerlinger can make the following claim,

> The day of tolerance of inadequate measurement has ended. The demands imposed by professionals, the theoretical and statistical tools available and rapidly being developed, and the increasing sophistication of graduate students of . . education have set new high standards that should be healthy stimulants both to the imaginations of research workers and to developers of scientific measurement (1986, pp. 431-432).

It is ironic, because the scales used in present-day empiricist research on teaching have *nothing* to do with *scientific measurement* .

Scientific Experiments are Radically Different from Empiricist Experiments

As mentioned in Chapter V, the accepted view of experimentation in empiricist research does not follow the experiments of Galileo or Lavoisier, but follows the Fisher statistical designs first suggested in the present century. We should recall the experiments described in Chapter IV. Experimenters go into classrooms and manipulate the observed atomistic variables of teaching in order to establish their relations with the variables of results on standardized achievement tests. Ideally, samples are drawn in such a way that every member of the target population of teachers and students has the same chance of being included. Ideally, its basis in chance allows for the application of the statistics of probability, with calculations being made to show the likelihood that the experimental and the control samples are drawn from the same population.

I used the word, "ideally" intentionally. For the problem is that whereas Fisher's design worked reasonably well in research on agriculture (and still does), the situation is radically different in teaching. Not only is the experimental approach of empiricist research not the experimental approach of scientific research, but in being used on teaching, it is being used in an empiricist area to which it is ill suited. Thus the obvious conceptual and practical problems are considerable.

In research on teaching, random sampling is not even closely approximated to. Random samples of schools, of students and of teachers would all need to be drawn separately. Given that one or more such samples are not so drawn, we are then faced with all the problems of so-called internal and external validity which arise therefrom, which Campbell and Stanley (1963), and Snow (1974), and all the empiricist research manuals so carefully, and properly (given the assumptions of their approach) note. *There are also the complications of content, context, and so on* [5]. There is the problem of assessing the "yield" (called "measuring" by empiricist researchers): the more specific and appropriate the methods of assessment of what has been learned, the more difficult are they to actually achieve.

Stenhouse (1979) has pointed to what are perhaps the most devastating negatives of all. Unlike in the agricultural case, experimental indications of supposed measures of gross yield, as a basis for standard procedures to be applied are completely unacceptable in teaching. First, because actions involve *meanings*, which logically cannot be distributed systematically, they cannot really be scaled or measured, and as a result cannot be controlled, or statistically sampled [6].

---

5. Which problems I discuss in the next chapter.
6. I shall say something about yield here and write in some detail about the impact of meaning in teaching, in the following chapter.

Second, and of singular significance for such research, experimental results as measures of gross yield are acceptable only on the assumption that *the procedure need not be appropriately varied within the population*. This assumption is acceptable in agriculture, but manifestly not in teaching. With a standard procedure, it does not matter if some or many plants fail or die, as long as the total yield is increased and/or improved. But in teaching, every pupil and student counts and we can *and ought*, for reasons of Normative Theory, Type 12, to vary "treatments"[7]. Given that this kind of empiricist experiment developed in agriculture, it may be pertinent to consider Stenhouse's analogy that, "The teacher is like a gardener who treats different plants differently and not like a large-scale farmer who administers standardized treatments to as-near-as-possible standardized plants" (1979, p. 77).

Drawing upon Heuristics from Empiricist Theory, the findings of this experimental design are a statistical form of empiricist generalization. It will be remembered that the results obtained are of the following kind: questions should be asked one at a time and be appropriate in level of difficulty (University of Canterbury); more time was spent in reading groups and in active instruction, and less time was spent dealing with misbehavior; reading groups should be organized for efficient, sustained focus on the content (Brophy and Evertson); teachers should move around the room a lot, monitoring pupils' seatwork and communicating to their pupils an awareness of their behavior, while also attending to their academic needs (Gage); teachers' diagnostic competence *per se*, was not related to progress in learning (Helmke and Schrader); the difference in classroom pass rates on the BSF curriculum-based tests is 5.2% in mathematics and over 10% in reading (Finn and Achilles). These findings are empiricist statistical propositions, or empiricist generalizations, or normative recommendations based upon empiricist generalizations. For concomitant reasons, they cannot be related one to another in a scheme of rational connection, which continually evolves and interrelates experimental results, developing the Scientific Theory even further. They can at best be related in a commonsense scheme of empiricist connection of emphases, of statistical correlations, and of general and normative claims.

Empiricist pedagogical researchers not uncommonly make assertions about scientific methods which are just incorrect. Thus the aforementioned Campbell and Stanley are famous for claiming that: "If the more advanced sciences use tests of significance less than do psychology and education, it is undoubtedly [sic] because the magnitude and clarity of the effects with which they deal are such as to render tests of significance unnecessary" (1963, p. 42). But scientific research is not concerned with what Campbell and Stanley call, "effects"; it is concerned with rational connection between abstract concepts embedded in Scientific Theory, which is tied deductively by abstractive connection to the empiricist concepts of the observational world. And in that which Campbell and Stanley call, "the more advanced sciences", presumably physics and chemistry,

---

7. See discussion of individuality in the next chapter.

tests of significance are not used. However, in contrast, say, in the medical research which contributes to what is itself a Practical Theory (Cluster 4) which draws upon various sciences, tests of significance are used in the comparison of, for example, new drugs; but this has little to do with the central concerns of scientific research which use and develop Scientific Theory. While belief in the possible efficacy of the new drug is based upon rigorous biological and chemical knowledge, the *comparisons of treatment* remain statistically empiricist. Scientific research proper produces powerful results *because* it does not deal in "effects", because it does not need tests of significance, and because it uses quite different methods. As already pointed out, scientific research and Scientific Theory became powerful centuries before tests of significance were invented.

Consider the conflation of these two fundamentally different forms of experiment, in the following quotation from the educational research manual by Borg and Gall:

> We stated above that the experiment is the most powerful research design currently available for testing hypotheses about causal relationships between variables. Yet it is also important to note that the experiment is not a perfect method. Even the findings of a well-designed experiment are potentially refutable. This point is well made by the philosopher Karl Popper: *But what then, are the sources of our knowledge? The answer, I think, is this: there are all kinds of sources of our knowledge but none has authority. . I do not, of course, deny that an experiment may also add to our knowledge, and in a most important manner. But it is not a source in any untimate sense.* As you read this and the next chapter, you will find that many factors can threaten the validity of an experiment. By controlling these factors, the power of an experiment to demonstrate a cause-and-effect relationship is strengthened. As Popper observed, however, it is not possible to do an experiment that provides an irrefutable demonstration of cause-and-effect. Therefore, replication of experiments-- especially ones that test alternative causal hypotheses-- are desirable (1989, pp. 641-642).

There is here a mischievous and gross "slide" in the meaning of the key word, "experiment". Whereas Popper is talking about true scientific experiments, such as those of Galileo and Lavoisier, Borg and Gall's book describes *empiricist statistical* experiments ! Popper, the world's greatest living philosopher of *science* is being erroneously brought into the argument to bolster the epistemological and social status of empiricist statistical experiment.

The experiments of Galileo and Lavoisier, on falling bodies and combustion were not Millian/Fisherian statistical experiments, but were guided by the abstract concepts of their evolving Scientific Theory, together with the conceptions provided by their prior thought experiments and this Scientific Theory. And the reason for varying the conditions in such an experiment is *not to test the probability, or degree of truth of the hypothesis* , as empiricist researchers and too many logicians believe, *but to discover the scope of the hypothesis* . As described in Chapter II, in Galileo's elegant experiment on accelerated motion, the theoretically derived relation between the distances and times was tested by letting the ball roll a quarter, a half, two-thirds, and so on, of the groove, and measuring the time taken in each case. The relations discovered by Galileo and Lavoisier were not judged by tests of significance, but by that which could be established between the rationally connected propositions of abstract concepts and the derivations to the observable world.

As implied in Chapter II, what empiricist pedagogical experimenters fail to realize is that in scientific experimentation there is already organization-- inbuilt Scientific Theory already at work in helping scientists to reach pertinent results by way of corroboration, elaboration, or falsification. This aspect is missing from empiricist experiments on teaching.

The powerful and ampliative results obtained by Galiieo and Lavoisier and the scientists since their time cannot be achieved by the present approach to experiment in pedagogical research. *Only by replacing the Mill-Fisher approach with experimental approaches which are driven by Scientific Theory will the kinds of results which have been achieved in scientific research become possible in research on teaching*.

This looks like a task of considerable complexity, and it is. Indeed, it is immensely difficult; and it may be impossible. It may be asked how experimenters can make use of Scientific Theory in their experiments, if there is as yet no such Scientific Theory in pedagogy? Yet surely, it may be said, there can be no such Scientific Theory in pedagogy until real scientific experimentation is carried out? But it was just such a situation which Galileo and Lavoisier faced in their respective areas. Consider Lavoisier. It will be recalled that, "Chemical analysis depends intimately upon the purity and identity of the substances under investigation. On the other hand, the purity and identity of the substances depend necessarily upon the correctness of the chemical analysis." But, together with the results of other researchers from which to draw, Lavoisier gradually fought his way through the confusion to develop the necessary abstractions and their interrelationships. For instance he laid it down explicitly that an element was the most basic stage of chemical decomposition and that in all experiments an equal quantity of matter exists both before and after the experiment, and nothing takes place beyond these changes and modifications of the combination of elements. These conceptions have become chemical presuppositions which are obvious to us, but were not obvious to the eighteenth century. Lavoisier was developing the Theory at the same time he was doing the experiments. Scientific experiments are made in order to test a true hypothesis (Cluster 5, Type 8), i.e. some implication of Scientific Theory, and the design depends upon the current content of the Theory and the kind of question being answered.

The strength in the empiricist experimental design is that it is a general one which can be used with any Heuristic. From the point of view of scientific research however, this is precisely its weakness. For the following reason, even this strength is something of an illusion. Because it derives from Scientific Theory, the result of a scientific experiment has much of its scientific significance written on its face. But even when an empiricist experiment is judged to be statistically significant, a further, and exceedingly important decision needs to be made: the decision whether this difference is of *practical pedagogical significance*. Because this decision is both philosophical and normative, and requires educated commonsense, and because the statistical methods themselves cannot make this decision, this crucial point has normally been ignored by researchers in reporting their results. In an effort to improve matters, results are now often reported in terms of the size of the difference found between the different treatments— called a "confidence interval", or "confidence limit". This is a change, but only a small improvement. For the issue of establishing pedagogical importance remains. Hence Scriven's apt claim that perhaps most pedagogical research has been trivial for most of this century (1988, p.134).

It may be worth remembering that Wegener, from outside traditional geology, began something which eventually turned the earth sciences on their heads. If there is ever to be scientific research on teaching, then I believe it will have to come by way of an original movement in theory (Clusters 3, or 6, or 9)— perhaps something like Macmillan and Garrison's erotetic theory (1988).

Serendipity occurs in Scientific Research

There are numerous instances of chance happenings in scientific research, which have resulted in important discoveries. We may cite such well-known examples as: Darwin's chance reading of Malthus when he was at the time puzzling over a "mechanism" to explain evolution, or Fleming's chance leaving of a mould onto which the penicillin bacillus fixed itself, or (less well-known) Watson's ideas about "cosy corners" in the DNA helix, which derived from his chance reading of a paper on the structure of metals, or his idea that the fact that adenine always paired with thymine and guanine with cytosine meant that the base sequences of the two intertwined chains were complementary. Of this episode, Crick says,

> The key discovery was Jim's determination of the exact nature of the two base pairs. . He did this not by logic but by serendipity. . The more important point is that Jim was looking for something significant and *immediately recognized the significance of the correct pairs when he hit upon them by chance* — "chance favours the prepared mind" (1988, pp. 65-66).

Perhaps the most famous serendipitous discovery was that of Roentgen's observing of the darkened platino-cyanide plate, which led him to X-rays. In relation to this discovery, Medawar makes the following telling and disturbing point. In the late nineteenth century, surgery had improved radically, but the surgeon was handicapped by not knowing in advance what was wrong inside. Suppose that it had been proposed to research into a way of making human flesh transparent? Suppose also that a project review system for the award of grants similar to that which prevails today in scientific and empiricist research applied at the turn of the century. We can imagine,". .the regretful, almost pitying shakings of the head that would accompany the verdict that such a proposal was idiotic in the extreme. Imagine, too, the meeting of a peer review committee as they judged their colleague to be in need of psychotherapy" (1981, p. xxi-xxii). Nevertheless it is the case that a method of making human flesh transparent was indeed discovered by Roentgen in the course of his normal approach to his scientific duties.

Such discoveries could not have been premeditated, or predicted.

In contrast to all the above, there seems never to have been any serendipity in empiricist research on teaching. To mimic Medawar's words, the empiricist researcher by the literature he reads (other empiricist studies) and even by the company he keeps (statisticians and other empiricist researchers) is preventing himself from winning a prize; he has made himself discovery-proof. Given that there is no explicit, interrelated, theoretical base (Cluster 9), and that empiricist research has such an inductive orientation, it is even difficult to imagine what serendipity or luck could *mean* in such research[8]. The delights and insights of serendipity in any field of endeavor are reserved for those who have an appropriate background; and the delights and insights of serendipity in scientific research are reserved for those who have a scientific background.

Scientific Research as an Enterprize has Social Aspects

Scientific research also has a social side. It applies in relations between scientists. Emphasis on the social is one of the strengths of Kuhn's argument. But Kuhn's point is also astray and has confused a whole generation of empiricist pedagogical researchers and social scientists. *For there is both a positive and a negative aspect to this social dimension.*

The positive aspect is that essence of scientific research as a social system, which is *its marvellous procedural values* — including its checks and replications and institutional criticism, which make for a general rationality and a continuing evolution of its accounts of the empirical world. Thus we can regard scientific research, "..not as a particular set of procedures or techniques, but rather as a general commitment to evidence" (Siegel, 1988, p. 52), logical, conceptual, and observational. The scientific community, advancing its scientific knowledge of the world is quite compatible with partiality, petulance, and passion in scientists as individuals. It is the enterprise or institution as a whole which censors and restricts partiality and corrects errors. Bronowski is precisely to the point when he writes,

> The values of science derive neither from the virtues of its members, nor from the finger-wagging codes of conduct, by which every profession re-

---

8. It would seem that so-called cognitive science may be developing theory and that this may begin to provide one kind of background against which there could be serendipitous findings. In my view however, cognitive science is not Scientific Theory, but theory in the Cluster 3 or 8 senses.

minds itself to be good. They have grown out of the practice of science, because they are the inseparable conditions for its practice (1965, p.60).

Bronowski is alluding to the pursuit of truth, the honest description of results, the self-destroying nature of fraud. Without these values, one scientist cannot rely upon the work of another. If such values are not followed in the overwhelming proportion of cases, scientific research just cannot continue. Truth must be the overwhelming *value*. "Truth is the drive at the centre of science; it must have the habit of truth, not as a dogma but as a process" (Bronowski, 1965, p.60). Such procedural and social values have been responsible in large measure for the immense success of scientific research over the years since Copernicus. It is such values which put the test to the claimed relationships between the abstract concepts produced by the creative imaginations of individual scientists.

The negative aspect certainly features in scientific research, but is not what makes it acceptable. Rather it is what distracts scientists from their search for the way the empirical world works. It consists in those social pressures to conform, and those fears for one's professional position which keep scientists from striking out in original, but empirically-valid ways. Indeed the empirical checks on scientific claims and the working applications in technology force scientific research to be, for most of the time, a rational matter. Social pressures can have negative effects, perhaps even long-term ones, but never permanent ones. One of the most long-term negative effects of social pressure occurred in the earth sciences, in the U.S.A. during the period 1926 to 1964, when continental drift was scorned [9].

As Marvin writes, "It is a rare scientist who is willing to be caught out favoring an hypothesis that others laugh at" (1974, p.95). And as Hallam reports: ". . for an American geologist to express sympathy for the idea of continental drift was for him to risk his career" (1976, p.15). Opdyke recalls: "I came from a department of geology at Columbia University that was dead set against continental drift. Marshall Kay, Walter Bucher, Joe Worzel, and those boys wouldn't hear of it; it was anathema" (reported in Glen, 1982, p.312). Pressures must really have been considerable. For as Hallam points out, "It would be quite reasonable . . . to expect a large number of earth scientists to remain somewhat sceptical or at least unconvinced, but why did not such an ex-

[9]. Continental drift became a source of comic relief, and taunting in geology and geophysics classrooms across the U.S.A. For instance, P.E.Raymond, Professor of Palaeontology at Harvard, relished telling the story of how he had found one half of a Devonian period pelecypod (a bivalve mollusc) in Newfoundland and another half in Ireland; and when he placed them together, voila, the two matched perfectly! They were actually the two halves of the same mollusc, which had been wrenched apart by continental drift in the late Pleistocene ! (Marvin,1974,p.106). "Although any loose statement denigrating or mocking continental dispersion got easy passage and approval for publication, anyone who was unwise enough to argue for displacement of continents was cold-shouldered by referees and editors, and became the butt for snide comments" (Carey, 1976, p.9).

citing new idea stimulate active research into testing it, rather than promoting such a vehement adverse reaction?" (1983, p.151). For opponents, drift theory may have been seen as a threat to their livelihood and to the accumulations of their life's work. Du Toit speaks of how in the history of geology, ". . it is interesting to observe how deeply conservatism appears to have become entrenched" (1937, p.1), and of the danger, ". . lest the science become stereotyped through too close an adherence to accepted beliefs." It is also germane to mention that Carey, Professor of Geology at the University of Tasmania, and a supporter of movement since the late 1930s, visited the United States in 1966. In addresses at a number of universities, he encountered deep hostility[10], but talked so persuasively, ". . that, much to the discomfort of geological faculties, he aroused the interest of many graduate students in continental drift, an idea totally new to them" (Marvin, 1974, p.152). In Chapter II, it was mentioned that a year after the large-scale change in beliefs, Harvard University Press published a book listing the main ideas of the first half of the twentieth century— which made no reference to continental drift! So some individual scientists can be bigots. But the point to notice is that despite all this negative pressure, eventually the new conceptions triumphed.

It is in the social sciences and empiricist pedagogical research that this negative social aspect of science has been powerfully and destructively paralleled and has often taken the place of the rational and truth-pursuing critique and falsification of scientific research. For, unfortunately, in empiricist research as currently pursued, it is even more difficult to make rigorous and definitive checks against experience. It was just this negative social aspect which restricted research on teaching for the eighty or so years in which evolving statistical-empiricist approaches of the Mill-Fisher sort were seen as *the* ways of doing research. It was the power of their advocates, and their social control in university departments and in powerful organizations such as the American Educational Research Association, which rigorously excluded alternative ideas. As significant evidence, it should be noted that it was only in 1988 that the *American Journal of Educational Research*, the house journal of the AERA began to publish research which was not observational, experimental, or statistical. And this occurred only after a long struggle by members who saw value and possibility in alternative approaches.

Conclusion

In any discussion of what makes scientific research, scientific research, there can be no "knock-down" arguments. But a case can be constructed by accretion. This is what has just been attempted in the above points. If sufficient criteria are marshalled, which empiricist research on teaching fails to meet, then it looks like a completely different enterprise.

If those who research on teaching wish to do scientific research, and to claim

---

10. Personal communication from Mr. D. Hannan.

its kudos, then they should try to do scientific research, and not something else under its name. If empiricist researchers on teaching really wish to begin doing scientific research, then an immense intellectual effort will be required– of a radically different kind from those achieved so far. The points made above have many implications for that which empiricist researchers would need to do to achieve this status.

One other thing should be said. Much of current scientific research is "big science"– physics, say with the particle accelerators of Fermilab, or CERN, or the International Thermonuclear Experimental Reactor (ITER), and biology, say with the huge teams engaged in the project which is mapping the human genome, or in research on nerve regeneration. But scientific research did not begin as big science. How many research assistants and collaborators did Copernicus, Galileo and Newton have? And if a scientific approach to pedagogical research is possible, it will not be big science either.

As a simpler, and, in my view, decidedly more realistic option, suggestions are made in the next two chapters, regarding what researchers might do to substantially improve the empiricist efficacy of their current approaches.

CHAPTER VIII

ACKNOWLEDGING THE COMPLEXITIES OF TEACHING WHICH ARE BEING IGNORED BY PRESENT-DAY EMPIRICIST RESEARCH

As a reminder of the complexity which must be taken into account if an adequate empiricist approach is to develop, the following four short accounts of teaching are provided. In a short space, the immense empirical and conceptual complexity of teaching, and by implication the inadequacy of typical empiricist approaches and findings are here demonstrated. In the first example Bush stresses *personal initiative*, *perseverance*, *standards of attainment*; he builds *self-confidence* at the same time as the boys create *a thing of beauty and utility*; he helps them to see that *school is a part of, not apart from life in general*. In the second example, Campbell has students *not merely reproduce scientific propositions, but gets them to understand them* and *come to see their relevance for their everyday lives*; he helps them grasp that *science is not a set of dogmas, but the striving for a continually evolving account of what might be the case*. In the third example Milz makes use of *subtle forms of motivation*, getting young children *not merely to read, but to want to read more*. In the last example, Harry ensures that students grasp the idea that *simple black and white categories are inadequate to deal with the complexities of actual historical situations*, as is a *too rapid moral judgment*. These examples also show how *depth of knowledge empowers teachers*. (All the italics are added.)

The first three examples come from Ken Macrorie's wonderful book, *Twenty Teachers* (1984). It shows even better than can the merely four examples provided here, just how complex teaching is, and how inappropriate, arrogant, perhaps perverse, it is for empiricist pedagogical researchers to believe that their present methods can capture anything of serious importance for the understanding of teaching.

Scriven has put my concerns nicely in the recent book on methods of research published by the American Educational Research Association:

> .. the same factors that lead to the preparation of conceptually incompetent researchers explain the rash tendency of researchers to rush into building a lifetime of research on a foundation of conceptual sand. You might, for example, build a lifetime of research on the teaching process upon a weak analysis of what teaching itself is. Teaching is, in fact, a very difficult no-

tion to define, and if you decide that you can define it in any way you like or if you think it is a very simple notion to analyze, you will finish up doing a great deal of research on a process which is definitely not teaching, and is only related to it in some obscure way. In short, your research will be useless for any practical concerns, such as the improvement of teaching, and for any theoretical concerns since, as you have defined teaching, its relevance to any understanding of the phenomena that people are trying to explain in working on a theory of teaching is not clear (1988, p.136).

Allowing for the Complexities: Sam Bush, Master Teacher of Woodwork

A boy comes to Bush wanting to build a chair. Right from the beginning Bush makes the boy decide. What kind of chair is it to be— period? modern? kitchen? black? The pupil is initially confused, but recovers. Bush insists on drawings, because to cut wood implies one knows what is required. Carpentry is not just playing about. Bush peruses the drawing, then asks the boy what he believes he is doing. Bush wants the boys to be able to justify their decisions, to realize that they are responsible, and that their, ". . *decisions are taken seriously, and because often [their] ideas are better than mine.*" Then he has them draw the object to scale, using proper drawing instruments. He wants to see the difficulties of construction made perfectly clear to them. At that point Bush may make suggestions re improvement, but without being overwhelming. He says that if he tells a boy he must do a particular thing, then the boy will do it, but, ". .he will also hate me for stealing his project." At this point Bush makes a rapid judgment of character, in order to encourage the boy, without allowing him to overestimate his own abilities.

Bush tries to make the boys believe in themselves, to realize that they can do things they never dreamed possible. He takes all their attempts seriously, but he is hard on the child who will not make up his mind. "I want him to draw up a personal expression with conviction, stick to its construction . . . and see in its completion a measure of his integrity." Bush is trying to introduce the boy to the boy's own self, and to deal with the work in a mature way.

There is nothing mysterious about woodworking says Bush; there are no perfect answers listed at the end of the teacher's copybook, which the boy must reproduce. "The unknown solutions are available through concentration, effort and study." After everything is decided in detail, only then do the boys begin sawing. Bush explains the use of every tool when the tool becomes required, and he expects the child to consider its function with care and to use it thereafter without further instruction, for he emphasizes that all the abilities required are within the boy himself. He continually tells them that, ". . *the release and development of these capacities— perseverance, imagination, courage, decisiveness— are the value of the work and the point of the course* ." Bush expects them to develop qualities of character, of permanent value to them.

He says that the pieces of furniture which visitors can see, ". . standing around waiting to be finished by last semester's pupils say more to the boys than

I can say. When they walk into this great room, they see that they are expected to do work of a very high quality." Bush does not allow any changes of intention in weak moments: for it is only their own selves which stand between themselves and an achievement of lasting importance. The boys have the examples of their predecessors to live up to. Bush points out that most furniture stores do not sell work of the standard the boys produce. "Do you see that mirror? Isn't that turning done beautifully? That piece means something special to me because it was done by a boy who was falling apart in life." Because he involves himself with every aspect of the work, Bush is aware when a boy is having trouble. Based upon his estimate of the boy's character, he may provide indirect help.

Such a manner of teaching is immensely time consuming, but Bush cannot see any other way of achieving such significant results. Unfortunately, such an approach restricts the time Bush has for his own work, but somehow he finds time because he believes it is important for him to be seen to be making objects, ". . . so that the boys' efforts are not so much in a school shop as in an active, creative studio. . . An opportunity is before the boys to make something excellent. It's a rare student who doesn't sense that immediately." When a problem arises, he speaks to the boy or to the class to emphasize that what they are doing in the shop is not something divorced from the other activities which they engage in, " . . or from their core as persons. . . I want them to dive into [their] idealism with high expectations and then measure up. . . All the work we do is hard. The problems. . . demand their full attention and effort. If they do a sloppy job, I usually ask them to do it over."

At the very beginning, when the boys are standing expectantly, waiting to start a project more challenging than they have ever encountered, Bush emphasizes that courage and hard work will succeed. (Macrorie, 1984, pp.4-8).

Allowing for the Complexities:
Don Campbell-- Teaching High School Physics

"I guess I've always thought students learned science best in the lab. I try to start from the lab and go to the classroom wherever possible. Let them do some observing, get a functional background to base their theories on." Campbell's students begin with a course on optics. Only after they have observed some of the behavior of light, he believes, is it appropriate for the question to be asked, "What is light?"

Campbell uses a film which begins with several simple experiments, difficult to perform in a classroom. In a demonstration of refraction, the film's viewers see events as though they themselves were in a swimming pool, below the waterline, looking up at a swimmer whose feet are dangling over the edge into the pool. At one point, the feet appear to separate from the rest of the person, who appears to float in the air above the water. What the film is then demonstrating are refraction and total internal reflection. Naturally, the students want to view that part of the film again, to check if they saw what they thought they saw. They then start to consider possible explanations. "I tell them that as we go along in this course, we're going to look at a lot of phenomena and then try to pull them together[1]...When they start to ask, "Why did that happen?" or "What's going on there?" I tell them to remember the question and we'll take a look at it later." Often the students can then answer it themselves.

Campbell's students do experiments with plane and parabolic mirrors. They then do an experiment with a mirror, a long strip of tape, a clear filament flashlight, a source of electricity, a 3 X 5 card, and a meter stick. The students are instructed to place the light bulb about one and a half focal lengths from the mirror, to turn on the light, and to move the card in the reflected light until an image, the real image of the light bulb appears on the card. "Oohs and aahs can be heard all over as they begin to see the real image for the first time. . Move the card and it is just light, but at that point, and only that point, there is the image! . . Until the students have hands-on experience with reflection, they don't have much real understanding of such a phenomenon."

Campbell avoids any cookbook approach. He does not tell them what light is. He tries to get them to develop their own ideas after they have observed enough for the phenomena to have meaning for them. The students have encountered science books before. So they soon suggest that light is a wave. Other students wonder about this. Someone will suggest that it may be a particle. "I ask, 'If light were a particle, how would it reflect? How can we test this?' and we talk about the bending of light as it passes from one medium to another, the refraction, and they learn that Newton predicted it would be a particle, and that light would travel faster in glass. When measured, light was found to travel more slowly in glass" (i.e. this theory, Type 8 was not corroborated). At that point of course, the students are confident that light is a wave! Cannily, Campbell then works through ideas of wave theory (Cluster 9) and the students begin considering the possibility of waves. What might waves do? How might waves be used in explaining the phenomena the students have observed? Many of the students remain convinced that light is a wave, but some still maintain that light is a particle. Campbell then introduces the photoelectric

---

1. There is an important difference between discovery and justification in original scientific research, and the pedagogy of teaching Scientific Theory, which has already been discovered and justified, e.g. Campbell talks about the students' "observing some of the behavior of light", but he can present this to them only because the developed Theory of Light is already in existence. It is however possible for a careful, committed, and imaginative teacher to bring pedagogy closer to the actual world of scientific discovery. Teachers can put students in problematic situations. As Freundlich argues, "The important thing is that students. . grapple wth phenomena that contradict their expectations, that they come to grips with problems not solvable within their own conceptual framework. They will then be aware of their own pictures of the world and of the possibility of others; and they will then be cognizant of the problems a specific picture attempts to solve" (1980, p. 120).

effect and says, ". . . that here was Albert Einstein looking at light somewhat like a particle again, calling light a "photon". Students then want to say, 'Well, it must be a particle that travels in waves.' " As a result the class evolves a supportable working theory of light (Cluster 9, Type 15).

Campbell uses a unit from the Project Physics Course of the Harvard Study Committee approach, "Where is the earth in the universe?" This begins with the Ancient Greek answer to that question and then works through Ptolemy, Copernicus, Tycho, Galileo, and Newton. The students begin to appreciate the cumulative nature of science (despite Kuhn !).

Campbell says that when he was in high school he memorized Kepler's three planetary laws, but had no idea of how they were applied. In his own course he has students apply some of the data from the Project Physics Course that allow them to plot the Mars orbit. They plot the data and then try to apply Kepler's law to the orbit they have drawn. At that point Kepler's Laws take on meaning. The students can see that the orbit of Mars is not a circle. In a rudimentary way they are able to grasp why Kepler had such difficulties.

In the lab, Campbell gives students sufficient information to be able to make some decisions and to attempt some solutions on their own, after which he asks some leading questions. "Let them work things out for themselves. . . Nine sets of students working on a problem get a straight line and the tenth set gets a curved line. Is the curved line wrong? I never deal with it as if it's wrong. If the technique were wrong, then we'll deal with that. But maybe they were the ones who found the thing that's right." Campbell tells the students about some of his own experience in research— that sometimes you expect nature to work one way, but it fails to do so. For example, we know that the calibration curve for a Geiger tube goes out in plateaus and then the line turns and goes up. But during a summer institute in which Campbell was studying, he found that the curve for his tube ran up, plateaued, and came down. He reported it in his class as a coming down, and all the other teachers laughed. So Campbell challenged the teachers to test it in the lab. When they did, they discovered that the line graph for their Geiger tubes indeed went down instead of up. They had all been unaware that the tubes with which they were working were flawed. So says Campbell, he tells his own students always to report what they have found not what they believe they ought to have found.

Campbell reports that some students have all sorts of neuroses about science and think it is boring because they have never had any positive encounters. He is certain something can be done about this. Thus he introduces a technique or idea only when it appears that it will shortly be needed. About six weeks from the end of semester the students perform an experiment where they calibrate the field strength of a coil in fundamental units, and measure the strength of the magnetic field. "We had a nice big equation on the board and I asked them, 'If I had put that on the board early in the course and asked how many of you would try it, what would you have done?' Most of the students said, 'I would have walked out.' But by the end of the year they can handle it pretty well."

Campbell also attempts to make science relevant to the students' lives. In discussing momentum and energy, he often uses cars as a frame of reference, e.g.

calculating the forces on someone who is wearing a seat belt during an accident. As a result the students see clearly why bones break (Macrorie, 1984, pp. 137-146).

Allowing for the Complexities:
Vera Milz and the Teaching of Reading

Milz says that as a beginning teacher in the early 1960s she was expected to make use of a basal reading text. It was Dick, Jane and Sally. The old "Look, look! Run, run! See, see!". In the workbooks the children had to search through the page for any words beginning with "b". Milz was comfortable enough during the first year trying things out, but by her third year could see that this approach failed to meet the needs of all the children. She began to explore alternatives.

She discovered Bill Martin, the author of the *Little Owl* books; the Grade One children loved a book called *Brown Bear, Brown Bear, What do You See?* They would grasp the rhyme and begin to read, "Brown Bear, Brown Bear, what do you see?" "I see a red bird looking at me." They used the pattern in other writing, e.g. at Halloween they produced, "Orange pumpkin, orange pumpkin, what do you see?" "I see a white ghost looking at me." Roach Van Allen, of the University of Arizona was another inspiration. He argued that children should write their own books, and then learn to read them. Leland Jacobs from Columbia Teachers College introduced her to the vast world of children's literature. Every two weeks he brought a new author to the class to talk to the teachers. As a result, Milz discovered that there were all sorts of exciting books available for children, but her school district mandated a set text for teaching reading, so that she could do these other inspirational things only in spare time. She visited other schools and saw children who created stories that were typed into books, which they then read aloud.

She has a mailbox in the room, where the children can receive notes from her, and send her letters. She always replies to them by the next morning. These notes provide a personal reading text for each pupil. After being absent or sick, she returns to many notes saying that they have missed her. Some children now in high school still communicate with her. The pupils also keep a daily journal, which they like to read to one another, and often share with the entire class.

As she empasizes, this is all very different from teaching children first a letter, then a sound, then a word, and expecting they will somehow put all the words together to make meaning. Reading she points out is not just connecting a number of words. For in talking we say things which create a meaning so that our communications are understood. "I want to help children to do this right from the beginning. . . They write because their notebook must have their name on it. They label their possessions and find that if they lose them, they get them back. . . In contrast, many teaching programs ask children to learn words such as 'the' or 'and'." But, she argues, such words by themselves lack meaning.

One year there was a nine year old boy in her second-grade class with a school record that said, "Cannot read *The Little Red Story Book* ." He had repeated the class several times. "I said, 'I think he's proven that he can't read it,' and I asked him to write about things that he liked. 'Tell me about your house. What's your bedroom like? What's your favorite toy?'. . until then he had shown no progress on any reading test. But by the end of the year he had shown six months' growth in reading. . . He [had] felt really bad about himself. Now he's a child who's excited about reading."

Milz invites parents to become involved. She has them write to their children, by putting notes in their lunch boxes. The children's stories are typed up and placed in covers. As we all know, there's something wonderfully exciting and encouraging when writers see their own work in any kind of published form, and children are equally encouraged. They write about happenings in their own lives, just as professionals do. When she reads the class a story by her favorite children's author, Tommy DePaaola, she points out that he is writing about his own grandmothers and about other relatives, and that of course the children can do the same.

Fifth and Sixth Graders come back to Milz's class and read the stories they wrote in Grade One to the children.

Unfortunately, Milz still finds that too many teachers in too many districts are being forced to use basal reading texts, with boring checklists of skills, with the ironic result that there is *no time for the pupils to do any reading of real books*.

Some publishers use professional authors, but in accordance with superficial curriculum theory (several Clusters) they nevertheless still simplify the vocabulary. But Milz has taken the simplified version of a story and also the original and had her pupils read both of them. The children, ". . had no difficulty handling the so-called 'hard words' of the original stories. In one basal reader one publisher changed 'fortunately' and 'unfortunately' to 'what good luck' and 'what bad luck'." Her children just love to play with long words, so when she presented them with the two books, they inevitably went to the one with "fortunately " and "unfortunately".

She found out that it was not pedagogically dangerous to let her children make mistakes, yet fail to correct them immediately. During one school year she retained everything the children wrote, and found no evidence that children continued to make the same mistakes. Milz argues that the continual new learning which the children were acquiring, had the concomitant effect of changing their spelling.

By the end of the first school week some of her first-graders are already going to the Media Center or Public Library saying," 'I want the books by so and so'-- the same author they've heard me read in class. *If they like the book they want to read more by that writer-- just as adults do* . . .How are my children doing in reading? *Standardized tests can't show that fully.* For one thing, they fail to take children's logic into consideration." Again, she says, ". .*there's no place for one of the girls in my class to put down that in one year she has read three of the Laura Ingalls Wilder books at an age when many children in conventionally taught classes are reading nothing but primers* . Likewise a child who loves books but may be at a point of struggling to learn to read may be labeled a failure by the test" (Macrorie, 1984, pp.79-91).

Allowing for the Complexities:
Harry and the Teaching of American History

Harry views the history of the United States as the history of the growth of opportunities for participation in the processes of democracy. He argues that US history should focus on the struggle of Americans to get their, "share of the American pie". *This view of history makes certain periods critical*, such as the drafting of the constitution, Jefferson, Jackson and the coming of the Civil War.

Thirty-seven years of teaching history have made Harry see more in US history than just the growth of opportunities. He realizes there are many different stories of his country. Harry knows what key events and ideas each story tells well, or leaves out. It is important to note that he has used many different stories in his long career.

One year he taught the history as the history of Black struggle. But although Black history highlights certain important events and ideas, there are other important ideas that are left out. "What you end up not doing are certain familiar things that maybe are important to a person outside the Black community, which is essentially a white context. The struggle of slaves for freedom in Jacksonian America does not reflect all the important issues of Jacksonian America."

The third way to teach American history is to focus on economic growth. He has never done it, but includes a heavy dose of economics in the approaches he currently uses. A fourth way is to teach it by topics, which Harry does not like and has never done. The disadvantage of topics is that they do not satisfactorily demonstrate causes and effects and the logic of events. A fifth way is the judicial approach, but he sees this as fundamentally political, though in a different sense. A sixth way is to see it as a course on American culture. This approach he has tried and liked.

This extensive knowledge of the potential of American history, and of segmenting and structuring that story, *has led Harry to realize that there is one period that is more imortant than others* . In Harry's view, students will not understand American history without understanding the Age of Jackson. The three key ideas in the Age of Jackson which must be made clear are the growth of democracy, the growth of industry, and the potential for conflict.

He looks at the way in which the Age of Jackson exclusively restricted civil rights to free white males. Then he looks at one group that was excluded– to see what American society and Andrew Jackson did about the Indian. They decided to move all Indians who lived east of the Mississippi, to the west, a situation most vividly dramatized in the fate of the Cherokee, who were moved out of Tennessee in the Trail of Tears to Oklahoma. Jackson was acting in direct violation of a decision of the Supreme Court. He was doing that which was unconstitutional. But he was so popular that no-one was about to impeach him. But Harry and the history class decided to impeach him.

So the class held a trial where the greatest number of students were the US Senate. Others acted as the prosecutor, the defense, and as witnesses. His idea was: "Let's dramatize in this all the issues of the rights of people, the rights to be involved in the interference of government and the dangers of government intervening in your lives, the dangers of excluding some people while you include others. Let's bring all these out in a trial of impeachment of Andrew Jackson." This is only one example of innovative teaching which Harry has used when teaching this critical period. In his view, such simulation captures best the complexities of the issues of the Age of Jackson.

In this account, it is evident how Harry is able to segment and structure a story that includes many important ideas: civil rights, growth of opportunity for participation in the democratic process, interference of government, potential for conflict, the constitution, and the Supreme Court. His knowledge of content is important to his teaching. *Harry believes that he cannot teach a subject that he does not know as a scholar.* (The year prior to the interview he taught anthropology, but found difficulties for he had no depth and no sense of the discipline and its present development.)

As an historian with expert knowledge, Harry believes that he has to compromise, to popularise, when teaching American history to high school students. Having to compromise historical integrity troubles him, but he settles for teaching a few things well, at the price of ignoring issues that may not be as important. He is a talented storyteller and an excellent discussion leader. He makes extensive use of these talents in the classroom to make students listen and think (Gudmundsdottir and Shulman, 1987, pp.62-64).

The wide variety of teaching and learning portrayed in merely four accounts should be noticed. If may be plausibly suggested that the addition of several more such accounts would continue to expand this diversity. *The variety and multifarious nature of teaching thus become revealed* . Such accounts should also make clear that research methods which attempt to study teaching and thereby to improve it, will have to be very much more sophisticated than empiricist research as presently constituted. For instance, the reader will not have failed to notice the irrelevance of much of the empiricist research on reading by Anderson, Evertson and Brophy, considered in Chapter IV, for a teacher of Vera Milz's ability. Indeed, given the variety of ways in which reading has been successfully taught in the last fifty years-- phonics, phonetics, look-and-say, I.T.A., whole language, and so on (at age four, my younger sister just taught herself to read), there is something perverse about the confidence of these recommendations on the basis of "research".

Specifics from some of the above descriptions of teaching will be drawn upon in discussions which follow.

Some Conceptual Matters: Distinctions between Reason and Cause, and Action and Movement

As suggested in earlier description and discussion, empiricist research ignores the necessity and significance of theory of various kinds, and has implicitly assumed that a behavioral theory of observables (Cluster 6) is sufficient to explain and improve human educational activities such as teaching. Its practitioners still work with such a theory, despite disclaimers from people such as Gage and Needles (1989). *If there are ever to be adequate empiricist methods, this assumption must be recognized, and the situation strongly modified.*

In the present section it is suggested that causal explanation is *external* to its effect, while in contrast, reason explanation is *internal* to the construal of and meaning of the context in which it makes sense (Melden, 1961; Taylor, 1964; Aune, 1977; Davis, 1979). It is further argued that (again despite disclaimers in recent defences of empiricist research) such features make the actions of teaching different from the movements upon which empiricist research concentrates. An improved empiricist approach, an approach which researches into the whole empirical world rather than into just part of it, must allow for *both* causes and reasons. Indeed, reasons will be central.

Explanations of much of what occurs in the lives of human beings can be provided by pointing to reasons: he likes to go out in wet weather; she is interested in the causes of disease; the problem on which they were working indicated to them that the solution was to be found in the tropics; the teacher dropped the book on the floor, to show the effects of gravity; the lecturer coughed to get attention. But explanations can also be provided by pointing to causes. He slipped on the wet pavement; she contracted measles; the hot weather made them both perspire; the book was bumped from the teacher's hand; the chalk dust caused the lecturer to cough.

Situations in schools cover the whole range between the two extremes of reason and action and cause and movement. For instance, because of what the science teacher says and does over a term of science teaching, a class may become aware that care, order, sequence, and precision are important for their understanding of scientific experiments, i.e. they come to *grasp the reasons for* such procedural values. But throughout the term the teacher may also be making use of particular *techniques of behavior modification to achieve* orderly procedures and behavior by the students in the laboratory. In their strict form, techniques of behavior modification are *causes* of change. In most instances it will be reasons that are at work; in others, it will be causes. And there will be gray regions where we are uncertain whether the students' learning of procedures results from reason or cause, or whether both apply at the same time. Such gray areas do not however affect the significance of the conceptual distinctions, or the argument that adequate research methods must make allowance for both kinds of effect. Causal explanation may be enough for some pedagogical matters, but with most teaching it is reasons that are chiefly significant; the causal processes assumed to be so significant by empiricist research on teaching are mere periphery.

It is all too easy for empiricist researchers to overlook this distinction between reasons and causes, or to act as though only explanations in terms of causes or correlations were acceptable, or rigorous, or scientific. Correlations may often be thought to be implicitly causal, and are too often made use of by empiricist researchers as though they were causal (see, for example, Gage and Giaconia, 1988). This is unfortunate.

Let me now probe the distinctions between reason and cause, by considering some aspects of the enterprize of teaching (Komisar,1968; Soltis,1978). Teachers indulge in a wide array of physical moves, but in most cases not because they are caused so to do. They make the moves because of their intentions for the students, and their *understanding* of the aims, goals, rules, procedures, and subtleties of the enterprize of teaching, and of their *meaning* and significance, together with how they construe the state of affairs at that moment.

Suppose that in his teaching of the Age of Jackson, "Harry" has just written the words, "Trail of Tears" on the board with a new piece of chalk, when the chalk snaps. The sudden change of pressure under his hand *causes* him to scratch a chalk line under the three words. But this situation is quite different from that where he, in wishing to *emphasize the significance* of that forced march of American Indians to Oklahoma, deliberately draws a similar line under the very same words. In the latter case, Harry's grasp of historical significance provides the *reason* why he draws the line, not the cause. Because Harry has been teaching American history for thirty-seven years, such an action may have become habitual, but that in no way destroys the basic point. Prior to its becoming habitual, he would consciously and deliberately have drawn the line. In the first case the breaking and change in pressure caused Harry's hand to move in a certain manner; in the second case Harry's grasp of the meaning of the situation as part of his interrelated understanding of the context of U.S. history during the presidency of Andrew Jackson indicates to him that he *should* draw the line.

The reason he should draw the line is connected to or constituted by certain meanings, e.g. he wants the students to construe the Trail of Tears as an example of the authority of an extremely popular president ignoring the Supreme Court. The breaking of the chalk (the cause) had the *separate* existent of the scratching of the line (the effect). In contrast, the grasp of the significance of the Trail of Tears did not cause Harry to draw the line. Indeed it is perhaps to misdescribe the situation to talk about Harry's "drawing a line", for he is actually making that movement *as part of the teaching* . The reason for his movement is that he has grasped conceptually that visible lines under particular words are construed by observers (students in this instance) *as having a meaning* . Also carefully to be noticed is that this action is indeed happening in a rule-governed and meaning-impregnated classroom, i.e. in a pedagogical context where there are actions similar to the rules of a game-- like chess or basketball. Just as the *contexts* and the rules of chess and basketball give meaning to the movements of the players, it is the understanding by Harry and his students of the pedagogical context, of what goes on in history classrooms, of what history teachers and students know and do, which explains why he draws the line. There is no causation here.

What is more, whereas the breaking (cause) is *conceptually separable* from the scratching of the line (effect), the underlining is not something conceptually separable from the meanings of the context of history teaching; the underlining *is* the emphasizing; there is a composite "underlining plus emphasizing of significance", understood as such by both teacher and taught. If Harry, say, began scratching lines in a random way all over the board, or underlining everything he wrote, his students would indeed begin looking for a cause, e.g. Harry had just had a brain hemorrhage, or Harry was a previously unrecognized split personality and his second personality suddenly took control. Of course Harry could have a reason— such as wishing to startle his students, or to demonstrate the self-defeating nature of underlining everything and thus of "emphasizing" everything. But these latter possibilities strengthen rather than undermine the general distinction between reason and cause, because as soon as we knew about Harry's reason, we would realize that our speculation about causes such as hemorrhage or split personality had been unnecessary. *A causal explanation is external to its effect. A reason explanation is internal to the construal of the meaning of the context in which it makes sense.*

Of course it is always possible to provide at least some kind of account in empirical, causal terms of a teacher's movements, by drawing upon the disciplines of physiology, biology, biochemistry, biophysics, and so forth. And this is in principle possible for everything physical which human beings do. It is also this fact which helps to confuse the issue, because such accounts are quite *irrelevant to an explanation of why Harry is drawing the line* . He is drawing because of his understanding of teaching. Empiricist research as presently constitued is not able to uncover such meaning. A more qualitative approach, perhaps of an interpretive or hermeneutic kind seems to be required here (Ericson and Ellett,1982 ; Palonsky, 1986). Or perhaps all we require is some educated commonsense?

Let me now introduce the parallel distinction between actions and movements. For the supplementary theoretical orientation (Cluster 6) of empiricist research is to concentrate on movement rather than on action. This is a conceptual distinction of considerable import: *overt movement is conceptually different from action* . Here I would also see the significance of the philosophical distinction between the empirical and the conceptual [2]. For the claim is that the crucial issue may often be not so much a matter of looking and seeing, but rather of (a) careful thinking about the situation, and/or of (b) making appropriate enquiries of the participants: of making distinctions in meaning as they relate to a particular context in which human beings are active. For a teacher such as Harry or Vera Milz, to praise a student is not just to utter such words as, "Well done", but is to say them in particular contexts of history or of reading. As mentioned in Chapter VI, unless the student believes that his own achievement is worthy of such words, the utterance will not act as praise, i.e. will not *be* praise, but will be something else, such as patronization, or condescension, or irony, or sarcasm.

At times such *rules and meanings will be instantiatory of, or constitutive of* the action. This is the case in actions such as the performative utterance, "I now pronounce you man and wife", in an action such as christening a ship, or when Harry or Don Campbell writes the marks "49%" on the student's paper, thereby failing him. But, in the last example it is not the movement of the hands of Harry or Campbell which are instantiatory or constitutive of the failing of the student; it is the *meaning shared by teacher, student, and society, of these sym-*

---

2. Despite Quine's subtle arguments to the contrary (1961, 1969).

bols, *together with the agreed societal rule* of the teacher's right to perform this action.

Consider again the example of Harry's drawing the line under the words, "Trail of Tears". In terms of bland movement, we may perhaps claim that what we actually observe is a man's hand holding a piece of chalk and moving towards the right. At even such a simple conceptual level, it should nevertheless be noted that we have still drawn distinctions between men and women, and moving and remaining still, and are aware of different descriptions of direction. In other words, even to produce such a simple description of overt movement, we require conceptualizations in the sense of presuppositions which are brought to the observation, conceptualizations which might in some special situations conceivably be disputed. The important point to note is that for us to grasp what Harry is *doing* is not to look for matters external to the action, not merely to take note of the movement, but to grasp the meaning of the action within the pedagogical context. Isolated from these meanings and that context, there is no history teacher, there is no blackboard, and there is no "underlining" or "emphasizing". Indeed, *there are not even any students* . Take away the context of the history classroom in that school and the same observable movement could be any number of things: conceptual art, testing the legibility of the chalk, drawing the hypoteneuse of a triangle, drawing a specific length, "doodling", and so on. Indeed, exactly opposite actions are quite compatible with the *same observable* movements. As Dearden says, "A man writing his signature on a piece of paper may either be issuing a death-warrant, or issuing a reprieve" (1968, p.132).

Unless observers *share* such understanding of meaning and context they cannot *observe* the action. At best, they can observe the movements. The meaning of what people do, the appropriate way of describing their activities, becomes clearer and clearer as we come to understand the surrounding social life, and as we come to learn its public language (Winch, 1958). Because I am interested in an improved form or forms of empiricist research on teaching, of central significance to me is the fact that teaching also covers the whole range between the two extremes of reason/action and cause/movement. There is a vast range of intentional and habitual actions, action on the basis of short and long-term planning, and immediate *ad hoc* reactions to unforseen circumstances. There are formal and explicit social structures to which teachers and students conform, which explain why teachers and students can be seen doing particular things. There are reason-based activities, which become habitual, as when a particular teacher always ends the lesson with a summary on the blackboard. There are socially-formed responses, like the raising of a hand in response to a question. And at the far pedagogical end we occasionally move into the region of causes. To grasp the nature of teaching, to explain teaching, to be able usefully to advise teachers on how they should teach, researchers must come to grips with the whole range.

These distinctions between reason and cause and action and movement are stressed because they point to the inadequacy of the present kinds of empiricist studies, as studies which are attempting to grasp hold of the real empirical world of teaching[3]. The operational definitions and statistical tests and correlations exclude *the richness of the total pedagogical situation* .

Some Seminal Features of Teaching which are Inadequately Construed

Over the last few decades there have been numerous detailed analyses of the concept of teaching, such as those by Smith (1961), Peters (1967), Dearden (1968), Macmillan and Nelson (1968), Langford (1968), Hirst and Peters (1970), Green (1971), Beck (1974), Soltis (1978), Derr (1979), Kerr (1981), Hintikka (1982), Rothstein (1983), Chambers (1986), Macmillan and Garrison (1987, 1988), Heslop (1991). Unfortunately, in precisely the manner criticized by Scriven above, empiricist researchers, though claiming to do research into effective methods of teaching, have almost entirely ignored such philosophical analysis[4]. They have assumed that if they can both pick out observable examples of teaching and provide some operational definitions of atomistic aspects of the presumed, observed behavior of teachers, then they can immediately begin relevant research. Such a view is at best naive. The following sections will show why this is so. They discuss intention, content, context, learners, and meaning (all of which have been highlighted by earlier analyses), and attempt both to make use of earlier critiques and to present a new justification.

Acting on the Intention to Achieve Learning

Teaching, either as action or as enterprize involves *acting on an intention*. Following the seminal ideas of the writers listed above, it may be argued that teaching is *any action or set of actions intended to achieve learning* . This conception is basic, relatively uncontroversial (for an another view see Freeman,

---

3. There is of course a vast philosophical literature surrounding such distinctions. It may be that by some form of supremely sophisticated reductionism, reasons will eventually be explained by causes. Though there is some interesting work going on in this area (see for example, Churchland,1988; Evers, 1990), it is manifest that no such reduction is as yet in place, and thus it is perfectly pedagogically proper to stress the importance of such conceptual distinctions for a presently appropriate research on teaching. Indeed, even were the reduction to be achieved, in my view it would still be appropriate to make use of these distinctions for pedagogical purposes.
4. Drawing on ideas similar to those of the Duhem-Quine thesis, some philosophers of education (e.g. Evers and Walker,1982; Walker, 1985) have challenged the legitimacy of such analyses. Given that conceptual analysis of teaching and related notions has been characterized by the extreme care of detailed and extended treatments of examples and contrasts, and the exploration of the whole vast context of usages, my brief reply is that such challenges are themselves illegitimate.

1973), but very important. It is not begging any questions about method. Furthermore it provides a firm foundation for a developing discussion because, (i) in a general way it describes all the actions of teaching both in schools and in other places, and (ii) it separates the actions of teaching from other actions also carried out in ordinary life, such as chatting, arguing, playing, entertaining, and so on, because these have a different intention. Equally, it sits squarely with such developments at the cutting-edge of pedagogical theory (Cluster 4), as the erotetic theory of Macmillan and Garrison (1988).

To count as teaching, an action or actions must indeed be actions, not just movements or happenings. The teacher must have an intention to achieve some learning by learners, whether vaguely formulated in his mind, or very specific. As argued earlier, what is significant is meaning, the meaning the teacher intends to pass on, or the meaning of common symbols shared by teacher and student. From the accounts of their work given above, it can be seen that Bush, Miltz, Campbell and Harry work with a variety of intentions.

Empiricist research ignores the fact that when we observe someone teaching, the intention (and thus the action) may not be obvious. But if the intention is misunderstood, the observer may well misdescribe what the teacher is doing. As a result, it logically follows that any generalizations derived from such observations may be faulty. However, if such generalizations do prove to be apt, then we may need to search for some other explanation of this fact. The matters being dealt with are so commonsensical, that even a misconstrual of the intention has little adverse effect on pedagogical conclusions; or the commonsense features which underpin the Heuristic being tested are such that they assert themselves despite the faulty construal of the intention; or the generalizations are apt by accident.

It is relevant to remember that Gage writes, "By teaching I mean any activity on the part of one person intended to facilitate learning on the part of another". If he had really made allowance for intention, his research would have been radically different. This may be especially so with respect to the second or later part of a compound intention (Soltis, 1978, p.47). For instance, suppose Don Campbell is observed to be telling a story about his own experience with a Geiger Counter (Macrorie, 1984, p. 143), and periodically writing words upon the blackboard. It is not that Campbell wants the students to memorize the words on the board, or even to remember the experience itself, but rather that he wants them to grasp that scientific research often does not turn out the way we may anticipate. The story-telling is aimed at a continuing insight into the subtlety and open-endedness of science. Further, if Campbell stops talking for ten seconds to let the meaning of his words become clear, he has not then stopped teaching, for the hiatus is itself a part of a whole *sequence of acts* which together form the activity of teaching physics. (He is still acting on his intention.) His immediate intention in pausing is, of course, to pause, but his longer-term intention is to do something further, which relates to the learning of physics. This is why writers such as Komisar (1968) and Soltis (1978) have pointed out that the word, "teaching" may apply to a particular act, or to a whole activity or enterprise (where the intention of the enterprise as a whole may or may not be obvious at a particular time). Thus, unless there is some detailed discussion between Campbell and the observer about that which Campbell is doing, the research is likely to be superficial.

There is a further problem. The intentions of teachers change— they even change during lessons. William Hare is making a point, the importance of which is much underestimated by curriculum planners, when he says that, "We may see at once in practice that an objective that had been given careful consideration simply cannot be achieved in the circumstances (italics added)" (1982, p.137). Effective teaching may involve a change in the immediate objectives of a lesson, for the teacher then judges that at that stage either a different objective, or a modification of the original objective is required, if the overall general goal of understanding is to be attained. Effective teachers are sensitive to the continually evolving situation. The approach must be "appropriate to the learners" (see below). There will therefore be a subtly-developing relationship between long-term intentions and goals, and more immediate objectives and actions.

Empiricist research as currently practised can hardly deal with such subtleties of intention. *And in failing to deal with intention it is failing to do research into teaching*. Macmillan and Garrison raise this problem in a more profound manner, in pointing out that Gage in his Process-Product empiricist research, ". .lets his ontology follow from his methodology: he starts with a method and finds the objects in the world that fit it. This in turn makes him unable to deal with the central problem noted by so many critics including ourselves— the intentionality of teaching" (1988, p.75).

Gage and Needles in their 1989 paper have attempted to answer this criticism. Their key response is:

> The "low-inference" process variables used in some Process-Product research are not no-inference variables. They do entail some inference as to the meaning of the behaviors. Whether the research is conducted in a laboratory or a natural setting, meaning must be attributed to the behaviors. . . A Process-Product researcher in the classroom infers meaning from events that are concurrent with and closely proximal to a given behavior. Suppose a teacher utters words grammatically identifiable as a question. The researcher must infer whether the question was intended to (a) call attention to a student's inappropriate behavior, (b) make certain that the student has the materials needed to complete an assignment, (c) check on the student's knowledge of the lesson content, (d) probe the reasoning etc. . . By considering the behavior surrounding the question, the researcher can attribute meaning to the teacher's question. Typically, researchers do so; they do not record mere "behavior" uninterpreted as to intention (p.256).

Superficially regarded, this reply looks reasonable. Thus for instance Brophy and Good's widely-used "Dyadic Interaction System" forces the observer to distinguish between Discipline, Direct, Open, Process, etc. questions. Clearly, judgment as to intention is needed here. But there are still at least half-a-dozen problems. First, there is no check against the observer's inference, e.g., by asking the teacher after the session is over. Second, when fitting the presumed action into the pre-established observation categories, the subtle differences between, or the multiple nature of, or the evolving nature of intentions, are ignored. Third, the inferences about intention are not drawn merely with respect to one teacher whom the observer has come to know well (e.g. studied ethnographically), but about many teachers, and, what is more serious, the actions of these very different persons are statistically summed. Fourth, there is no check on whether the observer's interpretation of the teacher's intention matches the effect on the pupil(s). Yet in many cases we need to know the mental state of the pupils, in order to know how we should *describe* the action. For example, as discussed in Chapter VI, that which is being coded as praise will not be praise unless it is construed as such by the pupil(s). And the pedagogical significance of classifications such as Brophy and Good's "Direct Question", defined as, "Teacher calls on a child who is not seeking a response opportunity" will depend upon why the child is seemingly not seeking to respond. He is deliberately giving some other child a turn? Still thinking about the answer? Has misunderstood the point? And so on. Fifth, meaning is attributed by the observer in order to place the words uttered into the pre-conceived category; *but for the remainder of the research, all meaning is ignored.* Finally, there is a radically truncated epistemology involved. Heron (1981) has pointed out, that in scientific research as an endeavor, there are at least, a product, a process of inquiry, and an experiential knowledge acquired through sustained encounters with that which is being researched into. If this is the case with scientific research, then, in the case of empiricist research on human beings there is an even greater need for this, "experiential knowledge acquired through sustained encounter with that which is being researched into". As Heron says,

> . . empirical research on persons involves a subtle, developing interdependence between propositional knowledge, practical knowledge, and experiential knowledge. The research conclusions . . necessarily rest on the researcher's experiential knowledge of the subjects. This knowledge of persons is most adequate as an empirical base when . . researcher and subject are fully present to each other in a relationship of reciprocal and open inquiry, and when each is open to construe how the other manifests as a presence (1981, p.31).

The gist of this problem is that if the research fails to make appropriate room for intention, then it cannot, logically cannot, be research on teaching. While such a critical dimension is largely ignored, the research must be radically unsatisfactory.

So a key problem is failure to address intention. Yet an adequately abstract notion of intention embedded in a meaning-giving theory (Cluster 3 as part of Cluster 4) may well be a possiblility. For instance, Garrison and Macmillan strive to produce such a theory in what they term, "an erotetic theory of teaching" (1988). Whether this contribution is a move towards a true Scientific

Theory of teaching, or is merely a basis for a better kind of Empiricist Theory (Cluster 8) is a matter for debate and for future development. They believe that it is a move towards a Scientific Theory proper, which also involves intention. Macmillan in a more recent paper has provided a hint which might be taken up. He writes,

> We should not be afraid of "inferences" in scientific contexts. In physics, no one seems afraid of going from cloud tracks to electrons and atoms, but in education we are afraid of the inference from behavior to belief or value. What is necessary is a clear recognition of the place of inference in such areas, of the types of evidence that are appropriate— not a fear of inference itself. And we should also recognize that many things that process-product researchers take to be inferences actually are not; they are just normal attributions of actions to individuals based upon the common understandings that are built into our ways of talking and observing. If we want to study teaching, we have to take intentionality (*in all its guises*) into account (italics added) (1989, p.4).

If, as argued in this book, scientific accounts relate both propositions involving abstract concepts and propositions involving concepts of observables, then Macmillan's point is important, and the traditional negative reaction of empiricist researchers to the issue of intention may be rendered unacceptable. *For it may be that to acknowledge the place of intention is to be scientific, and to ignore or reject its role is to be unscientific.* It may be that an abstract concept of "intention" can be embedded in Scientific Theory so that it relates by rational connection to other abstract concepts, and by way of abstractive connection can be interpreted to the observable world.

Indeed, the whole program of PBTE described in Chapter IV has been based upon an ignoring of intention. As a result, it is hardly to be wondered that researchers who support PBTE have tried to make their "modules" "teacher-proof": teachers' conceptualization has been considered irrelevant. PBTE construals have been more "teacher as robot" than "teacher as person". It is deeply ironic that many supporters of PBTE have at the same time characterized criticism of their program and proposals as unprofessional.

There is a "Content" to be Acquired or Learning to be Achieved and the Teacher Must Know It

If teachers intend to bring learning about, then there is a *content*, a subject-matter of the intention— that which is to be taught. The point derives from a simple, but profound logical truth, viz., that, as I have discussed elsewhere (1989, pp.47-58), there is always teaching or learning *of something* . This presents a problem for research in three separate but related ways. First, there is the effect on teaching and learning *of the teacher's own mastery* of his subject's content. Second, there is the *way in which different kinds of learning and different contents affect the manner* in which they should be taught. Third, there is the difficulty of capturing the complex varieties, and the more sophisticated kinds of learning in the *standardized tests of achievement* which are used for assessment within empiricist research.

Let us consider the teacher's own mastery of content first. The content to be acquired may be infinitely diverse and various. In the teaching described at the beginning of this chapter, all the following and much more could be content which is intended to be learned: the students' belief in their ability to construct fine wooden objects, or in their ability to finish what they have started, or to produce imaginative explanations for scientific phenomena; the concept of "refraction" or "electron"; the knowledge that light travels in straight lines, or that Andrew Jackson was the first American President from west of the Appalachians; a skill in the use of a carpenter's plane, or an ability to find the focal point of a lens; admiration for skillful carpentry, or for the early discoveries in inertia and gravity of Galileo and Newton; a liking for reading, or for books by Bill Martin; a propensity to be careful and circumspect when working with sharp tools; the habit of always storing tools carefully at end of a session in the workshop, or of always mentally summarizing the contents at the end of a chapter of the history textbook. We cannot conceive of teaching without some such content embedded in the teacher's intention. Empiricist research on teaching ignores almost all of the implications and effects of this diversity of content.

The details of the teaching of Bush, Milz, Campbell and Harry also point to the significance of the teacher's knowledge of content. For it should be clear that the successes they achieve would be impossible without their experience and their depth of knowledge. The need for this very subtle mastery of content has been acknowledged not merely by educators who stress the significance of direct instruction, but also by the most perceptive of all the progressive educators—John Dewey. Deeply concerned with the relationship between schooling and democracy, Dewey argued that particular approaches led to democratic virtues, while others negated them. In discussing what the teacher needs to know to be the democratic leader in the classroom, he writes:

> The first condition goes back to his own intellectual preparation in subject matter. This should be *abundant to the point of overflow*. It must be much wider than the ground laid out in textbook or in any mixed plan for teaching a lesson . . Some of the reasons why the teacher should have an excess supply of information and understanding are too obvious to need mention. The central reason is possibly not always recognized. The teacher must *have his mind free to observe the mental responses and movements of the student members* . . .Unless the teacher's mind has mastered the subject matter in advance, unless it is *thoroughly at home in it*, using it unconsciously without the need of express thought, he will not be free to give full time and attention to observation and interpretation of the pupils' intellectual reactions (italics added) (1933, pp. 274-275).

Dewey asserts that, "some of the reasons why the teacher should have an excess supply of information and understanding are too obvious to need mention." It may be assumed that Dewey had in mind such reasons as: being able to guide the learners to key issues and problems; being able to prevent learners from pursuing irrelevancies or matters of a peripheral nature; to make it possible for the teacher explicitly to summarize what lies ahead in the subject; to be able to give succinct, but accurate and intellectually respectable answers, conceptualized at the intellectual level of the learners; and so on. These requirements, together with Dewey's point that teachers need to be able to feel confident in their own depth of knowledge, so that they can give, ". . . attention to observation and interpretation of the pupils' intellectual reactions," and to channel the pupils' own purposes, show some of the subtleties of quality which can be introduced by a teacher who possesses knowledge in depth.

As study of the history of ideas demonstrates, and as current philosophers such as Toulmin (1972; 1982) and educationists such as Novak (1977) have reiterated, human understanding follows an evolutionary development, both in the bodies of knowledge themselves and in the individual human being's mastery of these. Teachers who themselves lack the grasp of their matter in such quality and quantity can only be aware of this evolutionary fact as a belief, not as something they themselves have gradually come to experience and know. *Not having themselves experienced so that they know, they cannot pass on this dimension in their teaching* : i.e. they lack an authoritative grasp. They can at best allude to it in a second-hand way. This is indeed a specific example of the general logical point that teachers cannot teach what they do not know. There is also the related problem of the mental empowerment given to the person who has knowledge in depth, an empowerment which is lost to the less knowledgable teacher. To treat these two cases as the same, as empiricist research has normally done, is to trivialize the significance of knowledge.

Because of the formidable variety of intention and content and the power of depth of knowledge, it is being argued here that teaching is a much more subtle and complicated set of actions than is commonly supposed by laymen, or allowed for by present-day empiricist research.

There is secondly the issue of the effect of particular content on *the way a subject should be taught and adjusted to the level of the learners*. An effective teacher of English for ten year-olds is not necessarily the same thing as an effective teacher of English for grade twelve; an effective high school teacher of English is not often an effective teacher of mathematics. This is because, ". . . the subject matter and the varying purposes for teaching different things at different levels themselves make demands on how the teacher should proceed" (Barrow, 1988, p.10). Indeed, as shown in Chapter V, only in a superficial manner has empiricist research addressed the significance of content. For a decade or so Margaret Buchman (1984) has been arguing for the significance of attention to content, and Shulman (1986a, 1986b, 1987), and other writers such as Winne (1987), and Smith and Neale (1989), in what has become known as "pedagogical content knowledge", have begun to point to the significance of knowledge of content and a teacher's ability to adjust it. This movement has also begun to consider these effects upon the success of teaching, and to suggest ways of making allowance for them in research. But the research methods they discuss are not traditional empiricist research methods, though they might in some sophisticated way be grafted onto empiricist methods as presently constituted.

There is thirdly, the problem of *assessing* that which has been learned as a result of the teaching. There is research evidence to show that there is insufficient overlap between the teaching and the assessment of the results of that teaching on standardized tests of achievement (Armbruster, Stevens and Rosenshine, 1977; Freeman, Kuhs, Knappen, and others 1983; Shulman, 1986). But commonsense and conceptual analysis show much the same things. Indeed, as many as thirty years ago, Banesh Hoffmann in *The Tyranny of Testing* (1962) showed how a keen intellect can expose all the inadequacies, subjectivites, ambiguities, dangers, and inanities of such tests. Thus at the beginning of this section, it was mentioned that what is to be learned may be principles, facts, concepts, laws, theories, skills, procedures, values, habits, attitudes, processes, and so on. Standardized achievement tests can at most test for the first few of these— and even then in a manner which must leave out much that is fundamental. Standardized achievement tests have three limited virtues. They can best cover a large amount of claimedly factual material in a short time. But of course, adequate learning even of facts involves additional aspects such as affect. They also make some allowance for that large and important part of our knowledge which is not easily available in response to direct questioning, but depends upon recognition: but even here, recognition in the situation of a test is scarcely recognition in living and meaningful contexts. Thirdly, they are easily marked— even by computers! But they can scarcely capture the vast variety and intelligence of learning. As Darling-Hammond so cogently puts it, standardized achievement tests :

> . . assess one's ability to find *what someone else has already decided is the one best answer* to a predetermined question. The tests do not measure the most important aspect of problem-solving ability— the ability to consider and evaluate alternatives, to speculate on the meaning of an idea based on first-hand knowledge of the world, to synthesize and interpret diverse kinds of information, to develop original solutions to problems. Moreover the tests do not really measure performance of any kind. Performance. . means the ability to do something; it is active and creative. Recognizing a correct answer out of a predetermined list of responses is fundamentally different from the act of reading, or writing, or speaking, or reasoning, or dancing, or anything else that human beings do in the real world (1984, p.56).

In short, there are problems with the validity of the key means of assessment used in empiricist research. To be valid, it must not just concern itself with a particular subject, such as mathematics, and with a particular level such as Grade IV , as a sort of given, as did Helmke and Schrader, above; but, at the very least such studies must differentiate between beginning teachers and knowledgable, experienced teachers. This point also relates back to that of teachers' intentions. For knowledgable teachers do things differently and continue to change them, because they are knowledgable. To study knowledgable teachers in the same manner, or to recommend that they use the same techniques as neophytes, or as the average, is otiose. Bush and Milz and Campbell and "Harry" are light-years away from the young teacher just out of college.

That in their more recent writing, leading empiricist researchers such as Gage (1989) have acknowledged the need for more subtle and epistemologically more complex forms of assessment is thus to be welcomed and profoundly encouraged.

There is a Particular Context Within Which the Teaching is Embedded

Because they change the meaning involved, different contexts change either the meaning intended by the actor, or at least how the action is construed by others. A child's avoiding eye contact in Vietnam is construed as a gesture of respect; in USA as a gesture of evasiveness, or insolence. Because different times of a week have different meanings for students, teaching on Friday afternoon may have different requirements from teaching on Monday morning. As already mentioned, appropriate learning takes place in a meaningful and evolving context, and actions are embedded in and take much of their meaning from the context.

We may note the words of two leading empiricist researchers in their summary of a wide range of empiricist research into teaching:

> The data reviewed here should make it clear that . . what constitutes effective instruction *varies with context*. What appears to be just the right amount of demandingness (or structuring of content, or praise, etc.) for one class might be too much for a second class but not enough for a third class. Even with the same class, what constitutes effective instruction will vary according to subject matter, group size, and the . . objectives being pursued (Brophy and Good, 1986, p.370).

These words are interesting because they appear to be making my present case: that context has a continuing complicating influence on how a teacher can and ought to teach, and thus has to be taken into account if there are to be rational research methods. Yet, except in the very limited sense of taking the grade level into account when deciding how to code the items observed, the empiricist approach ignores context.

The points made by Brophy and Good are not new, but they are correct-- indeed *necessarily* correct, i.e. we did not require empirical research of the empiricist or any other kind to establish their truth. For no matter how good or bad the teaching, *learning is something achieved by the learners, not something done to* them. The learners themselves, whether compliant or resistant to the teacher, in learning reluctantly or willingly, necessarily make the context vary. Furthermore, because of the unlimited complexities teachers face, and because of the unique combination of factors present at any one time in a classroom, unless there is to be robot-like implementation of a rigid, pre-conceived approach, the personal judgment of teachers just has continually to be used, with the resulting variation which Brophy and Good describe. Empiricist researchers also have consistently failed to realize just how radically a change of content or context (and concomitantly of intention) alters the appropriateness of a particular method or technique, and can immediately render inappropriate the ideas about how to teach, recommended by even the most recent piece of research.

The empiricist researchers, Good and Grouws have observed that the stability of context is an important issue, though one seldom discussed in (empiricist) research literature: "If the conditions of teaching are not stable, it is difficult to make inferences about teacher effectiveness. For example, if student populations are changing rapidly (changes in the school neighbourhood, bussing, etc.) then the lack of stability in teaching performance may be due to the fact that the conditions of teaching are changing" (1977, p.50). This is true and important, but reveals a lack of understanding of the deeper complexities. The authors are talking about merely the most superficial aspects of context. The most serious methodological error lies more deeply, being part of the point made in Chapter VI about general concepts such as "praise". My criticism there was that empiricist methodology assumes it can do something which is not logically possible, viz. separate the thing called "discipline question", or the thing called "praise" (or any of the other atomistically isolated features), from its integrated, interrelated context, and deal with it as an atomistic individual entity, analogous to an element in chemistry, or a force in physics. For this presumed separation involves a false epistemology. Of course the researchers believe they are making the separation, and they do indeed make observations and use check lists of presumed occurrences of the entity. But this is all merely surface method, which misses the essences and depths of the situation. We may recall the example of Sam Bush. His talk straddles both motivation more strictly interpreted, and praise. It is subtle praise, demonstrating Bush's confidence in his pupils. But it is not the kind of atomistic praise which has been the subject of empiricist research: it would not have fitted into the pre-conceived categories. Context is subtle, and its most significant effects and implications are glossed over in present empiricist approaches.

Consider another example— from Gage. Gage's findings may be some of the more acceptable principles which the Process-Product version of empiricist research has recommended. But most of them can be accepted only in *particular contexts judged to be suitable* by the teachers. For instance, one of the recommendations mentioned in Chapter IV advises that teachers should call on a child by name prior to, ". .asking the question, as a means of ensuring that all pupils are given an equal number of opportunities to answer questions" (1978, p.39). But this advice though perhaps sound in many instances, is hardly context-less. Is making sure that all pupils have an equal opportunity to answer questions always the overriding concern? It all depends upon the reasons for the questioning: encouraging one particular child? keeping them all alert? summarizing rapidly? Gage explicitly indicates that this work is relevant for the teaching of reading in Third Grade, but that is context of a rather different kind: a very general context of level and content. The more a context varies, the less can teachers apply any findings directly, and the more they have to use their judgment and their art. Gage of course acknowledges that art must be used in applying his findings. But this admission rather misses the more fundamental point— as will be shown in the next paragraph. For one thing, art always implies *judgment* by teachers. And it is at least conceivable that with particular classes, teachers' judgment may suggest that they *never* make use of any of Gage's supposedly scientifically-established recommendations.

But there is an even more significant aspect of context. Thus, as just mentioned, in claiming that he is producing the scientific basis of "the art of" teaching, Gage properly acknowledges that teachers need to use his findings *in an artful way,* and this includes making allowance for the context. But what he, like most other empiricist researchers fails to realize is that this admission of the necessity of artful practice invalidates his whole approach to research, and the very basis of his findings. For (unless they were being forced to teach in a particular, pre-conceived manner set up to suit the research) teachers will have been using what they know about teaching *in an artful way while they were being observed by Gage and his researchers.* (Surely he does not think they should be or are artful only when they make use of findings such as his.) They will have been making typical pedagogical moves, which they artfully applied with respect to the context (and the intention and content and learners). And of course, we would expect there to be some positive statistical correlations between such typical artful teaching and successful learning by the students, even with inadequate instruments of assessment. What Gage's observational methods have really captured are simple generalizations about some *typical artful moves in the art of teaching. But his actual research methods themselves have completely ignored the existence of this artfulness.*

So there are three interrelated problems of context: (a) There is the epistemological problem: if context is ignored in the *establishment* of the findings, *How much and what sort of credence* can we place on them? i.e. What is their epistemological status and what does this status tell us about them? (b) There is the practical problem: as contexts vary, we have to ask ourselves, even if we accept the findings as being candidates for pedgogical help, *Where and when ought teachers to apply* these contextually-superficial findings? Moreover, (c) (and even more problematic for Process Product research) *how do the teachers know* where and when ? (Certainly not from the empiricist research itself.)

There are Learners, and the Method or Manner Must be Appropriate for Them

A special aspect of context is the learners towards whom all the intentions and contents are aimed. There must be some kind of mental contact between a teacher who knows X and a learner who does not (yet) know X. Learners are conscious, self-conscious, and to a greater or lesser extent rational. They have hopes and fears and intentions and personalities and characters and schemes of their own and engage with and filter the content being taught. Teachers must metaphorically put themselves in the place of the learners, to try to conceive of the difficulties the learners are encountering. Teaching aims at the achievement of learning in students, but learning is something *that the learners have to achieve for themselves*, not the doing of something *to* the learners. Learners too have intentions: learners are not just bundles of reactions to external stimuli. They are unlike stars in their courses, or physical forces, or chemical reactions, or DNA replications, or the movements of tectonic plates. They live in particular places and are affected by their classmates and by what they have already learned.

A little reminder for empiricist researchers may be useful here, a reminder of the unpleasant complications of the real world of teaching as contributed by the learners. In her pungent autobiography about teaching in a New York City high school, Natkins describes the following incident during her final term of teaching— after twenty years experience. A knowledge of the high schools of New York City, or of those of many other large cities will show that this incident is by no means untypical. Mrs. Natkins has just sent "Pablo", a hulking six-foot student to the Dean because Pablo has refused to change his seat in class. (Throughout the previous few lessons he has been sitting next to his girl-friend and fondling her under her skirt.) On leaving, Pablo slams the door furiously. Mrs. Natkins is standing with her back to the desk, at last getting some discussion going in the disturbed class, when Pablo bursts back in. In a fury he races through the room and pins her against the desk. His clenched fist is just an inch below her jaw:

> "Whadja fill out that pink slip for?" He was trembling with rage. I was petrified. Disjointed thoughts tumbled into my mind. . "Pablo," I tried to coax, "let's talk about the pink slip later. It's not that important." I couldn't get the thought of a broken jaw out of my mind. It terrified me. "Whadja fill out that pink slip for?" God, he was relentless. "Come on Pablo, let's talk about this outside. Let's not make things worse." I couldn't let myself stay pinned against the desk with that damned iron fist ready to smash my jaw. . "Look, she's scared," one of the girls giggled. "Yeah, she's turning white." Another giggler (Natkins, 1986, pp. 65-66).

Finally one of Pablo's friends, himself rarely in class, pulls him away. Pablo lets fly with his punch, which rips in two the *Newsweek* from which Mrs. Natkins has been teaching and is holding in her hands. "For the second time that period I tried to change direction fast. I sent a messenger down to the Dean's Office and tried to teach as if I were in control. The rest of the term would be Hell if I didn't" (p.67).

But except for measuring their learning on standardized achievement tests, and except for the development of ALT (itself limited), empiricist research has largely ignored the variegated aspects of that which the learners bring to the situation. Qualitative research on teaching has tried so to allow for the actions and intentions of the learners, but empiricist research has not. Once again, closer attention to context ought to help, with empiricist approaches perhaps being applied alongside, or adapting the techniques of ethnographic studies, used so as to take careful notice of the mental states of the learners[5].

This section may be summarized in suggesting that empiricist research has been inadequate in its attention to the real empirical world of teaching, which is immensely more complex than the present approaches recognize.

The Lack of "Validity" of Empiricist Pedagogical Research

If the above criticisms are accepted, then empiricist research such as Process-Product deals with only limited features of teaching, fails to allow for the continual interconnections and reciprocity of the different features of teaching, and so cannot be performing research on *teaching*.

Teaching implements intention, makes allowance for a continual evolution of context, gages the reactions and provides for the individual differences of the learners, and continually modifies and adjusts the specific subject matter– all of which interrelate in complex, reflexive ways unallowed for in empiricist research. In fact, were such a researcher as Gage to more closely examine his own very perceptive *accounts* of teaching, quoted above, he would see that his actual approach is scarcely structured to capture such richness. And if it cannot capture the richness, it cannot deal with the teaching. But Nuthall and Church in the piece of research discussed in Chapter IV write that, ". .[teaching] occurs daily, in much the same way, in thousands of different classrooms. However good or bad the teaching may be, it has the kind of *regularity which makes it an ideal object for systematic behavioural analysis* (my italics)" (1973, p.10)– indeed this is the assumption which underpins all the research considered in Chapter IV. For the reasons outlined above, this claim begins to look radically unacceptable.

Indeed, Gage condemns his own Process-Product empiricist approach out of his own mouth. In the 1989 paper he writes, ". . .The essence of Process-product is the search for relations between process and product variables. . . *A flaw in the essence would be fatal to the enterprise* (italics added)" (1989,p.254). Certainly the relationship between the variables is the the essence of its method. But the essence of any appropriate research enterprise must be not just research into vari-

---

5. Fortunately, the significance of some of these ideas is beginning to be recognized, and implications for research are being drawn. Jean-Luc Patry has been talking about teaching as situation-specific for some years (1987; 1990). As I was putting the final touches to this manuscript, I encountered an article by W.S.Carlsen in the *Review of Educational Research* which argues that, ". . . research on questioning must acknowledge that the meaning of questions is dependent on their context in discourse, that the content of questions cannot be ignored. . . Research on questioning has generally failed to recognize that classroom questions are not simply the behavior of teachers but mutual constructions of teachers and students" (1991, p.157). And also: "In recent years process-product research has been increasingly careful to qualify and interpret its findings by context. Context, however, is usually defined in a static fashion [e.g. by age and/or grade]" (1991, p. 172).

ables which can be conveniently captured. It must be research into the entity as it really is, i.e. into *teaching* – not into some pale simulacrum of it.

Later in the same paper, Gage avers that Process-Product work can be scientific, ". . only in the methodological sense of using tools to minimize bias effects, error, and unreliability, and to ward off the many kinds of threats to internal and external validity identified by Campbell and Stanley" (1989, p.259). Ignoring the debatable claim that an enterprise ,"can be scientific only in the methodological sense", consider the claim that Process-Product research is warding off, ". . the many kinds of threats to internal and external validity". What Gage's research has "warded off" are found in the standard list in empiricist research manuals, viz. such threats as history, maturation, sensitizing the students to the purposes of the experiment, and so on. But we should notice how such features are irrelevant, if the research, in dealing with problematic variables, is not itself dealing with teaching.

Perhaps there is something similar with respect to the application of its findings. It may be that its findings are most applicable in the arguably artificial pedagogical world which empiricist research has created for itself, for example where teachers are being forced to teach reading in the specific ways advocated by Process-Product researchers such as those listed by Brophy and Good, in Chapter IV. But empiricist findings would seem to be much less applicable in the pedagogically more sophisticated worlds of the teachers described in the beginning of this chapter. For instance, someone such as Vera Milz, the teacher of reading, would seem to be precisely the kind of teacher who would be constrained rather than aided by being forced to implement the findings of the Brophy-Good research on the teaching of reading.

Moreover, as already argued, the method of assessing learning in empiricist research is itself extremely circumscribed, and ignores many of the most valuable dimensions of learning. Such research tends to achieve similar results with the same kinds of procedures, i.e. the research is reliable in the technical empiricist sense. But, for reasons already given, such technical reliability may be worthless.

I would suggest that all too often what empiricist researchers have achieved is actually the reverse of Gage's claims. *Despite the best of intentions*, they have made the results reflect as little as possible the way things are in classrooms and as much as possible their own preferences.

CHAPTER IX

EMPIRICIST RESEARCH SHOULD ACKNOWLEDGE ITS EMPIRICIST STATUS DROP ITS SCIENTIFIC PRETENSIONS AND BECOME EMBEDDED IN AN ADEQUATE PEDAGOGICAL THEORY

It has been suggested by various writers, for example Warren (1976, p.31) and Barrow (1984, p.7) and in several places in this book, that a key reason for the limitations of empiricist research is that it fails to be rooted in theory of various kinds, and that the kind of pedagogical theory in which it should be embedded, must, amongst other things, have both empirical and normative aspects, i.e. it must not merely be concerned with what is, but must also be concerned with what ought to be. In both its research procedures and its applications in the real world of teaching, empiricist research needs to be embedded in apt and intelligent, explanatory and normative pedagogical theory. This means that, it must be based squarely in the alternative traditions, conceptions, and implications of the humanities, *because it is only in the humanities that we find sufficiently powerful images of both the is and the ought.*

We should remember just how limited are the findings of present-day empiricist research, "..compared to the knowledge base necessary for a teacher to practice" effectively (Confrey,1987, p.389). The point is however, as already argued, not just that the findings are limited, but that they are also often distorting of the very enterprize they are supposed to be helping. As mentioned earlier, it is not uncommon for teachers to use the procedures recommended by empiricist research only while they are being supervised, or examined, and to revert to methods they prefer immediately the observer leaves the room. There is surely something deeply and distressingly amiss with a method of research which promotes such behavior.

Embedding Empiricist Research in an Appropriate Pedagogical Theory

This chapter argues that not merely is the empiricist contribution limited in effectiveness at present, but that even after it has radically improved, it always will be restricted, because of the logically-mixed nature of Pedagogical Theory as an evolving kind of normative Practical Theory. Intelligent teaching must always involve contributions from Normative Theory and from the sort of practical experience that is based upon," . . common sense, casual empiricism, [and] thoughtful speculation and analysis" (Lindblom and Cohen, 1979, p.12). Amongst other things, it is the ignoring of the importance of teachers' personal experience and depth of tacit and practical knowledge, which so invalidates the widespread legislation of various American states— legislation based upon the findings of Process-Product, empiricist research.

The description of the theory of Cluster IV now becomes significant. It will be recalled that Dearden described such theory as, ". . the product of . . the endeavour to achieve an intellectually deepended understanding of educational practice in all its aspects: both curricular and institutional, and both empirical and evaluative" (1984, p.8). Such theory, as was argued above, is a kind of Practical Theory, which makes use of precepts, professional wisdom, traditions, assumptions about the nature of learning, psychological claims, ideologies about the best way to teach, beliefs about the nature of persons, social norms and values, philosophical argument, personal experience, and so forth. Despite its conglomeration of different disciplines and logical kinds, its point is always to help teachers and administrators to decide two logically different matters (a) *what ought* to be done in their classrooms and (b) *how* it should be done. Question (a) is logically prior to question (b), and though these are both immensely complex questions, and the relationship between them perhaps more so, something will now be said about them.

Empiricist researchers have tried to provide answers to question (b) without any explicit address of question (a). *In such research, the methodology has begged the normative question.* Even if empiricist research were scientific, what it should be aiming at, what it should research into, and when its findings ought to be applied, would still need to be drawn from Normative Theory, and would involve a thorough understanding of the context of application. Present findings are as applicable to an authoritarian approach, or to indoctrination as they are to genuine education. (Indeed, it is conceivable that in some contexts where they are opposed to a teacher's personal style, they may well cause psychological harm to the teacher and educational harm to the children.) And although it is the case that the actual research has been carried out in classrooms where education rather than indoctrination was probably the aim, that has been contingent. This situation is morally unacceptable. An appropriate research must itself be embedded in Pedagogical Theory which amongst other things involves Normative Theory. To ignore norms does not make them go away, it merely misleads, for they are still there, but lurking unacknowledged.

Normative Theory in Pedagogical Theory

A few key aspects of such normative theory will now be sketched. Numerous philosophers and writers have made important suggestions. One may cite as fine examples, Arnstine (1967), Bantock (1981), Crittenden (1981), Dearden (1968, 1972, 1976), Degenhardt (1982), Dewey (1916/1950, 1973), Flew (1976, 1985),

Francis (1976), Gowin (1980), Green (1971), Greene (1973), Heron (1981), Highet (1950), Jackson (1989), Kaplan (1980), Oakeshott (1962), Passmore (1980), Reid (1961,1986), Soltis (1978,1987), Strike (1982), Siegel (1988), Warnock (1977), J.White (1973, 1982), but particularly Peters (1966, 1967, 1971, 1973, 1977) and Scheffler (1960, 1965, 1973). The following suggestions are intended to provide merely general orientation, rather than detailed argument.

The teaching stressed by Scheffler in his seminal writings passes on knowledge, but passes it on as provisional. The emphasis is on knowledge at the reflective and critical level. Such orientations are excellences indeed. But they do not come easily. The propensity to search rigorously for counter-evidence that may show our beliefs to be false or at least in need of amendment, and the desire to make certain we understand what we are talking about are not natural inheritances, but must be *taught to* and *learned by* students. As Raths, Wassermann, Jonas and Rothstein (1986) make perfectly clear, intellectual "skills" like physical skills, require effort and practice. Scheffler emphasizes that teachers ought to place their own claims and their reasons for claims in front of the learners and they must develop the ability of students to independently judge these. As he says, that which distinguishes this sort of teaching is its emphasis on *rational explanation and critical dialogue*, its attempts to give honest reasons and to welcome radical questions (1965, pp.11-12). It should, ". . develop a sort of learning in which the student will be capable of backing his beliefs by appropriate and sufficient means" (1965, p.10). Such teaching involves a rigorous attempt to get learners to think rigorously. The teachers themselves must, ". . exemplify the critical thinking they strive to develop in students, combining tough-minded instruction with a penchant for inquiry" (Holmes Group,1986, pp.28-29).

Wherever appropriate, independent judgment must be encouraged, for two reasons. First, it is itself a general virtue. Second, until independent judgment is practised, students' grasp of a particular discipline remains second-hand. But carefully to be noticed is that the independent judgment which learners are capable of will only be as good as (a) their current command of the content and methods of that particular discipline and (b) their current general ability to think critically[1]. Indeed there is a sense in which independent judgment, depth of knowledge of the methods and content of a discipline, and general critical ability are all interrelated and mutually empowering. The capacity for independent judgment itself rests upon standards. As the Holmes Group suggests, such teaching, ". . identifies students' misconceptions, and question[s] their surface responses that mask true learning" (1986, p.29). Genuine respect for the intellectual integrity of learners thus stresses the standards which make reasons cogent and criticisms sound. Respect for independent judgment is also of course part of general democratic behavior and as such is part of one widely-accepted normative theory of our society (Cluster 7). It is to be noticed just how little traditional empiricist research says about, has addressed itself to, or has even ac-

---

1. For considerations of the nature of critical thought, see for instance, McPeck, 1981; Ennis, 1987; Siegel, 1988.

knowledged the existence of these matters.

In having respect for genuine learning, and in showing respect for students as persons, the teacher's own personal life is inevitably involved. In such teaching, teachers are not just observing life, or preparing students for life, they are engaging in life with the students. This is one reason why views of teaching which stress the findings of empiricist research too strictly, seeing teachers as "technicians" who are merely to apply such findings, are so dangerously astray. (For Gage's point that all such findings should be used merely as, "the scientific basis" of pedagogic *art*, tends to get forgotten by school districts, administrators, and supervisors, who take notice of what he and other empiricist researchers have recommended. For they look for the teacher's application of the "rules" in all lessons.) Using their imagination and insight, teachers can and should apply a spectrum of approaches: after becoming clear about the content to be learned, they can at the very least demonstrate, cajole, challenge, question, indicate, feign doubt, model, orient, coax, coach, hint, suggest, lead and elicit. Peters's general educational point about involving the "wittingness" of the learners (1966) applies here. Surely there has also to be an equivalent "wittingness" of the teachers in what they are doing. Perhaps it is only to be expected that empiricist research, embedded as it is in presuppositional behaviorist theory (Cluster 6) will have ignored the wittingness of teachers, and other aspects of the morality and humanness of teaching, for behaviorist theory is a-moral. But when it comes to pedagogy, the a-moral is often equally the immoral.

Some Empiricist Research as Anti-Educational

Except in setting up the Heuristic being tested, empiricist research ignores teachers' personal experience and depth of knowledge. It also ignores the personalities of teachers– yet their personalities are at the heart of what teachers do. Such omissions do not just lessen the efficacy of any findings, or contribute to the immorality of the approach, they raise huge questions about its validity. For as Herron has pointed out (1981), researchers ought not to assume one theory of behavior (Cluster 3) for themselves and an entirely different one for their "subjects". If their own theory is intelligently self-directional and intentional, then to be consistent, presumably rational people should also apply this conception to the "subjects". The consistency Herron is alluding to is the methodological consistency of persons doing research on persons; but, for this same reason that the research has to do with persons, we might equally claim that moral or normative consistency argues for the very same point. The point is even stronger: it is the ignoring of personal experience, knowledge (and intentions), intelligent self-direction, differences of personality, and so forth, in the "subjects" which *exemplifies* the immorality.

For his part, Peters is arguing in a wide array of books and articles, but particularly in *Ethics and Education* (1966) and *Education and the Education of Teachers* (1977), much the same ideas as Scheffler, but emphasizing the concept of *education* : the criteria which activities must meet to count as education. I have already pointed out that he emphasizes "wittingness" and voluntariness on the part of the learners; he also rejects the concomitant indoctrination, and argues for the broadening effects or education, the development of "cognitive perspective", and the quality of educational activities– which are in fact self-transforming. The "wittingness" becomes part of a subtle commitment to the subject, as the learners slowly come to appreciate its intrinsic values. The cognitive perspective becomes part of the transformation of the person. Failing to become so committed in this sense is at the same time failing to grasp the nature of the subject. Failing to develop cognitive perspective is at the same time failing to transform oneself.

When these ideas are put together, it may be suggested that as part of the underpinning Normative Theory, what needs to be recognized is that *teaching and research on teaching ought to be educational*, and that research into teaching just has to take this into account as a continuing part of its methods.

I should wish to argue that unfortunately, some of the effects of such research are *not educative, not even neutral, but are anti-educational*, as Degenhardt (1984) has so successfully shown, for the research is implicitly stressing a technicist, truncated, causal view of teaching, rather than a view based on moral norms, on significant human values, on meaning, and on the application of intelligence and judgment. As Fenstermacher says, "The effect of presupposing exogenous causation is the idea that teachers become more effective by [behaving] differently towards students (emitting different stimuli), as opposed to the idea that teachers become effective by thinking and feeling differently about what makes the activity of teaching worthwhile" (1979, p.173). Moreover, empiricist research, especially its Process-Product version is so influential and imperialistic in its claims to science that it *diverts attention (and money) from alternative, more humane ways of studying and talking about teaching*.

As just one example of a more humane way, we may consider Ayres (1989). In his study of six teachers, it is their *lives* which are seen as significant, not some empiricist or PBTE so-called "competencies": importance lies in the relationships between what the teachers *are* and how and what they teach. The author and the teachers engage in a biographical dialogue, an exploration, a refinement of personal conception. Unlike present-day empiricist researchers, the writer does not see himself as the expert come to show teachers what to do, but as a collaborator in helping them to find their pedagogical selves. There is a combination of interviews and observations, in which the teachers are brought to reflect upon and analyze themselves and their approaches. As the collaborative analysis proceded, these teachers became more able consciously to grasp their pedagogical selves."This process led to an awareness of aspects of their own practices that had been obscure or unavailable to them" (Ayres, 1989, p.8). And we who read the book become ourselves more pedagogically self-reflective, and probably, better teachers. Such pedagogical theory (Cluster 3 as part of Cluster 4) is itself educative. Much the same is true of recent books such as Macrorie's *Twenty Teachers* (1984), from which I drew three of the examples considered earlier, of Tom's *Teaching as a Moral Craft* (1984), of Connell's *Teachers' Work* (1985), of Raphael's *The Teacher's Voice: A Sense of Who We Are* (1985), of Palonski's *800 Shows a Year* (1986), of Freedman's *Small Victories* (1991), and others, none of which, students of research will ever encounter in courses on how to do empiricist research on teaching.

From such considerations it may be concluded that any adequate Pedagogical Theory as Practical Theory will involve something radically different from the simple-minded conversion of empiricist findings into seemingly straight-forward recommendations for action. There ought to be a Popperian openness about the use of the findings of empiricist research: at present there is too often a Kuhnian dogmatism. *Researchers have to justify themselves to teachers, not teachers to researchers.* As Stenhouse said in a talk to teachers,

> What I am trying to do is to encourage the feeling that all the statistics can be thrown out if they don't accord with the reality as you know it, and when you look at statistical results, somehow the thing to do is to end up not talking about standard deviations but talking about experience (1978, p. 67).

The place of various Clusters of theory in pedagogical research is grotesquely underrated. My point is that while inadequate research is bad enough, there is something more important: in ignoring the many different Clusters and in holding inadequate notions of theory, *the mind-set of people such as Kerlinger and Gage restricts conceptions of what research might be.* It holds back progress because it ignores or disparages the very Clusters and Types of theory whose development is required. Thus, Kerlinger is noted in many university departments for the "rigor" of his research approach. But the only rigor he displays is the rigor of the *statistical techniques.* His approach is entirely lacking in *conceptual* rigor; yet statistical rigor performed without rigorous initial careful and caring conceptualization of the nature of the matters being investigated is otiose. And besides using conceptual analysis, subtle conceptualization draws upon at least theory from Cluster 3 and Cluster 7, Type 12.

An Adequate Conception of Teaching

Such theory and such research would also need to include a clear conception of *what it is to teach in general and also what it is to teach particular subject-matter in particular contexts to particular kinds of learners.* Empiricist research as presently constituted has not come even near to such an achievement.

Two immediate candidates for the core of the general account of teaching are first, the traditional philosophical analyses of the last thirty years of Scheffler, Soltis, Passmore, Peters, Hirst, and so on; and second the recent erotetic account of Garrison and Macmillan (1988). There is considerable overlap between these analyses.

The traditional account of teaching mentions many of the matters discussed earlier in this chapter and points to the same sort of complexities raised above. The account by Garrison and Macmillan also bases itself squarely in intention, and develops several of the same concerns, for instance about content, but has a somewhat different emphasis, and also attempts a more logic-based description—

erotetic logic, the logic of putting and answering questions. Amongst their account they make the following important points. "To teach someone is to answer that person's questions about some subject matter" (p.21). "It is not that the student does ask the question. . but rather that the teacher believes that in some sense, the student ought to ask the question"(p.28). "It is the intention of teaching acts to answer the questions that the auditor (student) epistemologically ought to ask, given his or her intellectual predicaments with regard to the subject matter"(p.32). On such an account, it will be seen immediately just how important the teacher's knowledge of content again becomes. For, "To know which questions to answer requires that the teacher have considerable knowledge of the students and of the subject matter; without this, the teacher cannot know where a particular question or answer is going to lead" (p.38). But such a logic does not foreclose on how a teacher ought to proceed. Nor does it make decisions about content: rather it clarifies the nature of dealing erotetically with a particular content. The actual content to be taught will derive from that which is seen as valuable by theory of Clusters 3, 4 and 7, and from the epistemological structure of the discipline. How a teacher ought to proceed will derive from Normative Theory and from various kinds of improved empiricist research. Moreover, on an erotetic account, aspects of teaching outside those answerable by erotetic logic will still require refined pedagogical judgment, that is, such a theory properly allows for intelligent and educated action on the part of the teacher.

Local Theories as Part of Pedagogical Theory

What have sometimes been called, "local theories" may also be significant. It is not clear what variety of theory is intended in the term, "local theories". Perhaps there are features of Clusters 3, 4 and 5, with hopes of eventually becoming Cluster 9. Cronbach (1975) was an early advocate of the cogency of local theories, when he claimed that we should not make generalizations, ". . the ruling consideration in our research," but that the researcher should, ". . look within his data for local effects" (1975, p.124). Though Cronbach was talking about psychology, his point applies equally in pedagogy. Snow (1977) also suggested that there be a shift from universal general to local theories. He advocated developing, "local instructional theory" applicable in specific places and subgroups, saying that, ". . such theories would be intended to generalize more across time in one place than across places" (1977, p.12). Early discussion of local theories such as that of Gehlbach (1979) saw them in the quantitative camp, but in a recent article, Schibeci and Grundy (1988) emphasize that local theories can be used in cooperation with more detailed qualitative research. They write that local theories, ". . and case study methodology can be mutually complementary within a given context" (1988, p.93). Moreover:

> Local theories, like case studies, speak to specific situations. Unlike the results of pieces of specific research (for instance, a survey of a particular school population), local theories provide possible explanations rather than a specific body of facts. Because the theory generated by this research relates

to a local context, propositions that follow from the theory are more readily identifiable than propositions from universal generalizations. Thus local theories provide a more immediate basis of action than do other forms of quantitative research. In this way local theories closely parallel case research (1988, p.94).

It is noticeable that all this is hopeful rather than substantive— in their article the writers provide no detailed example. Nevertheless, the notion of a "local theory" does at least conceptually point to the kind of research mode into which empiricist research as a part of a general Pedagogical Theory may be able to develop. It may be that if the process-product version is going to take context into account, it will need to be radically restricted to particular situations. Thus one example might be the study of the teaching of mathematics, in Grade IV, by experienced, knowledgable teachers, in schools of a specific socio-economic level, in a particular city, at this time. It is conceivable that the Helmke and Schrader study of Chapter IV is really local theory masquerading as a more general level of theory— which is why questions can be asked about their recommendation for, for instance, "discrete support".

Misunderstanding the Importance of a More Complex Empiricist Method

It is not just that present empiricist approaches are inadequate for their subject matter. It is also that in misunderstanding the nature of scientific endeavor, and in over-emphasizing the importance of science as a basis for teaching, the researchers and the policy-makers who implement the findings, actually *undervalue empiricist research*. What Nathaniel Gage had to say some years ago is typical of such undervaluing of empiricist approaches. Given the fact that his own Process-Product work, like that of other present-day empirical researchers is itself empiricist and statistically sophisticated (though conceptually, qualitatively and pedagogically unsophisticated), this is ironic.

Not content to argue the case for his notion of the scientific nature of his empiricist work, he also attacks other ways of understanding teaching. He writes,

> Since *Summerhill* appeared in 1960, we seem to have been more than ever at the mercy of powerful and passionate writers who shift educational thinking ever more erratically with their manifestos. The kind of research I have been describing is a plodding enterprise, the reports of which are seldom, I regret to say, as well written as the pronouncements of authors unburdened by scientific method. But, in the long run, the improvement of teaching— which is tantamount to the improvement of our children's lives— will come in large part from the continued search for a scientific basis for the art of teaching (1978, p.41).

I share Gage's concerns about such books as *Summerhill*, which as any guide to what should be going on in, and how people should teach in, public schools, is complete pedagogical nonsense. But Gage's view of the alternative to building, ". . a scientific basis for the art of teaching" (1978, p.41) itself involves several mischievous misunderstandings. He writes, "For most of our history, the main alternative has been a combination of logic, clinical insight, raw experience, common sense, and the writings of persuasive prose stylists" (1978, p.41). Such an attack on non-scientific research on teaching cannot be ignored, for it helps to provide an inflated evaluation of present-day empiricist approaches generally, and of Process-Product research in particular.

Contra Gage, what is so deficient about logic (in whatever sense he is using)? And what is so bad if the insight really is "clinical"? Medical doctors seem to have achieved a very great deal in that way. And why does experience have to be denigrated as "raw"? Certainly,"Practitioners are an important source, perhaps the most important source, of practical wisdom in education" (Garrison, 1986, p.18) (why "perhaps"?). So there is surely mature experience, which itself can be subjected to sophisticated, adult, contemplative consideration? As examples of the cogency of experience, may be cited the descriptions and arguments of almost any of the recipients of the Presidential awards for outstanding high school teachers of science and mathematics such as James Skrocky (see for example, Maugh II,1984). Or consider the accounts quoted earlier from virtuoso teachers such as Sam Bush and Don Campbell. And who better than "Harry" to tell us about and to demonstrate to us the profound pedagogical "payoff", which results from depth of experience and practical wisdom? We have all had teachers in our own schooldays who inspired us. How many courses on the results of Process-Product, or any other kind of systematic empiricist research did they take? Or consider the writing of the books listed earlier. In their own diverse ways, all these authors show the immense importance of deep pedagogical experience, accompanied by perceptive, reflexive thinking. Surely experience can be strengthened by drawing upon various alternative theoretical (several Clusters) perspectives? And what is so suspect about commonsense? It is the pedagogical commonsense of classroom teachers which is often required to counter the unrealities of university educationists and researchers. Such experience and thought seem to me to be of deep significance, and essentially empiricist– but empiricist work of a superior kind. Stenhouse is much more realistic than Gage when he says to teachers:

> I want to make it quite clear that in reporting research I am hoping to persuade you to review your experience critically and then test the research against your critical assessment of your own experience. I am not seeking to claim that research should override your judgment: it should supplement it and enrich it. All too often . . research is presented as if its results could only be criticized technically and by other researchers. But I am arguing that it should be subject to critical appraisal by those who have educational rather than research experience and who are prepared to consider it thoughtfully in the light of their experience (1978, p. 52).

From the teacher's point of view, what empiricist research on teaching has at best achieved so far are claims which, ". . are not conclusions to accept, but hypotheses that need testing" (Stenhouse, 1979), by teachers in real classrooms.

Moreover, it is inappropriate to dismiss non-statistical work as "the writings of persuasive prose stylists". If all we had were, say, Rousseau's *Emile* and A.S. Neill's above-mentioned *Summerhill* , Gage might have had a point. But we also have sophisticated philosophical critiques of such work, such as Robin Barrow's *Radical Education* (1978), which separates the grain from the chaff of the "persuasive prose stylists"; indeed we have the writing of a whole generation of analytic philosophers of education such as Peters and Scheffler, and many other fine philosophers too numerous to list, which distinguishes the worthwhile from the misleading. Furthermore, we have the mature work of people such as John Dewey in *Experience and Education*. Dewey may hardly be called a, "persuasive prose stylist", but his writing provides teachers with insights and understanding, which they can use discriminatingly throughout their careers, and in many different contexts. Indeed the point is both more general and more significant. The pedagogical classics of Kant, Locke, Pestalozzi, Plato, Dewey, and all the others provide essential ideas for *pedagogical conceptions of what might be* — all sorts of fertile theoretical insights (Clusters 3, 4, 7 and 8), which should form the basis for a more intelligent, evolving, reflexive, empiricist research.

And although he alludes to other kinds of research, Gage seems to be unaware of their exciting *empiricist* possibilities. For instance, drawing on the distinction between discovery and justification, he says that qualitative and quantitative research in pedagogy operate in two different domains. He sees the place of the qualitative researcher as providing hypotheses for the quantitative researcher to test and justify or reject (1978, p.83). (Others such as Howe (1988) see considerable overlap and compatibility.) But there are deep problems in this viewpoint. How do the statistical relationships produced by Process-Product research *justify* ? First, suppose that a statistical relationship is claimed to have been established between two variables, say, teacher "warmth" and achievement on a standardized test. How is such a statistical relationship a justification for a connection? Those who advocate and pursue such research assume that without understanding why one action or set of actions is better it can be proved by statistical treatment that it is. But as has been emphasized throughout this book, in the absence of any explanatory theory (Clusters 3 and 9, and others) which can account for the statistical relationship, that is all there is. As Carlsen recently put the specific point: "Even if one is strongly outcome-oriented, statistical correlations of questioning behaviors with student outcomes are meaningless-- no matter how strong-- in the absence of a plausible explanatory model [i.e. Cluster 3, and perhaps others]" (1991, p. 172). (Of course, statistical correlations may well seem more significant to a researcher who has as his basis a suppressed behaviorist Presuppositonal Theory.) The statistical relationship is itself merely some evidence which encourages us to search for a theoretical explanation— but only if we are sure that the initial conceptualization was itself inadequate: in the present instance, only if we are sure that "warmth" is a sufficiently precise notion with which to work, and only if it makes sense to isolate it in this way from its context.

Second, to say that the place of the qualitative research is to supply hypotheses, is to take a narrow view of such research. I should wish to argue that qualitative research performs both the task of discovery *and* the task of justification. The irony of the kind of view held by Gage is that as the qualitative account is made more subtle and more elaborated, it begins to look increasingly like a theory which can explain the claimed correlation. That is, as a variety of Cluster 3, it looks more and more like a qualitative analogue of Scientific Theory. So Gage and researchers like him may have the account the wrong way around. Qualitative research is needed to explain the results of quantitative, empiricist research. What is even more to the point— an adequate empiricist research must be (to keep to the present imprecise terminology) *both* quantitative and qualitative.

There are moreover, several other dimensions of the Pedagogical Theory in which the empiricist research must be embedded, dimensions which may be of deep significance, but which have been traditionally underestimated or ignored, because they are contributed largely by practitioners rather than by university scholars. When these are recognized and emphasized, the classroom teacher's contribution to Pedagogical Theory and the research which is supposed to inform such theory begins to look more and more important in contrast to that which can be contributed by professional researchers, empiricist or otherwise. Two such key dimensions which are ignored by the restricted pedagogical mind-set of present-day empiricist researchers are the theories (Clusters 3 and 7) discussed by Schon as "knowing in action" and "reflection in action" in *The Reflective Practitioner* (1983), and by Polanyi as "tacit knowledge" in *Personal Knowledge* (1958), and elsewhere (1962, 1966). As Polanyi points out, this dimension can never be captured by any explicit account, but only by a variety of apprenticeship. Given their great complexity, it is not intended to discuss these matters here, but they are noted because they are significant for an improved empiricist approach. And this is in a sense to return the discussion to Feyerabend's point about multiple theories. For the proliferation of theories (various Clusters) for the improvement of scientific research is equally apt for the improvement of empiricist research. Martin has made a similar point: "The rationale for the proliferation of theories approach is this: The more theories one is used to working with in a given domain, the less likely is it that one will be blinded by one's commitments to any one of them" (1972, p. 126).

An Example of an Improved Kind of Empiricist Research: System for the Classroom Observation of Teaching Strategies (SCOTS)

I now discuss some research, which though not an exemplar of the above suggestions about embedding research in a Practical Theory of Pedagogy, nevertheless, goes part of the way towards this state. This research is reported in the important book, *The Teacher's Craft* (1985) by John L. Powell, whose author attempts to meld his own experience, deep thinking about the issues, and the research evidence of the study. Of all pieces of actual empiricist research so far performed, it is the one which I believe comes closest to approaches advocated in the present book. Given that Scotland, the geographical location for the research, is ethnically homogeneous, and has about the size and population of an average American state, there may be some reason for construing the research as producing "local theory" as alluded to above. The author also makes explicit the reasons for the importance of several of the significant philosophical points I have discussed above.

The SCOTS research grew out of a concern to discover whether progressive schooling was having an adverse effect. The researchers however soon decided that because of problems both conceptual and empirical, little that was worthwhile would flow from such a project, and so they changed tack. Showing the kind of philosophical acumen, so lacking in the studies described in Chapter IV, Powell explains why:

> The ill-defined nature of the terms could have been overcome by arriving at some arbitrary definitions to be applied within the context of the research, *but there was no way in which so simplistic a dichotomisation of types of teacher could have been usefully pursued* . We were faced with a direct choice between ending all investigation there and then, and. . . attempting to record, and to seek to understand, the complex patterns of activity that we saw in the primary classroom. We chose the latter (italics added) (1985, p.7).

Earlier systems of observation such as those of Flanders (1970) and his followers and Stallings (1977) were rejected, for the following philosophical reasons.

> 1.The behaviours recorded and counted [in a Flanders type system] were ones that could be unambiguously defined and recognised— and were therefore fairly simple ones.
> 2.The number of behaviours that could be categorised and recorded were, *at any one time*, quite small in number, and thus what was recorded was a very small sub-set of the total range of behaviours occurring.
> 3.The behaviours recorded had little meaning in themselves and it was therefore necessary to interpret the data collected after the action was over. (Although the recording itself may have been 'objective', subjectivity was involved in the subsequent interpretation of the data.)
> 4.When the interpretation was carried out, all behaviours placed in a category had to be treated as identical. Thus, for example, all initiations of verbal interactions were treated as identical events unless another simultaneous recording permitted some degree of sub-categorisation (e.g. according to what was going on at the time), and even then was in practice limited by the amount of simultaneous recording practicable (1985, pp.8-9).

RESEARCH EMBEDDED IN PEDAGOGICAL THEORY

So Powell and his associates devised their own schedule for observations, which, they admit, was one, ". .wholly different from virtually every other classroom observation system currently in use" (p.10). The final system of forty-three items became known as System for the Classroom Observation of Teaching Strategies (SCOTS). The System evolved in the following way. Originally Powell and another researcher went into the same classrooms with pen and pad. Notes were compared, particular notice being taken of teaching which seemed different from that generally practised. Once such a dimension of teaching was noted, an endeavor was made to place it into a five point scale—commonly deciding on two extremes and interpolating three intermediate categories. They write,". .what was difficult was to convert this draft into something that could stand up to use . . . when further teachers were observed, difficulties were liable to arise" (p.11).

The researchers attempted to avoid the conceptual overlap (which they called "lack of unidimensionality"), which was discussed in Chapter VI. The point may be illustrated by their item 21 which refers to the continuum of extrinsic-intrinsic motivation. The researchers were correctly aware that this dimension ought not to be confused with the *degree* of motivation. They continue,

> Initially the aim . .was to define the five options in largely behavioural terms. The supposition was that if one described behaviours, it would be easy to see which teachers were alike in particular respects. *The fallacy was in supposing that actions very similar in terms of behaviour were necessarily alike in their significance, and also in supposing that fundamentally very similar practices could not have quite diverse manifestations in behavioural terms.* Thus it was that variation frequently came to be described in more abstract terms, while behavioural descriptions came increasingly to be used as exemplars rather than [as] definitions (italics added) (pp.11-12).

In Chapter VI, I claimed that if we want to know how best to make praise an effective tool in our teaching, then we must get to know as much as possible about the nature of human beings as human beings. This means going to the fertile studies of the novel, drama, history, ethnography, and so on, not to simplistic empiricist studies, i.e. we need to grasp the *meaning* of situations. Powell's study recognized this fact. In discussing the findings about rebuking children (the opposite of praising) he writes:

> The experience of receiving a sharp rebuke from a teacher seemed to differ in accordance with the basic, lasting relationship between rebuking teacher and class or child. Thus a teacher who was generally recognised as being well-disposed could, without causing offence, speak bluntly or sharply, whereas another, viewed as generally hostile, could arouse hostility with speech not a whit more blunt. The need was to see events and actions as having meaning and significance in, and in relation to, the context in which

they occurred (1985, p.7).

We see here that the researchers had rediscovered the hard way the philosophical point that action and movement are not the same thing, and that to act intelligently we need to be aware of the subtleties and meanings of situations. And they did not (indeed could not) discover this point by mere observation, but by observation followed by discussion and debate with the teachers and children who had been observed.

Also significant is that they had discovered the hard way that operational definitions cannot capture the complexity of teaching. They continue,

> . .the variables finally selected were not ones chosen to reflect any particular theoretical perspective relative to teaching. Rather the approach was essentially eclectic, one drawing on variables associated with a variety of theoretical perspectives . . The object was to include, so far as possible, all potentially relevant variables (p.13).

The original fifty-four items and codings were tried out on previously unobserved teachers, and appropriate modifications were made. Three researchers were trained in their use. For most of a term, the schedule was tried out in observing 138 teachers of Grades V, VI and VII, five times, in thirty schools in both city and country, in widely separated parts of Scotland. Each of the four quarters of the Scottish school day were observed, the fifth observation being in the morning. Attempts were made to observe the total range of styles rather than to observe something like a random sample of schools. Comparisons of codings were made.

Although reliability between observers was improved, the aim was not that, but to improve the schedule. Items were grouped in the following way: 1-5 matters requiring both observation and information from the teacher; 6-9 direction/control of work; 10-17 teaching/learning; 18-26 motivation, control, discipline; 27-36 organisation; 37-43 teacher's personality and relationships with pupils.

It was the primary aim of the schedule to allow the observer freedom to infer within defined limits. "He does moreover seek to record what happens rather than to fathom what the teacher intends should happen: something not only more practicable but also more relevant" (p.17). Powell also claims that his schedule is no more subjective than claimedly less-subjective schemes, writing with considerable insight,

> The very decision of what to record in a machine-like way at fixed time intervals-- as when, e.g. using a Flanders-type schedule-- itself involves subjectivity, and the decision as to how to use the frequency counts, even more. The essential difference between the SCOTS schedule . . and one of the allegedly more objective schedules lies not in whether subjective judgment is employed but in *when* it is employed. In the former case, it is at the time of observation; in the latter, predominantly when the analysis is

carried out or when the method of analysis is decided on. It is the contention of the present writer that subjective judgment, *subject to well-thought-out rules of application, is much more soundly based when used 'live' at the time of observation, when the whole range of evidence is before the observer's eyes.* And this advantage is over and above that of being able to record a great variety of factors simultaneously (italics added) (1985, p.18).

Moreover, it was realised that some data could be obtained *only through discussion with the teacher* : ". . although an observer can see whether pupils are operating in groups, he may be unable to infer with any certainty the basis on which the groups were formed," (p.18). Here we have an acknowledgement, however limited, of the importance of taking into account the teachers' intentions and the way in which teachers construe what is going on.

In the context of the present book, some of the general findings are worthy of note. The naivete of atomistic approaches to studying teaching was for instance clearly recognized. Powell writes,

> . . it was evident that teachers who used very similar approaches used them with markedly different levels of skill. Classes operating by group methods could be models of purposeful, self-generated activity, or a place of confusion in which pupils sought . . for guidance, while an overworked teacher sought to keep the show going somehow. Equally, teacher-centred classes could be stimulating places where virtually every pupil was eagerly engaged, or depressing ones characterised by rigidity, uniformity, and dullness. There were at least some grounds for thinking that what the teacher does may be less important than the way that he does it (1985, p.6).

The research has some nine key fndings, which in general look like a vindication of the claims made in Chapter VIII, that, in research, account should be taken of intention, content, the teacher's knowledge of the subject, the context, the learners, and methods suited to them. The following are typical: ". . teachers should have well thought-out operational procedures and they should ensure early in the school year that pupils are familiar with them," (p.163); "It is clearly essential that, by one means or another, every class broadly share the teacher's objectives," (p.164); "Much seemed to depend on the basic relationship between teacher and class: a teacher who was liked by the class or perceived as generally well-disposed to the class could criticize bluntly either work or conduct without causing offence and consequent demotivation," (p.166); Self-confident and able pupils were stirred to fresh effort by sharp criticism (p.166); While the extent of knowledge is not the major factor in quality, inadequate knowledge, ". . is something for which other skills and qualities cannot wholly compensate," (p.168); Because pupils differ in the way they comprehend, "There thus seems to be a strong case for providing teachers with specific training designed to enhance their skills in the structuring and restructuring of ideas and associated information," (p.169); "Fulfilling such a variety of objectives calls for balance and compromise; it also calls for awareness sensitivity and flexibility. It is more important that the teacher should succeed in these three respects than that any particular teaching style or practices be adopted. The writer is convinced on the basis of his observations that *there are many different ways of excelling in teaching* " (italics added) (p.171); ". . though the complexity of the teaching situation was appreciated early and the SCOTS schedule created in an attempt to reflect it, *the attempt to relate outcomes to particular teaching styles proved simplistic* " (italics added) (p.171).

In its own way, this research was paralleling several, though not all, of the more subtle empiricist approaches advocated in this book. That such a fine piece of research could result, is indeed encouraging.

CONCLUSION

This book argues that empiricist pedagogical researchers have confused the status of their work, and that scientific research is different from their empiricist research. This confusion has also prevented empiricist research from being truly effective in an empiricist manner. To become empiricistically cogent, these approaches must be improved: they must deal with the real world of pedagogy, not just with some aspects of it which the current statistical methods happen to be able to capture. Context, the intentions of the teacher, the state of mind and the intentions of the learners, the nature of the content which is being taught, are not merely accompaniments of the activities of teaching, they are the *very constituents* of such activities. They are equally the very constituents of sound empiricist study. Empiricist researchers have allowed the statistical methods to dictate how teaching will be studied, rather than ensuring that the nature of that which is to be studied will determine the methods. In so doing, they have distorted the nature of the very enterprize they claim to be researching. This alone would prevent their approach from being scientific.

As was argued in the Introduction, there is nothing inadequate about empiricist enterprizes as such, where such enterprizes are the most appropriate route to understanding. The inadequacy occurs when researchers fail to make due allowance for the complexity of their empirical subject matter. Researchers such as those in the Process-Product and other empiricist modes ought to recognize the empiricist status of their endeavors, realize what this means for future developments and modify their methodology in ways which make use of the wide range of humanistic and normative critiques. In particular, empiricist research should be embedded in a Pedagogical Theory of the Cluster 4 kind, one which takes explicit account of the several different kinds of theory currently ignored. It is noteworthy that something similar has been mentioned in a recent paper by Gage (1989), who in the second of three imaginative constructions of possible future pedagogical research between the years 1989 and 2009 is in much of what he writes, actually describing a more adequate empiricist approach:

> Processes in teaching were investigated in interpretive and cognitive terms as well as in terms of teachers' and students' actions[1]. Through the use of multiple perspectives, the teachers' pedagogical content knowledge was described in ever more valid ways. Products, or the outcomes of teaching, were investigated in ever more authentic terms– with essay tests, real-life performances, group processes, and concrete products, as well as with the multiple-choice tests that had been prevalent through the 1980s. Process-product relationships, in all the various grade levels, subject matters, student cultures, and economic levels (and combinations of these) were examined–

---

1. i.e. "movements", or "behavior".

sometimes through interpretive case studies, sometimes through correlational studies, sometimes through field experiments. . and sometimes through critical-theoretical analysis (p.144).

These ideas are much more satisfactory than Gage's views criticized earlier– though he persists in believing that statistical research is scientific. Though they are crucial, it will not be merely a matter of *adding* "multiple perspectives". If the observational-statistical-experimental approaches, the "correlational studies" and "field experiments" Gage alludes to here are to become adequate, and an acceptable part of such multiple perspectives, they will still have to deal with the real, complex nature of teaching ; and to achieve this, the empiricist entities so far correlated under such approaches will themselves have to become radically more theoretically (Clusters 3,4,5, and 7) sophisticated.

Moreover, as Gage suggests, the use of such techniques as essays, performances and case studies would certainly improve the realism of testing and would be much more subtle. Indeed, such approaches have traditionally been used by enlightened teachers, schools, and systems in the past. But it should be noted that such techniques can no longer be subjected to simple statistical analysis in the manner of the standardized achievement tests used in present-day approaches. The present reductionist scaling approach of empiricist research, which itself has rejected more subtle methods of assessing, will no longer be suitable. The reductionist approach will no longer be acceptable because essays, performances, and so on, as methods of assessment, themselves instantiate the more subtle empiricist approaches which this book recommends.

Perhaps the most debilitating traditional belief of empiricist pedagogical researchers has been that methods must be scientific to be adequate. They have pursued statistically-based empiricist methods, believing them to be scientific, thus over-valuing them and concomitantly undervaluing the importance of other kinds of empiricist methods. Yet I would argue that it is other kinds of sophisticated empiricist methods which are best suited to the study of a complex human enterprize such as teaching[2].

It was pointed out in Chapter II, that at the very beginning of the scientific revolution Francis Bacon said that knowledge is power. In his *Novum Organum* he said truly that, "We cannot command nature except by obeying her. We have to understand nature . . . we can at best wheedle her by following her own idiosyncracies." Scientific research is not wishing, or magic, or religion, or recipes, or the following of set statistical procedures. Neither is satisfactory empiricist research. It is because knowledge is power that scientific research has prospered. It is worth repeating: KNOWLEDGE is power.

---

2. To adapt Crick and Watson's words about DNA, at the conclusion of their famous paper in *Nature* : it has not escaped my notice that this specific critique of empiricist research on teaching may have wider implications.

REFERENCES

Aiken, H.E. (Ed.). 1966. *Philosophy of educational development*. Boston : Houghton Mifflin.

Allen, J. (1882). Have we a science of education? *Education*, 2 : 284-290.

Anderson, L., Evertson, C., & Brophy, J. 1979. An experimental study of effective teaching in first-grade reading groups. *Elementary School Journal*, 79: 193-223.

Anderson, L., Evertson, C., & Brophy, J. 1982. Principles of small-group instruction in elementary reading (Occasional Paper No.58). East Lansing, MI: Michigan State University, Institute for Research on Teaching.

Andreski, S. 1972. *Social science as sorcery*. Harmondsworth, England : Penguin.

Andrews, T.E. 1972. Certification. In, W.R. Houston, & Howsam, R.B. (Eds.), *Competency-based teacher education* (pp. 45-56). Chicago: Science Research Associates.

Arber, A. 1954. *The mind and the eye*. Cambridge, England : Cambridge University Press.

Aris, R., Davis, H.T., & Stuewer, R.H. (Eds.). 1983. *Springs of scientific creativity*. Minneapolis : University of Minnesota Press.

Aristotle.1961. *Physics*. (R. Hope.,Trans.) Lincoln, NE: University of Nebraska Press.

Ambruster, B.B. Stevens, R.J., & Rosenshine, B. 1977. *Analyzing content coverage and emphasis: a study of three curricula and two tests* (Technical Report No.26). Urbana-Champaign: University of Illinois, Centre for the Study of Reading.

Arnstein, D. 1967. *Philosophy of education: Learning and schooling*. New York: Harper & Row.

Austin, J. 1962. *How to do things with words*. Oxford: Oxford University Press.

Austin, J. 1970. *Philosophical papers*. Oxford: Oxford University Press.

The Australian Encyclopedia. 1958. Sydney: Angus & Robertson.

Ayers, W. 1989. *The good preschool teacher: Six teachers reflect on their lives*. New York: Teachers College Press.

Bacon, F. 1952. *Novum organum*. Chicago : Great Books of the Western World, Encyclopedia Britannica.

Bagley, W.C., Seashore, C.E. and Whipple, G.M. (1910). Editorial. *Journal of Educational Psychology*, 1 (1) : 1-3.

Bain, A. 1889. *Education as a science* (7th ed.). New York : Appelton; London: Kegan Paul, Trench.

Bally, A.W. (Ed.). 1980. *Dynamics of plate interiors*. Washington D.C.: American Geophysical Union; Boulder, Colorado : Geological Society of America.

Banks, M. R., Colhoun, E.A., & Hannan, D. 1986. Early discoveries of ice action in Australia. Unpublished manuscript.

Bantock, G.H. 1981. *The parochialism of the present*. London : Routledge and Kegan Paul.

Barker, S.F. 1971. Introduction. In E. Nagel, S. Bromberger, & A. Grunbaum. *Observation and theory in science*.

Barr, A.S. 1958. Problems associated with the measurement and prediction of teacher success. *Journal of Educational Research* , 51 : 695-699.

Barrow, R. 1978. *Radical education*. London : Martin Robertson.

Barrow, R.1984. *Giving teaching back to teachers*. Brighton, England: Wheatsheaf Books; London, Ontario : The Althouse Press.

Barrow, R. 1986. Empirical research into teaching: the conceptual factors. *Educational Research* , 28 (3) : 221-230.

Barrow, R. 1988. Context, concepts and content: Prescriptions for empirical research. *Canadian Journal of Education* , 13 (1) : 1-13.

Barzun, J. 1944. *Teacher in America*. Boston : Little, Brown.

Becher, R.A. 1980. Research into practice. In W.B.Dockrell, & D.Hamilton (Eds.), *Rethinking educational research,* London: Hodder and Stoughton.

Beer, A., (Ed.). 1975. *Kepler: Four hundred years. Proceedings of conferences held in honour of Johannes Kepler*. Oxford : Pergamon.

Bennett, N. 1988. The effective primary school teacher: the search for a theory of pedagogy. *Teaching and Teacher Education* , 4 (1) : 19-30.

Bennett, N., Jordan, J., Long , G., & Wade., B. (1976). *Teaching styles and pupil progress*. Cambridge, MA: Harvard University Press.

Bennett,W.J.1986.*What works*.Washington, D.C.:U.S.Department of Education.

Bergmann, Gustav. 1957. *Philosophy of science*. Madison, WI: The University of Wisconsin Press.

Bernstein, R. J. 1979. *The restructuring of social and political theory*. Philadelphia : University of Pennsylvania Press.

Bernstein, R. J. 1983. *Beyond objectivism and relativism*. Philadelphia : University of Pennsylvania Press.

Beveridge, W.I.B. 1957. *The art of scientific investigation*. London: Heinemann.

Best, J. W.1981. *Research in education*. Englewood -Cliffs, NJ : Prentice-Hall.

Biddle, B. J. &Donald S.Anderson, D. S. 1986. Theory, methods, knowledge and research on teaching. In M.C.Wittrock (Ed.). *Handbook of research on teaching* (3rd ed.). New York: Macmillan Publishing Company; London: Collier Macmillan.

Biringuccio, V. 1540/1990. *The pirotechnia* . C.S.Smith and M.T.Gnudi (Eds. & Trans.) New York : Dover.

Blalock,H.M.,Jr.1966.*Theory Construction* .Englewood Cliffs,NJ:Prentice Hall.

Blease, D., & Bryman, A.1986. Research in schools and the case for methodological integration. *Quality and Quantity* 20: 157-168.

Bleicher, J. 1982. *The hermeneutic imagination*. London : Routledge and Kegan Paul.

Blum, A. 1971. Social construction of knowledge. In, M.Young (Ed.). *Knowledge and control*. London: Collier-Macmillan.

Bobbitt, F. 1922. *Curriculum making in Los Angeles* (Supplementary Educa-

tional Monographs No.20). Chicago: University of Chicago Press.
Bobbitt, F.1924. Discovering and formulating the objectives of teacher-training institutions. *Journal of Educational Research*, 10 (3) : 3-11.
Bogdan, R. C. & Biklen, S.K. 1982. *Qualitative research in education.* Boston: Allyn & Bacon.
Bondi, Sir H. Preface. In, R. Harre, (Ed.), 1975. *Problems of scientific revolutions.* Oxford: Clarendon Press.
Boorstin, D. J. 1984. *The discoverers.* London and Melbourne: J.M.Dent.
Borg, W. R. 1981. *Applying educational research.* New York: Longman.
Borg, W.R. & Gall, M. 1989. *Educational research : An introduction* (5th ed.). New York: Longman.
Brandshard, B. 1973. *The uses of a liberal education.* La Salle, IL : Open Court.
Broad, W. & Wade, N. 1985. *Betrayers of the truth.* Oxford : Oxford University Press.
Brockman, J. (Ed.). 1991. *Doing science: The Reality Club.* New York: Prentice Hall Press.
Brodbeck, M. (Ed.). 1968. *Readings in the philosophy of the social sciences .* London: the Macmillan Co.; New York: Collier-Macmillan.
Bronowski, J. 1951.*The common sense of science.* London : Heinemann.
Bronowski, J. 1952. The creative process. *Scientific American* , V : 67-73.
Bronowski, J. 1965a. *Science and human values.* New York : Harper and Row.
Bronowski, J. 1965b. *The identity of man.* New York : The Natural History Press, for the American Museum of Natural History.
Bronowski, J. 1973. *The ascent of man .* London : B.B.C. Publications.
Bronowski, J. 1977. *A sense of the future.* Cambridge, MA : The M.I.T. Press.
Bronowski, J. 1978a. *The origins of knowledge and imagination.* New Haven and London : Yale University Press.
Bronowski, J. 1978b. *Magic, science and civilization.* New York : Columbia University Press.
Brophy, J.E. 1979. Teacher behavior and its effects. *Journal of educational research* , 71 : 733-750.
Brophy, J.E. & Evertson, C. 1974a. *Process-product Correlations in the Texas Teacher Effectiveness Study: Final Report* (Research report 74-4). Austin: University of Texas R & D Center for Teacher Education (ERIC No. ED 091 094).
Brophy, J.E. & Evertson, C.M. 1974b.*The Texas teacher Effectiveness project: Presentation of non-linear relationships and discussion* (Research report No. 74-6). Austin:University of Texas R & D Center for Teacher Education (ERIC No. ED 099345).
Brophy, J. E. & Good., T.L. 1986. Teacher behavior an student achievement. In M. C.Wittrock, (Ed.), *Handbook of research on teaching* (3rd ed.). New York: Macmillan Publishing Company; London: Collier Macmillan.
Broudy, H.S., Ennis, R.H., & Krimmermann, M., (Eds.). 1973. *Philosophy of educational research.* New York: John Wiley.
Browning, O. 1891. *Educational theories.* New York & Chicago: E.L.Kellog.
Bruner, J. 1976. Foreword. In N.Bennett, J. Jordan, G.Long, & B.Wade, *Teach-*

ing styles and pupil progress. Cambridge, MA: Harvard University Press.

Buchmann, M. 1984. The priority of knowledge and understanding in teaching. In L.G.Katz & J. D.Raths, (Eds.), *Advances in teacher education.* Norwood NJ: Ablex.

Buckingham,B.R.1920. Announcement.*Journal of Educational Research,* 1 : 1-3.

Bullen, K.E. 1976. Wegener: Alfred Lothar. In, *Dictionary of scientific bibliography,* Vol.XIV. New York: Charles Scribner's Sons.

Burgess, R. G. (Ed.). 1985. *Issues in educational research: Qualitative methods.* London : The Falmer Press.

Butterfield, H. 1957. *Origins of modern science.* New York : The Free Press.

Callahan, R. E. 1962. *Education and the cult of efficiency.* Chicago : The University of Chicago Press.

Campbell, D. & Stanley, J.C. 1963. *Experimental and quasi-experimental designs for research .* Chicago: Rand McNally College Publishing Co.

Carey, S.W. 1958. *The tectonic approach to continental drift.* Hobart, Australia: The University of Tasmania.

Carey, S.W. 1974. Transcript of Non-uniformitarianism: Fourth Bertrand Russell memorial lecture in philosophy and science. Adelaide, Australia: The Flinders University of South Australia.

Carey, S.W. (Ed.). 1976. *The expanding earth.* Amsterdam: Elsevier.

Carlsen, W.S. 1991. Questioning in classrooms: A sociolinguistic perspective. *Review of Educational Research ,* 61 (2) , 157-178.

Chalmers, A. 1982. *What is this thing called science?* (2nd ed.). Brisbane, Australia: University of Queensland Press.

Chamberlin, T.C. 1897. The method of multiple working hypotheses. *Journal of Geology,* 5 : 837-848.

Chamberlin, R.T. 1928. Some objections to Wegener's theory. In W.A.J.M. Van der Gracht, (Ed.), *Theory of continental drift* (pp.232-240). Tulsa OK: American Association of Petroleum Geologists.

Chambers, J.H. 1986. Is teaching teaching in Timbucktu and Tomsk? *Journal of Abstracts in International Education,* 15, (2) : 56-70.

Chambers, J.H. 1983/1989. *The achievement of education.* Lanham,MD: University Press of America.

Charters,W.W., & Waples, D.1929. *The commonwealth teacher-training study.* Chicago: The University of Chicago Press.

China Science and Technology Museum and China Reconstructs. 1983. *China: 7000 years of discovery: China's ancient technology.* Beijing: China Reconstructs Magazine.

Churchland, P.S. 1986. *Neurophilosophy.* Cambridge, MA: M.I.T.Press.

Churchland, P.S. 1988. *Matter and consciousness.* Cambridge, MA: M.I.T Press.

Clark, C. M. & Peterson, P.L. 1986. Teachers' thought Processes. In M.C. Wittrock, (Ed.), *Handbook of research on teaching* (3rd ed.) (pp.255-296). New York: Macmillan Publishing Company; London: Collier Macmillan.

Cohen, B. 1982. *Means and ends in education.* London: George Allen and Unwin.

Cohen, I. B. 1980. *The Newtonian Revolution.* Cambridge, England : Cambridge University Press.

Cohen, I B. 1985. *The birth of the new physics.* New York : W.W. Norton.

Cohen, L., & Manion, L. 1985. *Research methods in education.* London : Croom Helm.

Coleman, J.S., Campbell, E.Q., Hobson, C.J., McPartland, J., Mood, A.M., Weinfeld, F.D. & York, R.L. 1966. *Equality of educational opportunity.* Washington, DC : U.S. Government Printing Office.

Coley, N. G. & Hall, V.M.D. (eds.). 1980. *Darwin to Einstein .* Harlow, England: Longman & The Open University Press.

Commission on Teacher Education.1966.*The improvement of teacher education.* Washington, D C : American Council on Education.

Conant, J. B. 1947. *On understanding science.* New Haven : Yale University Press.

Conant, J. B. (ed.).1967. *The overthrow of the phlogiston theory: the chemical revolution of 1775-1789.* Cambridge, MA : Harvard University Press.

Confrey, J. 1981. Subject-matter specialists. *The Elementary School Journal ,* 82 : 87-91.

Confrey, J. 1987. Bridging research and practice. *Educational Theory ,* 37: 383-394.

Connell, R.W. 1985. *Teachers' Work.* Sydney: George Allen and Unwin.

Coode, A.M. 1965, A note on oceanic transcurrent faults. *Canadian Journal of Earth Science.* 2 : 400-401.

Copernicus, N. 1543/1952. *De revolutionibus* (G.C.Wallis,Trans.). Chicago: University of Chicago Press.

Courtis, S.A. & Packer, P.C. 1920. Educational research. *Journal of EducatonalResearch,* 1 , 5-14.

Cox, A (Ed.) 1973. *Plate tectonics and geomagnetic reversals.* San Francisco: Freeman.

Crawford, J. & Gage, N.L. 1977. Development of a research-based teacher training program. *California Journal of Teacher Education ,* 4 : 105-123.

Crick, F. 1974. The double helix: A personal view. *Nature,* April 26, 766-771.

Crick, F. 1988. *What mad pursuit.* Cambridge, England: Cambridge University Press.

Crittenden, B.C. 1981. *Education for rational understanding.* Melbourne: Australian Council for Educational Research.

Cronbach, L.J. 1975. Beyond the two disciplines of scientific psychology. *American Psychologist ,* 30 : 116-127.

Cruickshank, D.R. 1979. Clear teaching: What is it? *British Journal of Teacher Education* 5 , 1 : 27-33.

Darling-Hammond, L. 1984. Mad hatter tests of good teaching. *The New York Times,* January 8.

Darwin, C. 1958. *The Autobiography of Charles Darwin.* N.Barlow (Ed.). New York: W.W.Norton.

Davis, L. H. 1979. *Theory of action.* Englewood Cliffs, New Jersey : Prentice-Hall.

Dearden, R.F. 1968. *The philosophy of primary education*. London: Routledge and Kegan Paul.

Dearden, R.F. 1976. *Problems in primary education*. London: Routledge and Kegan Paul.

Dearden, R.F. 1984. *Theory and practice in education*. London, Boston & Melbourne: Routledge & Kegan Paul.

Degenhardt, M.A.B.1982. *Education and the value of knowledge*. London: George Allen and Unwin.

Degenhardt, M.A.B. 1984. Educational research as a source of educational harm. *Universities Quarterly*, 38, (3) : 234-244.

Degenhardt, M.A.B. 1988. 'Curriculum' as fraud. *Unicorn* (Australia), 15, (2) : 96-99.

Denham, C. & Lieberman, A. (Eds.). *Time to learn*. Washington DC: National Institute of Education.

Derr, R. L. 1979. The logical outcomes of teaching. *Educational Theory*, 29 : 139-148.

Dewey, J. 1938/1973. *Experience and education*. New York: Macmillian; London: Collier-Macmillan.

Dewey, J. 1950. *Democracy and education*. New York: Teachers College Press.

Dietl, P. 1973. Teaching, learning and knowing. *Educational Philosophy and Theory*, 5 : 1-25.

Dixon, K. 1973. *Sociological theory: Pretence and possibility*. London and Boston: Routledge & Kegan Paul.

Dockrell, W.B., & Hamilton, D. (Eds.). 1980. *Rethinking educational research*. London: Hodder and Stoughton.

Donahue, W.H. (Ed.). 1990. *Astronomia nova*. Cambridge: Cambridge University Press.

Doyle, W. 1986. Classroom organization and management. In M.C.Wittrock (Ed.). *Handbook of research on teaching* (3rd ed.). New York: Macmillan Publishing Company ; London: Collier Macmillan.

Driver, R. 1983. *The pupil as scientist?* Milton Keynes, England: The Open University Press.

Duschl, R.A. 1990. *Restructuring science education*. New York: Teachers College Press.

Duhem, P. 1906/1962. *The aim and structure of physical theory*. New York: Atheneum.

Dunkin, M.J. & Biddle, B.J. 1974. *The study of teaching*. New York: Holt, Rinehart & Winston.

Du Toit, A.1937. *Our wandering continents*. Edinburgh: Oliver and Boyd.

Egan, K. 1983. *Education and psychology*. New York: Teachers College Press.

Egan, K. 1988. The analytic and the arbitrary in educational research. *Canadian Journal of Education*, 13 (1) : 69-81.

Elvin, L. 1977. *The place of commonsense in educational thought*. London : George Allen & Unwin.

Entwistle, H.1971. The relationship between theory and practice. In J.W.Tibble, (Ed.). *An introduction to the study of education*. London: Routledge.

Entwistle, H. 1969. Practical and theoretical learning. *British Journal of Educational Studies,* 17 (2).

Epstein, J. (Ed.) 1981. *Portraits of great teachers.* New York: Basic Books.

Ericson, D.P. & Ellett, F.S. Jr. Interpretation, understanding and educational research. *Teachers College Record,* 83 : 497-513.

Erickson, F. 1986. Qualitative methods in research on teaching. In M.C.Wittrock (ed.). *Handbook of research on teaching* (3rd ed.). New York: Macmillan Publishing Company; London: Collier Macmillan.

Evans, K.M. 1984. *Planning small scale research (*3rd ed.). Windsor, England : NFER-Nelson.

Evers, C.W. and Walker, J.C. 1982. Epistemology and justifying the curriculum of educational studies. *British Journal of Educational Studies,* 30 : 213-229.

Evers, C.W. 1990. Educating the brain. *Educational Philosophy and Theory,* 22, 2 : 65-80.

Fenstermacher, G. D. 1979. A philosophical consideration of recent research on teacher effectiveness. In *Review of Research in Education 6* (pp. 157-185). Itaska, IL : Peacock.

Fenstermacher, G. D. 1986. Philosophy of research on teaching: Three aspects. In M.C.Wittrock (Ed.). *Handbook of research on teaching* (3rd ed.). New York: Macmillan Publishing Company; London: Collier Macmillan.

Feyerabend, P.K. 1960. Patterns of discovery. *Philosophical Review* 59 : 247-252.

Feyerabend, P.K. 1968. How to be a good empiricist. In P.H.Nidditch, (Ed.), *The philosophy of science.* Oxford: Oxford University Press.

Feyerabend, P.K. 1970a. Explanation, reduction and empiricism." In H.Feigl & G. Maxwell, (Eds.), *Minnesota Studies in the Philosophy of Science, IV .* Minneapolis: University of Minnesota Press.

Feyerabend,P.K. 1970b. Consolations for the specialist. In I.Lakatos, & A.Musgrave, (Eds.) *The methodology of scientific research programmes* Cambridge, England: Cambridge University Press.

Feyerabend, P.K. 1976. *Against method.* London : New Left Books.

Feyerabend, P.K. 1978. *Science in a free society.* London : New Left Books.

Feyerabend, P.K.1985. *Realism, rationalism and the scientific method.* Cambridge, England: Cambridge University Pres.

Findlay, J.J. 1897. The scope of a science of education. *Educational Review,* 14: 236-247.

Finn, C.E. 1991. What ails education research? In, D.S. Anderson and B.J. Biddel (Eds.). *Knowledge for policy.* London Falmer Press.

Finn, J.D. and Achilles, C.M. 1990. Answers and questions about class size: a statewide experiment. *American Educational Research Journal,* 27, (3) : 557-577.

Fisher, R.A. 1935. *The design of experiments.* Edinburgh: Oliver and Boyd.

Fisher-Box, J. 1964. Fisher, Sir Ronald Aylmer. In, *The encyclopaedia Britannica* V. 17 : 803-806. Chicago: Encyclopaedia Britannica Inc.

Flanders, N.A. 1970. *Analyzing Teacher Behavior.* Reading, MA : Addison-Wesley.

Flew, A. 1976. *Sociology, equality and education.* London: Macmillan.

Flew, A. 1985. *Thinking about social thinking*. Oxford : Basil Blackwell.
Fraas, J. W. 1983. *Basic concepts in educational research*. Lanham, MD: University Press of America.
Francis, P. 1976. *Beyond control*. London : George Allen and Unwin.
Frankel, H. 1980. Hess's development of his seafloor spreading hypothesis. In T.Nickles, (Ed.), *Scientific case studies* (pp.345-366). Dordrecht, Boston & London: D.Reidel.
Freeman, H. 1973. Teaching and intention. *Proceedings of the Philosophy of Education Society of Great Britain*, 7 , (1) : 10-23.
Freundlich, Y. 1980. Philosophy of science and the teaching of scientific thinking. *Teachers College Record* , 82 : 117-124.
Frick, T. & Semmel, M.I. 1978. Observer agreement and reliabilities of classroom observational measures.*Review of Educational Research* , 48 : 157-184.
Froebel, F. 1887. *The education of man* (W.N.Hailmann, Trans.). New York: Appleton.
Gage, N.L., (Ed.). 1963. *Handbook of Research on Teaching*. Chicago: Rand McNally.
Gage, N.L. 1969. Teaching methods. In R. L. Ebel, (Ed.), *Encyclopedia of educational research* (4th ed.). London: Collier-Macmillan; New York: The Macmillan Company.
Gage, N.L. 1972. *Teacher effectiveness and teacher education: the search for a scientific basis* . Palo Alto, CA : Pacific Books.
Gage, N.L. 1978. *The scientific basis of the art of teaching*. New York : Teachers College Press.
Gage, N.L. & Giaconia, R. 1981. Teaching practices and student achievement: Causal connections. *New York University Quarterly*, 12 : 2-9.
Gage, N.L.1983. When does research on teaching yield implications for practice? *The Elementary School Journal* , 83 : 492-493.
Gage, N.L. 1984. What do we know about teaching effectiveness? *Phi Delta Kappan* , 43 : 87-93.
Gage, N.L.1985. *Hard gains in the soft sciences: the case of pedagogy*. Bloomington, IN : Phi Delta Kappa Centre on Evaluation, Development and Research.
Gage, N.L. 1989. The paradigm wars and their aftermath. *Teachers College Record*, 91 : 135-150.
Gage, N.L. & Needles, M. 1989. Process-product research on teaching : a review of criticisms. *The Elementary School Journal*, 89 : 253-300.
Gale, G. 1979. *Theory of science*. New York : McGraw-Hill.
Galfo, A. J. 1983. *Educational research design and data analysis*. Lanham, MD: University Press of America.
Galilei, G. 1950. *Dialogues concerning two new sciences* (H.Crew & A. De Salvio, Trans.). New York: Dover Publications.
Galton, F. 1869. *Hereditary genius: an inquiry into its laws and consequences*. London: Macmillan.
Galton, F. 1883. *Inquiries into human faculty and its development*. London:

Macmillan.

Garman, N.B. & Hazi, H.M. 1988. Teachers ask : Is there life after Madeline Hunter? *Phi Delta Kappan*, 69, (9).

Gardiner, M. 1981. *Science: good, bad and bogus.* Oxford : Oxford University Press.

Garrison, J. W. 1985. Some principles of postpositivistic philosophy of science. *Educational Researcher*, 15, (7) : 12-18.

Garrison, J. W. 1986. Husserl, Galileo, and the processes of idealization. *Synthese*, 66 : 329-338.

Garrison, J. W. 1988. The impossibility of atheoretical research. *The Journal of Educational Thought*, 22 : 21-25.

Garrison, J.W. 1989. Democracy, scientific knowledge, and teacher empowerment. *Teachers College Record*, 89 : 487-504.

Garrison, J.W. & Macmillan, C.J.B. 1987. Teaching research to teaching practice. *Journal of Research and Development in Education*, 20 : 38-43.

George, F.H. 1977. The use of models in science. In, R.J.Chorley & P.Haggett. (Eds.). *Models in geography.* London: Methuen.

Georgi, J. 1962. Memories of Alfred Wegener. In S.K.Runcorn, (Ed.), *Continental drift.* New York: Academic Press.

Giere, R.N. 1988. *Explaining scientific change.* Chicago: University of Chicago Press.

Gimpel, J. 1980. *The cathedral builders.* New York : Harper and Row.

Gingerich, O. 1975. Does science have a future? In O Gingerich, (Ed.), *The nature of scientific discovery.* Washington, D.C. : Smithsonian Institution Press.

Giroux, H. A.1988. Schrag speaks: Spinning the wheel of misfortune. *Educational Theory*, 38, (1) : 145-146.

Glass, G.V. 1976. Primary, secondary and meta-analysis of research. *Educational Researcher*, 5 : 3-8.

Glass, G.V. 1977. Integrating findings: the meta-analysis of research. *Review of Research in Education*, 5 : 351-379.

Glass, G.V., Cahn, L.S., Smith, M.L.,& Filby, N. 1982. *School class size: Research on policy.* Beverley Hills CA: Sage.

Gleick, J. 1987. *Chaos : Making a new science.* New York: Penguin.

Good, T.L. 1979. Teacher effectiveness in the elementary school. *Journal of Teacher Education*, 30.

Good, T. L. & Brophy, J.E. 1982. *Looking in Classrooms* (3rd ed.). New York: Harper & Row.

Good, T. L. & Brophy, J.E. 1986. School Effects. In M.C.Wittrock (Ed.). *Handbook of research on teaching* (3rd ed.). New York: Macmillan Publishing Company; London : Collier Macmillan.

Good, T.L. & Grouws, D.1977. Teaching effects: a process-product study in fourth grade mathematics classrooms. *Journal of Teacher Education*, 28 : 49-54.

Gowin, D.B 1980. *Educating.* Ithaca, N.Y. : Cornell University Press.

Green, T. F. 1971. *The activities of teaching.* New York : McGraw-Hill.

Greene, M. 1978. *Teacher as stranger.* Belmont, CA : Wadsworth.
Greene, M. 1986. Philosophy and teaching. In M. C. Wittrock (Ed.). *Handbook of research on teaching* (3rd ed.). New York: Macmillan Publishing Company; London: Collier Macmillan.
Grimmett, P. P. & Erickson, G.L. 1988. *Reflection in teacher education.* New York and London: Teachers College Press.
Grosser, M.1976. *The discovery of Neptune.* New York : Dover.
Gudmundssdottir, S. & Shulman, L.J. 1987. Pedagogical content knowledge in social studies. *Scandinavian Journal of Educational Research*, 31 (2) : 59-70.
Gurwitsch, A. 1970. Toward a theory of intentionality. *Philosophy and Phenomenological Research*, 30 (3) : 62-75.
Hacking, I. (Ed.).1981.*Scientific revolutions.* Oxford: Oxford University Press.
Hacking, I. 1983. *Representing and intervening.* Cambridge, England: Cambridge University Press.
Haggerty, M.E.1929. Review of *The commonwealth teacher training study*, by W.W.Charters & D. Waples. *Elementary School Journal*, 29 , 627-628.
Hallam, A. 1973. *A revolution in the earth sciences.* Oxford: Clarendon Press.
Hallam, A. 1976. Alfred Wegener and the hypothesis of continental drift. In, J.T. Wilson, (Ed.), *Scientific American: Continents adrift and continents aground.* San Francisco: W.H.Freeman.
Hallam, A. 1983. *Great geological controversies.* Oxford: Oxford University Press.
Hamilton, D. 1980a. Educational research and the shadow of John Stuart Mill. In, W.B.Dockrell & D. Hamilton, (Eds.), *Rethinking educational research* (pp.3-14). London: Hodder and Stoughton.
Hamilton, D. 1980b. Educational research and the shadows of Francis Galton and Ronald Fisher. In W.B.Dockrell & D. Hamilton, (Eds.), *Rethinking educational research* (pp.153-168). London: Hodder and Stoughton.
Hanson, N.R. 1958. The logic of discovery. *The Journal of Philosophy*, LV, 1073-1087.
Hanson, N.R. 1972a *Patterns of discovery.* Cambridge, England : Cambridge University Press.
Hanson, N.R. 1972b. *Observation and explanation.* London : George Allen and Unwin.
Harap, H. 1929. Review of, *The commonwealth teacher-training study*, by W.W.Charters and D.Waples. *School Review* , 37 : 467-470.
Hardie, C.D. 1968. *Truth and fallacy in educational theory.* New York : Teachers College Press.
Hare, W. 1979. *Open-mindedness and education.* Kingston and Montreal: Mc Gill- Queen's University Press.
Hare, W. 1985. *In defence of open-mindedness.* Kingston and Montreal: Mc Gill-Queen's University Press.
Harre, R. 1970. *The principles of scientific thinking.* London : Macmillan.
Harre, R. 1972. *The philosophies of science.* Oxford: Oxford University Press.
Harre, R. 1983. *Great scientific experiments.* Oxford & New York: Oxford University Press.

Harris, C. W. 1969. Statistical methods. In R.L.Ebel (Ed.), *Encyclopedia of educational research* (4th ed.). London: Collier-Macmillan; New York: The Macmillan Company.

Harris, L. S. 1985. Drugs and drug action: Types of drugs affecting the physiological system. In *The encyclopaedia Britannica* (Vol.17, pp 502-504). Chicago: Encyclopaedia Britannica Inc.

Hartnett, A. & Naish, M.(Eds.). 1976. *Theory and the practice of education.* London: Heinemann.

Harvey, D. 1967. *Explanation in geography.* London: Edward Arnold.

Heather, D.C. 1979. *Plate tectonics.* London: Edward Arnold.

Hedges, L.V., Giaconia, R.M. & Gage., N.L. 1981. *Meta-analysis of the effects of open and traditional instruction.* Stanford, CA: Stanford University Program on Teaching Effectiveness.

Heikkinen, M. 1988. The academic preparation of Idaho science teachers. *Science Education*, 72, (1) : 63-71.

Helmke,A. & Schrader,F.W. 1988. Successful student practice during seatwork. *The Journal of Educational Research*, 82 (2): 70-75.

Hernandez-Cela, C. 1972. Probability in empiricism and science. In, D.Willer and J.Willer, *Systematic empiricism*. Englewood Cliffs,NJ:Prentice-Hall.

Heron, J. 1981. Philosophical basis for a new paradigm. In, P.Reason & J.Rowan (Eds.). *Human inquiry*. New York: John Wiley.

Heslep, R.D.1991.Review article— Questions and erotetic teaching. *Educational Theory,* 41, (1) : 89-97.

Hesse, M. B. 1963. *Models and analogies in science.* London: Sheed and Ward.

Highet, G. 1950. *The art of teaching.* New York : Vintage Books.

Hilsum, S. 1972. *The teacher at work.* Slough, England: N.F.E.R..

Hilsum, S. & Cane, B.S. 1975. *The teacher's day.* Slough, England: NFER.

Hintikka, J. 1981. A dialogical model of teaching. *Synthese,* 51, (1) : 39-61.

Hirst, P.H. 1966. Educational theory. In J.W.Tibble (Ed.), *The study of education.* London: Routledge and Kegan Paul.

Hirst, P.H. 1972. The nature of educational theory. *Proceedings of the philosophy of education society of Great Britain,* VI (1): 110-118.

Hirst, P.H. 1974. *Knowledge and the curriculum.* London: Routledge.

Hirst, P.H. (Ed.). 1983. *Educational theory and its foundation disciplines.* London: Routledge and Kegan Paul.

Hoffman, B. 1962. *The tyranny of testing.* Los Angeles : Crowell-Collier.

Hogben, D. 1980. Research on Teaching, Teaching, and Teacher Training, *Australian Journal of Education,* 42, (1) : 107-117.

Hollis, M. 1982. The social destruction of reality. In M.Hollis & S.Lukes (Eds.) *Rationality and relativism.* Cambridge, MA: The M.I.T.Press.

Holmes Group. 1986. *Tomorrow's teachers.* East Lansing MI: Holmes Group.

Holton, G. 1973. *Thematic origins of scientific thought.* Cambridge, MA, and London : Cambridge University Press.

Holton, G. 1978. *The scientific imagination: Case studies.* Cambridge, England: Cambridge University Press.

House, E.R. 1982. *No simple answer: Critique of the "Follow through" evaluation.* Urbana: University of Illinois, Center for Instructional Research

and Curriculum Evaluation.
Howe, K.R. 1988. Against the quantitative-qualitative incompatibility thesis. *Educational Researcher* , 17 (2) : 10-16.
Huberman, M. 1984. How well does educational research really travel? *Educational Researcher,* 16 : 5-13.
Huck, S.N. & Sandler, H. (1979). *Rival hypotheses: Alternative interpretations of data-based conclusions.* New York: Harper and Row.
Hughes, D. 1973. An experimental investigation of the effects of pupil responding and teacher reacting on pupil achievement. *American Educational Research Journal* , 10 : 21-37.
Hume, D. 1956. *An enquiry concerning human understanding.* Chicago : Henry Regnery.
Humes, W.M. 1980. Alexander Bain and the development of educational theory. In J.V.Smith & D.Hamilton (Eds.), *The meritocratic intellect* (pp.15-27). Aberdeen: Aberdeen University Press.
Jackson, P.W. 1989. *The practice of teaching.* New York: Teachers College Press.
Jeffreys, H. 1929. *The earth, its origin, history and physical constitution.* Cambridge, England : Cambridge University Press.
Jones, B. 1974. Plate tectonics: a Kuhnian case? *New Scientist* , 63 : 912-914.
Jones, R. S. 1982. *Physics as metaphor.* New York: New American Library.
Judson, H. F. 1980. *The search for solutions.* London: Hutchinson.
Kaplan, M. (Ed.). 1980. *What is an educated person?* New York: Praeger.
Kerr, D.H. 1981. The structure of quality in teaching. In J.F.Soltis (Ed.), *Philosophy of Education: NSSE Eightieth Yearbook.* Chicago: NSSE. 61-93.
Klug, A. 1968. Rosalind Franklin and the discovery of the structure of DNA. *Nature* , August 24 : 808-810 and 843-844.
Kounin, J.S. 1970. *Discipline and group management in classrooms.* New York: Holt.
Kuhn, T.S. 1957. *The Copernican revolution.* Cambridge, MA: Harvard University Press.
Kuhn, T.S. 1970. *The structure of scientific revolutions* (2nd ed.). Chicago: University of Chicago Press.
Kuhn, T.S. 1977. The essential tension: Selected studies in scientific thought and change. Chicago: Chicago University Press.
Lakatos, I. & Musgrave, A. (Eds.). 1971. *Criticism and the growth of knowledge.* Cambridge, England: Cambridge University Press.
Lakatos, I. 1971. Falsification and the methodology of scientific research programmes. In I.Lakatos, & A. Musgrave (Eds.), *Criticism and the growth of knowledge.* Cambridge, England: Cambridge University Press.
Lakatos, I. 1980. *The methodology of scientific research programmes.* Cambridge, England: Cambridge University Press.
Lanier, J. E. & Little, J.W. 1986. Research on teacher education. In M.C.Wittrock (Ed.). *Handbook of research on teaching* (3rd ed.). New York: Macmillan Publishing Company; London: Collier Macmillan.
Laudan, L. 1968. Theories of Scientific Method from Plato to Mach. *History of Science* , 7 : 1-63.

Laudan, L. 1977. *Progress and its problems*. Berkley & Los Angeles : University of California Press.
Laudan, L. and others. 1986. Scientific change: Philosophical models and Historical research. *Synthese*, 69 : 141-223.
Laudan, R. 1978. The method of multiple working hypotheses and plate tectonic theory. In T.Nickles (Ed.), *Scientific Case Studies*. Boston , Dordrecht & London: D. Reidel.
Lindblom, C.E. 1959. The science of "muddling through".*Public Administration Review* , 19 (2) : 79-88.
Lincoln, Y. S. & Guba, E.G. 1985. *Naturalistic inquiry*. Beverly Hills: Sage.
Losee, J. 1980. *A historical introduction to the philosophy of science* (2nd ed.). Oxford & London: Oxford University Press.
Lilley, I.M. 1967. *Friedrich Froebel, a selection from his writings*. Cambridge, England: Cambridge University Press.
Macmillan, C.J.B. 1989. Reply to Chambers. *Philosophy of education 1989: Proceedings of the 45th Annual meeting of the Philosophy of Education Society*. Normal, IL: Philosophy of Education Society, 96-100.
Macmillan, C.J.B. 1990. Telling stories about teaching. *Philosophy of Education 1990. Proceedings of the 46th Annual meeting of the Philosophy of Education Society*. Normal, IL: Philosophy of Education Society, 198-201.
Macmillan, C.J.B. & Garrison, J.W. 1988. *A Logical theory of teaching: Erotetics and intentionality*. Dordrecht, Boston & London: Kluwer Academic Publishers.
Macmillan, C.J.B. and Thomas Nelson,T. (Eds.). 1968. *Concepts of Teaching: Philosophical Essays*. Chicago: Rand McNally.
Macrorie, K. 1984. *Twenty teachers*. New York: Oxford University Press.
Makarenko, A.S. 1951. *The road to life*. Moscow: Foreign Languages Publishing House.
Makarenko, A.S. 1954. *A book for parents*. Moscow: Foreign Languages Publishing House.
Manicas, P. T. & Secord, P. F. 1983. Implications for psychology of the new philosophy of science. *American Psychologist* , 38 : 399-413.
Maritain, J. 1955. Thomist views on education. In N.B.Henry (Ed.). *Modern philosophies of education*. Chicago: Chicago University Press.
Maritain, J. 1963. *Education at the crossroads*. New Haven, CT: Yale University Press.
Martin, M. 1972. *Concepts of science education*. Glenview, IL: Scott Foresman.
Marvin, M.B. 1973. *Continental Drift*. Washington,D.C.: Smithsonian Institution Press.
Mason, S. F. 1962. *A history of the sciences*. New York: Collier.
Mather, K.F. (Ed.). 1950. *Source book in geology:1900-1950* . Cambridge MA: Harvard University Press.
Mathews, J. 1988. *Escalante: the best teacher in America*. New York :Henry Holt.

Mayr, E. 1982. *The growth of biological thought.* Cambridge, MA: The Belknap Press of Harvard University Press.

Maugh II, K. 1984. Presidential awards for science teaching. In, B.Miss (Ed.), *Science year 1984.* 234-238. Chicago: World Book.

McDonald, F. & Elias, P.1976. *The effects of teacher performance on pupil learning. Beginning teacher evaluation study.* Phase II, final report, Vol.1. Princeton, NJ: Educational Testing Service.

McEwan, H. 1990. Teaching acts: an unfinished story. *Philosophy of Education 1990. Proceedings of the 46th Annual meeting of the Philosophy of Education Society.* Normal, IL: Philosophy of Education Society, 189-197.

McKie, D. 1962. *Antoine Lavoisier.* New York: Collier.

McMillan, J.H. & Schumacher, S. 1984. *Research in education.* Glenview London: Scott, Foresman.

McNeill, L.M. 1988. Contradictions of control. *Phi Delta Kappan.* JanFebMar.

McPeck, J.E. 1981. *Critical thinking and education.* New York: St. Martins Press.

Medawar, P. 1968a. Lucky Jim. *New York Review of Books*, 28 March,

Medawar, P.B. 1968b. *Induction and intuition in scientific thought.* London: Methuen.

Medawar, P.B. 1980. *Pluto's republic.* Oxford: Oxford University Press.

Medawar, P.B. 1981. *Advice to a young scientist.* London & Sydney: Pan.

Medley, D.M. 1973. Closing the gap between research in teaching effectiveness and the teacher education curriculum. *Journal of Research and Development in Education,* 7 (1) : 39-46.

Medley, D.M. 1977. *Teacher competence and teacher effectiveness: a review of the process-product research.* Washington, DC. : American Association of Colleges for Teacher Education.

Medley, D.M. 1982. Teacher effectiveness. In H.E.Mitzel (Ed.), *The encyclopedia of educational research* (5th Ed.). New York: The Free Press; London: Collier Macmillan.

Meehl, P.E. 1967. Theory-testing in psychology and physics: a methodological paradox. *Philosophy of Science* , 34 : 103-115.

Meehl, P.E. 1978. Theoretical risks and tabular asterisks: Sir Karl, Sir Ronald, and the slow progress of soft psychology. *Journal of Consulting and Clinical Psychology* , 46 : 806-834.

Menard, H.W. 1986. *The ocean of truth.* Princeton, NJ: Princeton University Press.

Mendelsohn, K. 1977. *Science and western domination.* London: Thames and Hudson & Readers' Union.

Mill, J. S. 1843/1967. *A system of logic.* London: Longmans Green.

Millar, R. 1983. The relationship between theory and experiment in science education. In, R. Millar (Ed.) *Doing science: Images of science in education.* London: Falmer Press.

Monastery, R. 1991. Married to Antarctica. *Science News*, 139 (17): 266-267.

Monroe, W. S. & Engelhart, M.D. 1936. *The scientific study of educational problems.* New York: The Macmillan Company.

Nagel, E. 1962. *The structure of science*. London: Routledge.
Natkins, L.G. 1986. *Our last term*. Lanham, MD: University Press of America.
Neill, A.S. 1944. *Hearts not heads in school*. London : Herbert Jenkins.
Neill, A.S. 1962. *Summerhill: A radical approach to education*. London: Gollancz.
Novak, I. 1977. *A theory of education*. Ithaca, NY : Cornell University Press.
Nozick, R. 1965. *Anarchy, state and utopia*. Oxford: Blackwell.
Nutall, G & Church, J. 1973. Experimental studies of teaching behaviour. In G. Chanan, (Ed.), *Towards a science of teaching*. Windsor, England: NFER.
O'Connor, D.J. 1972. The nature of educational theory. *Proceedings of the Philosophy of Education society of Great Britain* VI (1) : 97-109.
O'Hear, A. 1981. *Education, society and human nature*. London: Routledge and Kegan Paul.
Olby, R.C. 1974. *The path to the double helix*. London: Macmillan.
Oldroyd, D.1986.*The arch of knowledge*. Sydney: University of New South Wales Press.
O'Neil, W.M. 1969. *Fact and theory*. Sydney : Sydney University Press.
Packer, M.J. 1985. Hermeneutic inquiry in the study of conduct. *American Psychologist* 40 : 1081-1093.
Paslonsky, S.B. 1986. *900 shows a year*. New York : Random House.
Passmore, J.1975. *Science and its critics*. London: Duckworth.
Passmore, J. 1980. *The philosophy of teaching*. London: Duckworth.
Patry, J-L.1987. Cross-situational consistency of behavior: Triple relevance for research in education. Technical Report, University of Freibourg Pedagogical Institute.
Patry, J.L. 1990. Teaching is situation-specific but theory is not toward a higher impact of research on practice. Paper presented at the Annual Meeting of AERA. San Francisco.
Payne, J. 1886. Science of education. *Barnard's American Journal of Education*, 26 : 465-468.
Payne, W.H. 1886. *Contributions to the science of education*. New York: American Book Company.
Pearson, K. 1911/1957.*The grammar of science*. New York: Meridan Books.
Peters, R.S. 1966. *Ethics and education*. London: George Allen and Unwin.
Peters, R.S. (Ed.). 1967. *The concept of education*. London: Routledge and Kegan Paul.
Peters, R.S. (Ed.). 1973. *The philosophy of education*. Oxford: Oxford University Press.
Peters, R.S.1972. Education and human development. In, R.F.Dearden, P.H.Hirst & R.S.Peters (Eds.) *Education and reason* (pp.111-130). London: Routledge and Kegan Paul.
Peters, R.S. 1977. *Education and the education of teachers*. London: Routledge & Kegan Paul.
Peterson, H. (Ed.). 1946. *Great teachers*. New Brunswick, NJ : Rutgers University Press.
Phillips, D.C. 1971. *Theories values and education*. Melbourne : Melbourne.

University Press.

Phillips, D.C. 1981. Perspectives on teaching as an intentional act. *Australian Journal of Education*, 25 (2) : 99-105.

Phillips, D.C. 1987. *Philosophy, science and social inquiry.* Oxford and NewYork: Pergamon.

Polanyi, M. 1958. *Personal Knowledge.* London: Routledge and Kegan Paul.

Polanyi, M. 1962. The unaccountable element in science. *Philosophy*, 37 : 1-15.

Polanyi, M. 1966. *The tacit dimension.* New York: Anchor Press.

Popp, J. A. 1980. Philosophical analysis, research on teaching and aim-oriented empiricism. *Educational Theory*, 30 : 321-334.

Popper, K.R. 1957.*The poverty of historicism.* London: Routledge & Kegan Paul.

Popper, K.R. 1968. *The logic of scientific discovery.* London: Hutchinson.

Popper, K.R. 1969. *Conjectures and refutations.* London: Routledge and Kegan Paul.

Porter, A.C. & Brophy, J. 1984. Synthesis of research on good teaching: Insights from the work of the Institute for Research on Teaching. *Educational Leadership*, 45 (9) : 74-85.

Powell, J.L. 1985. *The teacher's craft.* Edinburgh: Scottish Council for Educational Research.

Poynting, J.H. 1910. Gravitation. In, *The Encyclopedia Britannica*, 11th ed.. New York: The Encyclopedia Britannica Company.

Quine, W.V. 1951. Two dogmas of empiricism. *The Philosophical Review*, 60, 20-43.

Quine, W.V. 1960. *Word and object.* Cambridge, MA : The M.I.T.Press.

Quine, W.V. 1961. *From a logical point of view.* New York: Harper & Row.

Raths, L.E., Wassermann, A.J., Jonas, A. & Rothstein, A.M. 1986. *Teaching for thinking* (2nd ed.). New York: Teachers College Press.

Ravitch, D. 1977. The revisionists revised: Studies in the historiography of American education. *Proceedings of the National Academy of Education*, 4 : 1-84.

Ravitch, D. 1983. *The troubled crusade.* New York: Basic Books.

Ravitch, D. 1984. The continuing crisis : Fashions in education. *The American Scholar*, 53 : 183-196.

Reichenbach, H. 1938. *Experience and prediction.* Chicago: University of Chicago Press.

Reid, L.A.1961.*Ways of knowing and experience.*London : George Allen and Unwin.

Reid, L.A. 1986. *Ways of understanding and education.* London: University of London & Heinemann Educational Books.

Rice, J.M. 1914. *Scientific management in education.* New York: Holt.

Ripley, J.A. & Whitten,R.C. 1969. *The elements and structure of the physical sciences* (2nd ed.). New York: John Wiley and Sons.

Rogers, A. L. 1924. Discussion of the foregoing papers. *Journal of Educational Research*, 10 : 3.

Rosenshine, B. & Stevens, R. 1986. Teaching functions. In M.C.Wittrock (Ed.). *Handbook of research on teaching* (3rd ed.) (pp.376-391). New York: Macmillan Publishing Company; London: Collier Macmillan.

Ross, S. 1985. Priestley, Joseph. In *The New Encyclopaedia Britannica*, Vol.9, pp. 696-698. Chicago: Encyclopaedia Britannica Inc.

Rothstein, A.M. 1983. The element of success in transactions. *Journal of Educational Thought.* 17, (1) : 29-35.

Rothstein, A.M. 1989. The educative process and the search for complementarity in intercultural relations. *The Sun Journal of Educational Studies* , (Korea), 4 : 1.

Rousseau, J.J. 1762/1966. *Emile.* London: J.M.Dent.

Runcorn, S.K. 1962a. Towards a theory of continental drift. *Nature* 193, 311-314.

Runcorn, S.K. (Ed.). 1962b. *Continental drift.* New York: Academic Press.

Russell, T. L. 1987. Research, practical knowledge and the conduct of teacher education. *Educational Theory*, 37 : 369-376.

Scates, D.E. 1947. Fifty years of objective measurement and research in education." *Journal of Educational Research* , 41, (4) : 24-33.

Scheffler, I. 1960. *The language of education.* Springfield, IL: Charles C. Thomas.

Scheffler, I. 1965. *Conditions of knowledge.* Glenview. IL: Scott-Foresman.

Scheffler, I. 1967. *Science and subjectivity.* Indianapolis: The Bobbs-Merrill Company.

Scheffler I. 1973. *Reason and teaching.* London and Boston: Routledge and Kegan Paul.

Schibeci, R.A. & Grundy, S. 1987. Local theories. *Journal of Educational Research* , 81, (2) : 91-95.

Schiller, F.C.S. 1917. Scientific discovery and logical proof. In, C. Singer (Ed.). Studies in the history amd methods of the sciences. V.1. Oxford: The Clarendon Press. 235-289

Schoeck, H. & Wiggins, J.W. (Eds.). 1985. *Scientism and values.* Princeton: D. Van Nostrand.

Schon, D.A. 1983. *The reflective practitioner.* New York: Basic Books.

Scripture, E.W. 1882. Education as a science. *Pedagogical Seminary* , 2 : 111-114 .

Sellers, W. 1961. The language of theories. In H.Feigl & G.Maxwell, (Eds.), *Current Issues in philosophy of science.* New York: Holt, Reinhart & Winston.

Sergiovanni, T. J. 1989. Science and scientism in supervision and teaching. *Journal of Curriculum and Supervision,* 4, (2) : 93-105.

Shapere, D. 1983. *Reason and the search for knowledge: Investigations in the philosophy of science.* Boston: Kluwer.

Shavelson, R. J., Webb, N.M. & Burstein, L. 1986. Measurement of teaching. In M.C.Wittrock (Ed.). *Handbook of research on teaching* (3rd ed.). New York: Macmillan Publishing Company; London: Collier Macmillan.

Shelby, L. R. 1972. The geometrical knowledge of medieval master masons.

Speculum, XLVII (4) : 395-421.
Sherlock, A.J. 1964. *Probability and Statistics*. London: Edward Arnold.
Shipman, M. (Ed.). 1985. *Educational research: Principles, policies and practices*. London & Philadelphia: Falmer Press.
Shulman, L. S. 1986a. Those who understand: Knowledge growth in teaching. *Educational Researcher*, V : 4-14.
Shulman, L. S.1986b. Paradigms and research programs in the study of teaching: a contemporary perspective. In M.C.Wittrock (Ed.). *Handbook of research on teaching* (3rd ed.) (pp.3-36). New York: Macmillan Publishing Company; London: Collier Macmillan.
Shulman, L.S. 1987. Knowledge and teaching: Foundations of the new reform. *Harvard Educational Review*, 57 , (1) : 1-22.
Siegel, H. 1980. Objectivity, rationality, incommensurability, and more. *British Journal for the Philosophy of Science*, 31, (4) : 359-375.
Siegel, H. 1985. Relativism, rationality, and science education. *Journal of College Science Teaching*, XV, 2, 95-101.
Siegel, H. 1986. Relativism, truth and incoherence. Synthese, 68, (2) : 225-259.
Siegel, H. 1987. *Relativism refuted* : A critique of contemporary epistmeological relativism. Boston: D. Reidel.
Siegel, H. 1988. *Educating reason*. New York and London: Routledge.
Siegel, H. 1991. The rationality of science, critical thinking, and science education. In M.Matthews (Ed.). *History, philosophy, and science teaching*. Toronto: OISE Press; New York: Teachers College Press.
Slavin, R. E. & Karweit., N.L. 1985. Effects of whole class, ability grouped, and individualized instruction on mathematics achievement. *American Educational Research Journal*, 22 : 351-367.
Simon, H. W. 1938. *Preface to teaching*. Oxford: Oxford University Press.
Simons, H. (Ed.). 1980.*Towards a science of the singular*. University of East Anglia, Norwich, England: Centre for Applied Research in Education.
Sizer, T.R. 1985. *Horace's compromise*. Boston: Houghton Mifflin.
Slavin, R.E. 1984. *Research methods in education*. Englewood Cliffs, NJ: Prentice Hall.
Smith, B.O. 1961. A concept of teaching. In, B.O.Smith & R.H.Ennis (Eds.). *Langauge and concepts in education*. Chicago: Rand McNally.
Smith, J.K. 1983. Quantitative versus qualitative research: an attempt to clarify the issue. *Educational Researcher*, 12 (7) : 6-13.
Sockett, H.T. 1989. Has Shulman got the strategy right? *Harvard Educational Review*, 57 : 208-219.
Soltis, J.F. *An introduction to the analysis of educational concepts* (2nd ed.). Reading, MA : Addison-Wesley.
Soltis, J. F. (Ed.). 1981a. *Philosophy of education: NSSE eightieth yearbook*. Chicago: NSSE.
Soltis, J.F. (Ed.). 1981b. *Philosophy of education in mid-Century*. New York: Teachers College Press.
Soltis, J. F. 1984. On the nature of educational research. *Educational Researcher* 13 (4) : 5-10.

Soltis, J. 1987. The virtues of teaching. *Journal of Thought*, 22 (3) : 61-67.
Sowell, E. J. & Casey, R.J. 1982. *Analyzing educational research*. Belmont, CA: Wadsworth.
Sperry, R. 1983. *Science and moral priority*. New York : Praeger.
Stake, R.E. 1988. Case study methods in educational research. In, R.M.Jaeger (Ed.). *Complementary methods for research in education* (pp.251-279). Washington: American Educational Research Association.
Stallings, J. & Kaskowitz, D. 1974. *Follow through classroom observation evaluation 1972-73*. SRI Project URU-73. Menlo Park, CA: Stanford Research Institute.
Stallings, J. 1977. *Learning to look: a handbook on classroom observation and teaching models*. Belmont, CA: Wadsworth.
Stemple, F.W. 1961. Qualities of the master teacher.*The Educational Forum*, 25 : 37-43.
Stenhouse, L. 1977. Case study and case records: Towards a contemporary history of education. *British Educational Research Journal*, 4, (3) : 265-91.
Stenhouse, L. 1979. Using research means doing research. In, H. Dahl, A. Lysne and P. Rand (Eds.). *A spotlight on educational problems* (Festskrift for Johannes Sandven). Oslo: Oslo University Press.
Stenhouse, L. 1983. *Authority, education and emancipation*. London: Heinemann.
Stenhouse, L. 1985. J.Ruddick & M.Hopkins (Eds.) *Research as a basis for teaching: Readings from the work of Laurence Stenhouse*. London, & Portsmouth, NH: Heinemann.
Strike, K. A. 1982. *Educational Policy and the Just Society*. Chicago: University of Illinois Press.
Suppe, F. (Ed.). 1977. *The structure of scientific theories*. Chicago: University of Illinois Press.
Tatsuoka, M. M. 1969. Experimental methods. In Robert L. Ebel (Ed.), *Encyclopedia of educational research* (4th ed.). London: Collier-Macmillan; New York: The Macmillan Company.
Tatsuoka, Maurice M. Statistical Methods. In, H. E. Mitzel (Ed.), *The Encyclopedia of educational research*, (5th ed.). New York: The Free Press; London: Collier Macmillan.
Taylor, F. 1911. *The principles of scientific management*. New York: Harper Brothers.
Taylor, F.B. 1910. Bearing of the tertiary mountain belt on the origin of the earth's plan. *Bulletin of the Geological Society of America*, 21 : 179-226.
Theobald, D.W. 1964. Models and method. *Philosophy*, 39 : 260-267.
Thomson, W. 1891. *Popular lectures and addresses*. London: Macmillan.
Thorndike, E.L. 1911. The measurement of educational products.*School Review*, 20 : 289-299.
Thorndike, E.L. 1940/1969. Human nature and the social order. (G.J.Clifford, Ed. & abridger). Cambridge, MA: MIT Press.
Tibble, J.W. (Ed.). 1966. *The Study of education*. London: Routledge and Kegan Paul.

Tidyman, W.F. 1915. A critical study of Rice's investigation of spelling efficiency. *Pedagogical Seminary,* 22 : 391-400.
Tom, A. R. 1984. *Teaching as a moral craft.* New York: Longman.
Totten, S. M. 1981. Frank B.Taylor, plate tectonics and continental drift. *Journal of Geological Education* , 29, (5) : 3-11.
Toulmin, S. 1953.*The philosophy of science.* London: Hutchinson.
Toulmin, S. 1961. *Foresight and understanding.* New York: Harper and Row.
Toulmin, S. 1970. Does the distinction between revolutionary and normal science hold water? In, I. Lakatos and A.Musgrave (Eds.) *Criticism and the growth of knowledge.* Cambridge, England : Cambridge University Press.
Toulmin, S.1972. *Human understanding.* Princeton: Princeton University Press.
Toulmin, S. & Goodfield. J. 1961. *The fabric of the heavens.* London: Hutchinson.
Travers, R. M. W. 1969. *An introduction to educational research.* (3rd ed.). London: The Macmillan Company, Collier-Macmillan.
Travers, R.M.W. (Ed.). 1973. *Handbook of research on teaching* (2nd ed.). Chicago: Rand McNally.
Travers, R.M.W. 1983. *How research has changed American schools: a history from 1840 to the present.* Kalamazoo, MI: Mythos Press.
Tuckman, Bruce W. 1978. *Conducting educational research* (2nd ed.). New York: Harcourt Brace Jovanovich.
Tuthill, D. & Ashton, P. 1983. Improving educational research through the development of educational paradigms.*Educational Researcher*, 12, (3) : 6-14.
Tyler, R.W. 1953. The leader of major educational projects. *Educational Research Bulletin,* 32 : 42-52.
United states Office of Education. 1987. *What works.* Research about teaching and learning. Washington DC: U.S.Government Printing Office.
Urmson, J.O. (Ed.). 1960. *Western philosophy and philosophers.* London: Methuen.
Uyeda, S. 1978. *The new view of the earth* (M.Ohnuki, Trans.). San Francisco: W.H.Freeman.
Van der Gracht, W.A.J.M. (Ed.). (no date). *Theory of continental drift.* Tulsa, OK: American Association of Petroleum Geologists.
Vine, F.J. 1966. Spreading of the ocean floor: New evidence. *Science,* 154 : 1405-1415.
Vine, F.J. 1985. Plate tectonics. In, *The new encyclopaedia Britannica*, 25 : 871-880.
Vine, F.J. & Matthews, D.H. 1963. Magnetic anomalies over ocean ridges. *Nature*, 199 : 947-949.
Walberg, H.J. & Fredrick, W.C. 1982. Instructional time and learning. In H.E.Mitzel, (Ed.), *The encyclopedia of educational research* (5th ed.). New York: The Free Press; London: Collier Macmillan.
Walberg, H. J. 1982. Educational climates. In H. J. Walberg, (Ed.) (pp.289-202) *Improving educational standards and productivity.* Berkeley, CA: McCutchan.
Walberg, H. J. 1986. Syntheses of research on teaching. In M.C.Wittrock (Ed.). *Handbook of research on teaching* (3rd ed.) (pp.214-229). New York:

Macmillan Publishing Company; London: Collier Macmillan.

Walker, J.C. 1985. Philosophy and the study of education : A critique of the commonsense analysis. *Australian Journal of Education*, 29, (1) : 101-114.

Warnock, G.J. (Ed.). 1967. *The philosophy of perception.* Oxford: Oxford University Press.

Warnock, M. 1977. *Schools of thought.* London: Faber & Faber.

Warren, D. R. 1976. What we need is more research. *Educational Studies* 7, (2) : 132-139.

Watson, J.D. 1968/1980. *The double helix.* New York: W.W.Norton.

Weaver, R. M. 1948. *Ideas have consequences.* Chicago: The University of Chicago Press.

Wegener, A. 1929/1966. *The origins of continents and oceans* (4th ed.) (J.Biram, Trans.). London: Methuen.

Welsh, P. 1987. *Tales out of school: a teacher's candid account from the front lines of the American high school today.* New York : Sifton/Viking.

Westfall, R. S. 1977. *The construction of modern science.* Cambridge, England: Cambridge University Press.

Westfall, R. S. 1980. *Never at rest: a biography of Isaac Newton.* Cambridge, England: Cambridge University Press.

White, F.C.1983. *Knowledge and relativism.* Assen, The Netherlands: Van Gorcum.

White, J. 1973. *Towards a compulsory curriculum.* London: Routledge.

White, J. 1982.*The aims of education restated.* London: Routledge and Kegan Paul.

Whiteside, D.T. 1970. *The mathematical principles underlying Newton's "Principia mathematica".* Glasgow : University of Glasgow.

Whyte, L. 1975. Medieval engineering and the sociology of knowledge. *Pacific Historical Review* 44 (1) : 1-21.

Wiersma, W. 1985. *Research methods in education* (3rd. ed.). Boston: Allyn and Bacon.

Willer, D. 1967. *Scientific sociology: Theory and method.* Englewood Cliffs, NJ: Prentice-Hall.

Willer, D.1987. *Theory and the experimental investigation of social structures.* New York: Gordon and Breach.

Willer, J. 1971. *The social determination of knowledge.* Englewood Cliffs, NJ: Prentice-Hall.

Willer, D., & Willer, J. 1972. *Systematic empiricism: Critique of a pseudo science.* Englewood Cliffs, NJ : Prentice-Hall.

Willis, B. 1910. Principles of palaeography. *Science* , 31 : 241-260.

Wilson, J. 1972. *Philosophy and educational research.* Slough, England: NFER.

Wilson, J. 1975. *Educational theory and the preparation of teachers.* Slough, England: NFER.

Wilson, J. 1977. *Philosophy and practical education.* London: Routledge and Kegan Paul.

Wilson, J. 1979. *Fantasy and common sense in education.* London : Martin Robertson.
Wilson, J. 1983. Philosophy and the study of education. *Australian Journal of Education*, 27 : 3.
Wilson, J. 1988. What philosophy can do for education. *Canadian Journal of Education*, 13, (1) : 83-91.
Wilson, J.T. 1965. A new class of faults and their bearing on continental drift. *Nature*, 207, : 343-347.
Wise, A.E. 1979. *Legislated learning: the bureaucratization of the American classroom.* Berkeley, CA: University of California Press.
Wright, W.B. 1923. The Wegener hypothesis. (Discussion at the meeting of the British Association, Hull, Sept.11th 1922). *Nature*, 111 : 30-31.
Wright, C.J. & Nuthall, G. 1970. Relationships between teacher behaviors and pupil achievement in three experimental elementary science lessons. American Educational Research Journal. 7 , (4) : 477-491.
Wrightsman, B. 1980. The legitimation of scientific belief: Theory justification by Copernicus. In, T. Nickles. (Ed.). *Scientific discovery: Case studies.* Dordrecht, Holland: D.Reidel.
Wolcott, H.E. 1988. Ethnographic research in education. In, R.M.Jaeger (Ed.). *Complementary methods for research in education* (pp.185-221). Washington: American Educational Research Association.
Woody, C. 1920. Application of scientific method in evaluating the subject matter of spellers. *Journal of Educational Research* 1 , 2.
Ziman, J. 1968. *Public knowledge.* Cambridge: Cambridge University Press.
Zumwalt, K.K. 1986. Working together to improve teaching. In K.K.Zumwalt (Ed.), *Improving teaching.* Alexandra VA: Association for Supervision and Curriculum Development.

INDEX OF NAMES

Achilles, C.M., 127-129, 131, 134, 136, 137, 164-166, 194,
Adams, J.C., 42,
Alvarez, W., 79,
Arnstein, D., 230,
Anderson, J.R., 116-118
Anderson, L., 116-118, 136, 210,
Aristotle, 29, 36, 37,
Austin, J.L., 26n, 157n,
Ayer, A.J., 23,
Ayres, W., 100, 233,
Bacon, F., 76, 84, 183, 246,
Bain, A., 96-97, 167,
Bakker, R., 79,
Barr, A.S., 108-109, 125, 167,
Barrow, R., 3, 158, 164, 167, 221, 229,
Bennett, N., 113-114, 158,
Bentham, J., 94,
Berliner, D.C., 4n, 109, 126-127,
Bernstein, R.J., 53,
Biringuccio, V., 44, 44n, 45,
Blum, A., 78,
Bobbitt, F., 102-193, 106, 111,
Boccaccio, 17,
Bohr, N., 174, 185,
Boorstin, D., 31n,
Borg, W.R., 80-81, 82, 169-170, 173, 182, 195,
Bragg, W.L., 50, 51, 172,
Bronowski, J., 21, 48, 174, 198-199,
Brophy, J.E., 113, 116-118, 119, 131, 133-134, 135, 136, 137, 162-163, 210, 217-218,
Bruner, J.S., 12, 114,
Buckingham, B.R., 103,
Bush, S., 202, 203-204, 223, 237,
Butterfield, H., 36, 45-46,
Callahan, R.E., 101, 167,
Campbell, D (1)., 70, 140, 193, 194,
Campbell, D. (2), 203, 204-207, 213, 216-217, 223, 237,
Capen, S.P., 106, 107,
Carey, S.W., 64, 199n, 200,

Carlsen, W.S., 134
Carnap, R., 23
Cavendish, H., 44
Cervantes, 17
Challis, J., 42
Chamberlin, R.T., 61
Chamberlin, T.C., 61, 83, 176-177, 185
Charters, W.W., 103, 106-108, 111, 125
Church, J., 112-113, 131, 137-138, 140, 227
Coode, A., 66
Confrey, J., 154, 229
Copernicus, N., 29-32, 83, 167, 171, 172, 177, 179, 180, 201
Crick, F.C.S, 15, 16, 49-59, 84, 145, 148, 159, 172-173, 177, 180, 183, 186, 197, 246
Crittenden, B., 230, 238
Cronbach, L.J., 104-105, 235
Darwin, C., 14, 83, 197
Darling-Hammond, L., 222,
Dearden, R.F., 8, 75, 214, 230, 238
deCasper, H.S., 115-116
Degenhardt, M.A.B., 4, 230, 233
Denham, C., 127
Dewey, J., 14, 19, 20, 220-221, 230, 238, 269
Donahue, W.H., 34n
Donohue, J., 57
Dostoevski, 17
Doyle, W., 119
Duhem, P.M., 53, 54
Duschl, 181,
Du Toit, A., 62, 62n, 200
Eccles, J., 54
Egan, K., 164
Einstein, A., 42, 174, 185
Eisner, E., 80, 82
Ellett, F.S., 213
Ennis, R.H., 238
Erickson, F., 213
Evertson, C., 116-118, 131, 136, 137, 210
Ewing, M., 90
Fenstermacher, G., 130n, 233
Feyerabend, P., 62, 78, 85, 177, 185
Finn, C., 4
Finn, J.D., 127-129, 131, 134, 136, 137, 164-166, 194
Fisher, R.A., 23, 96, 103-106, 111, 193, 196
Flamsteed, J., 72
Flanders, N., 240

INDEX OF NAMES

Flew, A., 230
Franklin, R., 50, 54, 55, 56, 58
Frobel, F., 19
Gage, N., 2, 7, 14, 109, 118-121, 133, 134, 137, 140-141, 144-156, 158, 166, 194, 211, 217, 224-225, 227-228, 236-237, 238-239, 245-246
Gale, G., 45, 47
Galileo, 29, 35-38, 43-44, 62, 68, 72, 96, 136, 142, 145, 159, 171, 174, 175, 179, 195-196, 201
Gall, M., 80-81, 82, 126, 169-170, 173, 182, 195
Galle, J.G., 42
Galton, F., 23, 97-98, 101, 104
Galvani, L., 18
Garrison, J.W., 4, 12, 15, 53, 121, 146, 148, 152, 197, 217, 218, 234-235, 238
Giere, 12, 60, 63
Giroux, H., 20n,
Glass, G.V., 119, 129
Godel, K., 52
Good, T.L., 113, 119, 135, 162-163, 217-218, 224
Green, T.F., 132, 231, 238
Greene, M., 130n, 231
Grundy, S., 235-236
Hallam, A., 61, 199-200
Hamilton, D., 93, 94, 98, 104
Hannan, D., 59, 200n
Hare, W., 217
Hanson, N.R., 18, 78
"Harry", 202, 209-210, 213, 214, 223, 237
Hernandez-Cela, C., 189
Helmke, 22, 121-125, 131, 134, 137, 164, 194
Heron, J., 90, 218, 232
Hess, H.H., 64-65, 172, 186
Hilbert, D., 52
Hirst, P.H., 13, 234, 238
Hoffmann, B., 222
Hollis, M., 53
Holmes, A., 63-64
Holmes Group, 231
Huberman, M., 187
Hume, D., 23
Hughes, D., 112-113
Jackson, P.W., 231
Kelvin, Lord (W.Thomson), 17
Kendrew, 50
Kepler, J., 32-35, 40, 159, 188, 206
Kerlinger, N., 3, 7-8, 26-27, 137, 141-142, 157, 179, 192
King, L., 64

Kohlberg, L., 12
Komisar, B.P., 216
Kuhn, T.F., 12, 29-31, 35, 49, 51, 64, 75, 78, 85, 88-89, 91, 135, 171, 174, 175, 198
Lakatos, I., 40-42
Lamarck, 83
Laudan, L., 63, 183
Lavoisier, A., 44-49, 83, 89, 136, 142, 145, 148, 159, 177, 180, 186, 196
Lescarbault, J., 42,
Le Verrier, U.J.J., 42
Lieberman, A., 127
Losee, J., 44
Macmillan, C.J.B., 4, 12, 121, 146, 148, 197, 217, 218, 219, 234-235, 238
Macrorie, K., 202
Malthus, T.R., 14, 197
Martin, M., 239
Marvin, M.B., 61, 66, 199n, 200
Maxwell, J.C. 14, 64
Mayr, E., 14
McNeill, L.M., 161
Medawar, P., 49, 54, 95, 99, 173-174, 176, 197-198
Meehl, P. 188
Menard, W., 66
Mill, J.S., 23, 35n, 93-96, 196
Mill, J., 19, 20, 94
Millar, 169
Milz, V., 202, 207-208, 210, 213, 223, 228
Moore, T.W., 10
Morgan, J., 65
Nagel, E., 68, 174, 188, 190
Natkins, M., 226
Neill, A.S., 19
Newsome, G.
Newton, I., 17, 29, 35, 38-44, 68, 72, 96, 145, 167, 171, 172, 174, 175, 177, 186, 201
Newton-Smith, 87, 90
Nickles, T., 31n
Nisbett, 139-140
Nozick, R., 19
Nuthall, G., 112-113, 131, 137-138, 140, 227
Oldroyd, D., 14
O'Hare, A., 80,
O'Neill, W. M., 79
Pasteur, L., 84
Pearson, K., 23, 96, 98-100, 101, 111, 189
Palonsky, S., 213, 233

INDEX OF NAMES

Passmore, J., 231, 234
Pauling, L., 50, 55, 172, 187
Pestalozzi, J., 238
Peters, R.S., 231, 232, 234, 238
Phillips, D.C., 53, 77, 170, 183, 238
Piaget, J., 12, 182
Polanyi, M., 239,
Popper, K.R., 43, 51-54, 146, 171, 179, 183, 195
Porter, A.C., 133-134
Powell, J. L., 131, 239-244
Pratte, R., 238
Priestley, J., 46, 47, 48, 83, 148, 172, 177
Ptolemy, 29-31, 32, 83, 177, 179
Quine, W.V.O., 52-53, 54, 85, 183
Rawls, J., 19
Reichenbach, H., 35n, 87
Rice, J.M., 100, 127, 145, 167
Richards, I.A., 85
Ripley, J.A., 35, 38
Roentgen, W.K., 18
Rothstein, A.M., 231
Rousseau, J.J., 19, 20
Runcorn, K., 61, 65
Russell, B., 23
Scheffler, I., 53, 78, 80, 83, 85, 86, 90, 231, 234, 238
Schibeci, R.A., 235-236
Schiller, F.C.S., 87
Schon, D., 13, 239
Schrader, F.W., 22, 121-125, 131, 134, 137, 164, 194
Schrag, F., 20
Scriven, M., 4, 164, 197, 202-203
Shakespeare, 17
Shapere, D., 87
Shavelson, R.J., 4
Shulman, L.J., 126, 127, 131, 132, 221
Siegel, H., 53, 87, 178, 180, 187, 198, 231, 238
Slavin, R.E., 15, 128
Smith, J., 19
Snell, W., 74, 96
Soltis, J.F., 211, 216, 234, 238
Stahl, G.E., 44, 89
Stallings, J., 122, 135-136, 240
Stenhouse, L., 193, 234, 237
Strike, K., 238
Sully, J., 96-97
Taski, A., 52

Taylor, F., 101
Taylor, F.B., 59, 60, 64, 186
Thorndike, E.L., 17, 100-101, 167
Tom, A.R., 3, 108, 109, 125, 127
Travers, R.M.W., 93, 125
Toulmin, S., 29, 73, 221
Ts'ao, H-c, 17
Turing, A., 52
Tycho (Brahe), 32, 35, 83, 175
Urmson, J.O., 94n
Vine, F.J., 65, 79, 172
Voltaire, 22
Walberg, H.J., 109, 118-121, 137, 138-139, 141, 180-181, 182, 192
Wassinger, W.W., 10
Watson, J., 15, 16, 49-59, 145, 148, 159, 172, 179, 180, 183, 186, 197, 246
Wiesma, W., 181-182
Wegener, A., 59-67, 83
White, F.C., 53, 82, 83, 238
Whitten, R.C., 35, 38
Wilkins, M., 50, 55, 58, 187
Willer, D., 24, 69, 100, 104, 105, 106
Willer, J., 24, 69, 100, 104, 105, 106, 143
Wilson, J., 164, 238
Wilson, T., 66, 145, 172, 186
Wrightsman, B., 31n, 179

INDEX OF TOPICS

Abstract concepts, 22, 34, 37, 38, 40-44 *passim*, 48, 66, 69, 70, 71, 75, 156-157, 187-188,
Academic learning time (ALT), 120, 126-127
   Bobbitt's views as precursor of, 101-102, 101,
   critique of, 153-155,
Action cf. movement, 213-215
A Logical Theory of Teaching (Macmillan and Garrison), 4, 234-235
Alternatives
   importance of in scientific and empiricist research, 185
   Feyerabend's views, 62, 185,
American Educational Research Association, 2
American Educational Research Journal, 2
American opposition to continental drift, 61
Anti-educational nature of some empiricist research, 232-234
Appropriateness of manner, its importance in teaching and in research on teaching, 225-227
"Atomism" as scientific research approach,
   its embedding in Scientific Theory, 159-160
   unsuited to research on teaching, 160-164
Australian Council for Educational Research, 1
Barr, A.S., 108-109
   as precursor of contemporary approaches to research, 108, 109
   belated admission of the significance of context, 109
Biringuccio, V., and early conceptions of chemistry, 44
Bronowski, J.
   science as a search for hidden likenesses, 29, 174
   social values of scientific research, 198-199
British Journal of Educational Studies, 2
Carey, S.W. and continental displacement, 64, 200
Charters, W.W. and the *Commonwealth Teacher Training Study*, 106-108
   contemporary criticism of, 108
Class size, 119, 127-129
   research by Finn and Achilles, 127-129
   randomized experiments and, 128
   benefits of smaller classes, 128-129
   unnecessary empiricist research on, 164-166
Commonsense observation, misconceptions of, 36, 45-46
Communist schooling, 19
Concepts
   abstract concepts, 22, 37, 38, 40-44 *passim*, 48, 75, 156-157, 187-188
   abstract concepts cf. general concepts, 156-164
   abstract concepts of science, 69, 70

abstract concepts and ratio measurement, 71
   conceptual insight makes phenomena seem more simple, 75
   context-based meaning of empiricist, general concepts, 163
   conceptual distinctions between reason and cause and action and movement, 210-215
   examples of abstract concepts: "ellipse", 34; "oxygen", 48; "plate", 66; "transform fault", 66; "point-mass", 40; "angular momentum", 41; "element", "mixture", "compound", 47
   precision of meaning of abstract concepts, 75
   empiricist concepts, 22,
   general concepts, 20, 189
   empiricist concepts as general concepts, 156-160
   kinds of connection between concepts, 69-71
   "praise" as an example of a general, empiricist concept, 160-163
   stability of meaning of abstract concepts, 75
Conceptions of scientific research in empiricist research on teaching, 130-143
   atomistic assumptions, 131
   assumptions about context, 131
   assumptions about similarities and differences of teachers, 131-132
Confusion between the scientific and the empiricist, 5, 145-164 *passim* 168-170
Confusion of "scientific" and "disciplined", 168, 182
Context,
   significance of contextual understanding in understanding teaching, 162, 164,
   appropriate use of praise as an example of contextual understanding, 160-163,
Connection between concepts, 69-71,
   abstractive connection, 69-71, 188, 190
   empiricist connection, 69-71,
   rational connection, 69-71,
   refraction as showing different connections, 70-71,
Content, its importance in teaching and in research on teaching, 219-223
Context, its importance in teaching and in research on teaching, 223-225
Context of discovery, 179-180
Continental drift, 59-64, 199-200
   sarcastic and hostile reactions towards, 61, 199, 200
Copernicus and the heliocentric theory, 29-32
Correlations, 4, 99-100, 111-129 *passim*
Crick and Watson and discovery of structure of DNA, 15, 16, 49-59, 84, 91, 145, 148, 159, 172-173, 177, 180, 183, 186, 197, 246
Discovery cf. verification/justification, 78, 87
   difference between actual discovery and teaching of, 205n
Discovery in scientific research, 179-180, 197-198
DNA, discovery by Crick and Watson, 15, 16, 49-59, 84, 91, 145, 148, 159, 172-173, 177, 180, 183, 186, 197, 246
Du Toit's fieldwork support for continental drift, 62

INDEX OF TOPICS

Educational Research: an Introduction (5th ed.) (Borg and Gall), 3
    confusions in, 80-83, 195
"Educational Research as a Source of Educational Harm" (Degenhardt), 4, 233
"Empiricist" cf. "empirical", 5
Empiricist research, 20-26, 111-129
  carried on where it is unnecessary, 164-166
  confused with conceptual, logical and commonsense analysis, 164-166
  cf. scientific research, 5, 98, 167-201, 193-197
  confused with scientific research, 188-192
  imperialist assumptions of, 188
  inductivist orientation of, 170-173
  thought experiments and, 178-179
Empiricist research on teaching, 111-129, 130-143
  actions, neglect of importance of in empiricist research on teaching, 213-215
  alleged non-theoretical nature of, 140-141
  lack of validity of, 227-228
  narrowness of, 165, 165n
  power of its advocates, 200
  precursors of empiricist research on teaching, 94-110
  reasons, neglect of importance of in empiricist research on teaching, 210-213
  theoretical nature of, 144-165 *passim*, 166
  summary statement about, by J.Willer, 143
  confusion with medical research, 146-148
  ways of improving, 5, 76, 196-197, 202-246 *passim*
Empiricist Theory, 20-26, 69-74
Empiricist generalizations, 20, 24, 22, 99, 105, 111-129 *passim*,
Erotetic theory, of Macmillan and Garrison, 12, 197,
Euler's Theorem and plate tectonics, 66,
Expenditures on educational research, 2, 4
Experience, importance of in teaching, 237
Experiment in scientific research, 37, 48, 193-197
Experiment in empiricist research, 136-137, 142-143, 193-197,
  statement about, by J.Willer, 143
Fads and fashions in teaching, as Normative Theory, 20
Falsification, as a scientific method, 43, 51-54, 54-58 *passim*, 91, 146, 171, 179, 183-184, 195
Feyerabend and importance of competing theories, 62
Foundations of Behavioral Research (Kerlinger), 3, 7-8, 26-27, 141-143
Fisher, R., statistical methods of, 103-106, 193
    "rigorous uncertainty" of these, 104
    testing Heuristics, 105
    place of suppressed Scientific Theory in his work, 105-106
Franklin, R. and discovery of structure of DNA, 50, 51, 54, 55, 56, 58
"Fudging" in research, 34n
Gage, N.
    critique of his defense of Process-Product research, 144-156, 236-239

his more acceptable recent views, 246
Galileo
   theory of circular inertia, 17, 35-37, 44-45
   thought experiments, 35-37, 174
   law of falling bodies, 35-38, 44-45
Galton, F., his statistical presuppositions as largely unexamined basis of modern empiricist research, 97-98
Giving Teaching back to Teachers (Barrow), 4
Gunpowder, empiricist discovery of, 21
Handbook of Research on Teaching (3rd ed.) (Wittrock), 2, 4
Hirst, P.H., views of Practical Theory, 13
Hume, D., empiricist conceptions of, 23
Induction, as logically faulty, 51, 51n
   Mill's views of, 95
Intention, its importance in teaching and in research on teaching, 215-219
Kantian-type argument for existence of common observation, 79-80
Kepler, discovery of the three planetary laws, 32-35
   Erroneous claim that he, "fudged his data", 34
   First Law not discovered by inductive methods, 35
   teaching his laws, 206
Kuhn's stage theory of the development of science, 12, 29-31, 32, 51, 64, 88-92, 187, 198,
Lavoisier's oxygen theory, 17, 44-49
Learners, their importance in teaching and in research on teaching, 225-227
Local theories, 235
Magnetic anomalies and seafloor spreading, 65
Maxwell, J.C., equations of electro-magnetic radiation, 14-15
Meaning in teaching, 161-163, 213-214
Medawar, P., views of scientific discovery, 49, 54, 95, 99, 173-174, 176, 197-198
Meta-analysis, 109, 118-121, 137, 138-139, 141, 180-181, 182, 187, 192
Metaphysical beginnings of scientific research and theory, 33, 179-180,
Method of multiple working hypotheses (T.C.Chamberlin), 61, 176-177
Micro-teaching, 11
Mill, J.S.
   his empiricist views, 23, 93-96, 173
   confusion of different kinds of concepts, 95
   views of induction, 95
National Foundation for Educational Research (England and Wales), 1
Neptune, discovery of, by Adams, le Verrier, and Galle, 41-42, 183-184
Newton, I.
   mechanics, 190-191
   theory of rectilinear inertia, 17
   laws of motion and gravitation, 17, 38-44, 29, 35, 38-44, 68, 72, 96, 145, 167, 171, 172, 174, 175, 177, 186, 201
   Poynting's questioning of precision of law of gravity, 42, 42n

Normative Theory, 9, 19-20
  in Pedagogical Theory, 230-234
  Normative Theory as doctrine and dogma, 19-20
  Normative Theory as rational normative argument, 19-20
  presupposed in research on teaching, 152-153
Paradigm (Kuhn), 29-31, 64, 78
Pearson, K.
  correlation, 99-100
  his confusion of scientific and empiricist research, 98-99, 189
  views about "facts" in scientific research, 99
Pedagogical Theory,
  as a necessary basis for adequate empiricist research on teaching, 230-246
  as Practical Theory, 12-14
  as emerging, it is hoped, from (empiricist) research, 137
  local theories as part of, 235-236
Performance based teacher education (PBTE), 125-126
  precursor of PBTE, 102
Philosophia Naturalis Principia Mathematica (Newton), 38-39, 40,
Phlogiston theory of combustion, 44-46
Plate tectonics, 64-67
Popper's theory of scientific research, 43, 51-54, 146, 171, 179, 183, 195
Polanyi, M., and tacit knowledge, 239
Precession of Mercury at perihelion, 16, 42, 183-184
Presupposition, 16-18
  presupposition about the significance of measurement, 16-17
Probability and statistics
  different purposes in empiricist research and in scientific research, 188-192
  different uses in empiricist research and in quantum mechanics, 191
  in classical mechanics, 190-191
  in statistical mechanics, 191
Process-Product research on teaching, 11
  Academic Learning Time, (ALT), 126-127
  critique of apologia by N.Gage, 144-156
  effects of size of class, study by Finn and Achilles, 127-129
  experimental studies of reading, by Anderson, Brophy and Evertson, 116-118
  findings of, 113, 117-118,120, 123-125,127, 128-129, 194
  meta-analysis and , 118-121
  PBTE and, 125-126
  Helmke and Schrader's study of mathematics and seatwork, 121-125
  techniques of, 111, 130-143
  University of Canterbury studies, 112-113
  views of scientific research assumed in, 130-143
Ptolemaic astronomy, 29-31, 32, 83, 177, 179
Questionnaire research by H.S. De Casper, 115-116
  results of, 116
Quantum mechanics

probability as relating the abstract concepts within the theory, 191
Questions
　Process question, 135
Quine, W.V.O., and the complexity of falsification, 52-54, 183
　exaggerations of the Quine-Duhem thesis, 53, 81
Ratio measurement, 71, 188, 192-193
Reason cf. cause, 210-213
Research on teaching, 111-129, 130-143
　alleged non-theoretical nature of, 140-141
　narrowness of present day empiricist, 165, 165n
　theoretical nature of, 144-165 *passim* , 166
　statement about, by J.Willer, 143
　confusion with medical research, 146-148
　ways of improving, 5, 76, 196-197, 202-246 *passim*
Research program (Lakatos), 40-41, 43, 185, 187
Rice, J.M., early empiricist study by, 100
Roentgen's glowing barium platino-cyanide screen, 18
Roman Catholic schooling, 19
Rules as constitutive of actions, 213
Scientific measurement, 191, 192-193
　difference between it and scaling, 72
Scientific research, 28-92, 167-201
　"atomism" in, 160
　experiments in, 136-137, 142-143, 193-197,
　views of Sir Karl Popper, 51-54
　not a matter of generalization, 75-76
　openness of, 186-187
　procedural values of, 198-199
　significant developments as large-scale and conceptual, 171
　views of K. Pearson, 98-100
　on teaching
　　not "big science", 201
　　views of H.Walberg, 138-139
　　views of N. Gage, 140
　no scientific method, or art of discovery, as such, 173-179
Scientific Theory, 9, 67-92
　abstract concepts, 187-188
　an European invention, 74
　classical statistical mechanics as an example, 191
　continues to improve, 76-92
　"incommensurability" challenge rejected, 87-92
　"meaning" challenge rejected, 84-87
　observational checks can be objective, 77-84
　"theory-laden" challenge re-described, 80-84
　measurement in, 192-193
　not a matter of generalization, 75-76

INDEX OF TOPICS

Scientific theory of pedagogy
  views of N.Gage, 144-155
  hopes for, 108
  long history of hopes for, 97
Scientific laws, 72
  confusion of by J.S.Mill, 95-96
S.C.O.T.S. (System for the Classroom Observation of Teaching Strategies); reported in *The Teacher's Craft* (Powell), an example of improved kind of empiricist research on teaching, 239-244
Seafloor spreading hypothesis (SSH) of H.H.Hess, 64-65, 91
Serendipitous discovery, 18, 197-198
Snell's Laws as examples of First-Level theory, 68-71, 73-74
Statistics and probability
  different purposes in empiricist research and scientific research, 188-192
  different uses in empiricist research and in quantum mechanics, 191
  in classical mechanics, 190-191
  in empiricist research on teaching, 111-129 *passim*, 130-143 *passim*
  in statistical mechanics, 191
Social dimension in scientific research, 61, 198-201
Stenhouse, L., views of problematic nature of statistical experiments on teaching, 193-194
Taylor, F.B., and continental drift, 59, 60, 64, 186
Teaching
  complexity of concept, 202-203, 234-235
  complexities of teaching 202-210
  concept of, 215-217
  concept of in empiricist research
    as indicated by explicit statements, 133
    as indicated by actual procedures, 130-133
  teaching history, 209-210
  teaching reading, 117-118, 207-208
  teaching science, 204-207
  teaching woodwork, 203-204
Teaching Styles and Pupil Progress, questionnaire, observation, factor and cluster analysis study by N.Bennett
  empiricist concepts of, 158-159
  results of, 114
Theory
  background to discovery, 180-182
  confusions of varieties, 7-8, 26-27, 76, 182
  Dewey's theory of democratic educaton, 19
  Empiricist Theory, 9, 20-26, 72
  First-Level theory, 9, 67-92 *passim*
  necessity in scientific research, 145-155 *passim*
  Normative Theory, 9, 19-20
  Normative Theory as doctrine and dogma, 19-20

Normative Theory as rational normative argument, 19-20
   presupposed in research on teaching, 152-153
   Scientific Theory, 9, 67-92
   Second-Level theory, 9, 68-69
   shifting meanings of, 7-8, 26-27
   Theory as Evolving Explanation, 9, 11-12
   Theory as Heuristic, 15, 105, 194
   Theory as Hypothesis, 9, 14-16
   Theory as Model, 14, 63-64
   Theory as Observational Presupposition, 18
   Theory as Ontological Presupposition, 16-18
   Theory as Presupposition, 9, 16-18
   Theory Contrasted with Fact, 9, 10
   Theory Contrasted with Practice, 9, 10-11
   Practical Theory, 9, 12-14
   varieties of theory, 7-27
Theory-laden (Hanson), 18
   observations not laden with all theories (White), 82-83
   theory-laden nature of observation, 18n, 67, 80-82, 186
   theory-laden nature not equivalent to subjectivity, 18n
   theory-ladenness of falsifying observations, not the problem often assumed, 80-84
   theory-ladenness in the earth sciences, 67
The Scientific Basis of the Art of Teaching (Gage), 2, 14, 15
   critique of, 145-155
Thorndike, E.L.
   conceptions of scientific research in education, 17, 100-101, 167
Thought experiments, 17-18, 35-37, 174-175
   importance for research on teaching, 178-179
Tradition, custom and rule of thumb, as part of Empiricist Theory, 21
Tycho (Brahe), 32, 35, 83, 175
United States Office of Educational Research, 1
Validity, lack of in empiricist research on teaching, 227-228
Vine, F.J., magnetic anomalies and and sea-floor spreading, 65, 79, 172
   conflicting views of Lamont Observatory scientists, 79
"Vulcan", 16, 183-184
Walberg, H.J.
   conceptions of meta-analysis as a basis for research on teaching, 109, 118-121, 137, 138-139, 141, 180-181, 182, 192
   kudos-seeking claims to falsifiable theory, 184
Wegener and continental drift, 59-65
   controversy with Taylor over priority, 59n
Wilkins, M., and the discovery of structure of DNA, 50, 55, 58, 187
Willer, D., scientific-empiricist distinction 24, 69, 100, 104, 105, 106
Willer, J., scientific-empiricist distinction 24, 69, 100, 104, 105, 106, 143

Other books by John H. Chambers:

Knowledge Authority and the Administration of Tertiary Education (1982)

The Achievement of Education (1983/1989)

Philosophy and Education

1. C.J.B. Macmillan and J.W. Garrison: *A Logical Theory of Teaching.* Erotetics and Intentionality. 1988   ISBN 90-277-2813-5
2. J. Watt: *Individualism and Educational Theory.* 1989   ISBN 0-7923-0446-2
3. W. Brezinka: *Philosophy of Educational Knowledge.* An Introduction to the Foundations of Science of Education, Philosophy of Education and Practical Pedagogics. 1992   ISBN 0-7923-1522-7
4. J.H. Chambers: *Empiricist Research on Teaching.* A Philosophical and Practical Critique of its Scientific Pretensions. 1992   ISBN 0-7923-1848-X

KLUWER ACADEMIC PUBLISHERS – DORDRECHT / BOSTON / LONDON